Lecture Notes in Mechanical Engineering

For further volumes:
http://www.springer.com/series/11236

About this Series

Lecture Notes in Mechanical Engineering (LNME) publishes the latest developments in Mechanical Engineering - quickly, informally and with high quality. Original research reported in proceedings and post-proceedings represents the core of LNME. Also considered for publication are monographs, contributed volumes and lecture notes of exceptionally high quality and interest. Volumes published in LNME embrace all aspects, subfields and new challenges of mechanical engineering. Topics in the series include:

- Engineering Design
- Machinery and Machine Elements
- Mechanical Structures and Stress Analysis
- Automotive Engineering
- Engine Technology
- Aerospace Technology and Astronautics
- Nanotechnology and Microengineering
- Control, Robotics, Mechatronics
- MEMS
- Theoretical and Applied Mechanics
- Dynamical Systems, Control
- Fluid Mechanics
- Engineering Thermodynamics, Heat, and Mass Transfer
- Manufacturing
- Precision Engineering, Instrumentation, Measurement
- Materials Engineering
- Tribology and Surface Technology

Ram P. Bajpai · U. Chandrasekhar
Avinash R. Arankalle
Editors

Innovative Design, Analysis and Development Practices in Aerospace and Automotive Engineering

I-DAD 2014, February 22–24, 2014

Springer

Editors
Ram P. Bajpai
Avinash R. Arankalle
Vel Tech Dr. RR and Dr. SR Technical University
Chennai
Tamil Nadu
India

U. Chandrasekhar
The Engineering Staff College of India (ESCI)
Hyderabad
Andhra Pradesh
India

ISSN 2195-4356 ISSN 2195-4364 (electronic)
ISBN 978-81-322-1870-8 ISBN 978-81-322-1871-5 (eBook)
DOI 10.1007/978-81-322-1871-5
Springer New Delhi Heidelberg New York Dordrecht London

Library of Congress Control Number: 2014937310

Printed on acid-free paper

Springer is part of Springer Science+Business Media (www.springer.com)

Preface

Designs and developments are the aspirations of tomorrow's technologies for the aero as well as the auto industry to be alive in the competitive world where cost-effective solutions, improvements in greenhouse environment, longevity/life cycle, eco-friendly materials and manufacturing, certification, and government legislation demands are becoming stringent. Whether it is aerospace or automotive, the pulse and echo are similar in meeting the expected performances in air or on the road, respectively. Both the industries have come to symbolize the essence of a modern industrial society. Perhaps more than any other single icon, it is associated with a desire for independence and freedom of movement, an expression of economic status. For the next decades, they are marching toward new concept designs, analysis and manufacturing technologies, where more swing is for improved performance through specific and/or multi-functional linguistic design aspects to downsize the system, improve weight-to-strength ratio, fuel efficiency, better operational capability at room and elevated temperatures, reduced wear and tear, NVH aspects while balancing the challenges of beyond Euro IV emission norms, greenhouse effects, and recyclable materials.

The conference aims to address these issues of tomorrow where academia—industry—R&D partnerships, and collaborative programs can be shared and implemented.

The organizers of the *International Conference on Innovative Design, Analysis and Development Practices in Aerospace and Automotive Engineering* (IDAD 2014) wish to provide a platform for deliberations on Design Engineering, Numerical Methods, Analysis/Optimization Techniques, Life Cycle Engineering, System Engineering, Configuration Managements, Advanced Materials, Novel Manufacturing/Prototyping, Vibration and Health Monitoring, Propulsion System and Quality and Reliability in the Aerospace and Automotive Field. The response to the conference was overwhelming on both national and international fronts. The submitted papers were reviewed by renowned experts in the field.

Logos of Organizers

Organizers

VEL TECH

Dr.RR & Dr.SR

TECHNICAL UNIVERSITY

VEL TECH RANGARAJAN Dr.SAGUNTHALA
R&D INSTITUTE OF SCIENCE AND TECHNOLOGY
University u/s 3 of UGC Act, 1956
#42, Avadi-Vel Tech Road, Avadi, Chennai - 600 062, Tamil Nadu, INDIA.
Phone: 044-26840249, 26840869, 26840896, 26841601, 26841993, 24334845, 45030136
E-mail : admission@veltechuniv.edu.in, admission@vel-tech.org, veltech@vsnl.com
Website : www.veltechuniv.edu.in, www.vel-tech.org

Supported by

सत्यमेव जयते
Department of Science and Technology
Ministry of Science and Technology
Government of India

Knowledge Partners

Armaments Research Board

Technische Universität Braunschweig
GERMANY

Institut für Verbundwerkstoffe
Institute of Composite Materials
IVW - GERMANY

NFF NIEDERSÄCHSISCHES FORSCHUNGSZENTRUM FAHRZEUGTECHNIK
GERMANY

NAFEMS

NATIONAL CHENG KUNG UNIVERSITY
TAIWAN

THE INSTITUTION OF ENGINEERS

Organizing Committee

Prof. Dr. R. Rangarajan, Chief Patron, Founder—President, Vel Tech RR and SR Technical University, India
Dr. Ram P. Bajpai, Convenor, Chancellor, Vel Tech RR and SR Technical University, India
Dr. P. Sarasu, Co-Convenor, IDAD 2014 Director Research, Vel Tech RR and SR Technical University, India
Prof. Avinash R. Arankalle, Organizing Secretary, Professor, Vel Tech RR and SR Technical University, India
Mr. S. Sivaperumal, Organizing Secretary, Assistant Professor, Vel Tech RR and SR Technical University, India

Advisory Committee

Shri. Ajay Verma, VP—Business Transformation, Dassault Systèmes, India
Dr. Arvind Chandrasekaran, Sr. Product Development Er. Alberta Centre for Advanced Micro-Nanotech Products, Canada
Dr. O. P. Bajpai, Director Academics, Vel Tech Dr. RR and Dr. SR Technical University, India
Dr. E. Balasubramanian, Associate Professor, Vel Tech Dr. RR and Dr. SR Technical University, India
Prof. Beela Satyanarayana, Vice Chancellor, Vel Tech Dr. RR and Dr. SR Technical University, India
Dr. U. Chandrasekhar, Director, ESCI, India
Dr. Chin-Hsiang Cheng, Professor, National Cheng Kung University, Taiwan
Dr.-Ing. Christian Boller, Director, IZFP, Germany
Dr.-Ing. Christoph Herrmann, Head of Institute, Technische Universität Braunschweig, Germany
Dr. Chun-Yi Su, Professor, CONCORDIA University, Canada
Dr. Deepak Srivastava, Founder and Lead, CANEUS, USA
Dr. Dhananjay Kumar, Professor, North Carolina A&T State University, USA
Shri. Eckhardt Schneider, Business Unit Manager, IZFP, Germany
Dr. François Falempin, MBDA, France

Dr. S. Gopalakrishnan, Professor, Indian Institute of Science, Bangalore
Dr. N. Gowrishankar, Whole Time Director, IP Rings Ltd, India
Dr. Gowtham Swaminathan, Associate Professor, Vel Tech Dr. RR and Dr. SR Technical University, India
Dr. Gregory Odegard, Associate Professor, Michigan Technological University, Michigan
Dr. Indira Narayanaswamy, Technology Director (ARD&P), ADA, India
Dr. Jagannathan Sankar, Director—(CAMSS), North Carolina A&T State University, USA
Dr. Jaiswal, Professor, Vel Tech Dr. RR and Dr. SR Technical University, India
Dr. K. Jaya Kumar, IAS, Joint Secretary, CSIR, India
Dr. K. Jeyaraman, Associate Professor, Vel Tech Dr. RR and Dr. SR Technical University, India
Dr. Jitendra Tate, Associate Professor, Texas State University, USA
Dr. Joseph Koo, Sr. Research Scientist, University of Texas at Austin, USA
Dr. D. K. Kharat, Director Armaments and Secretary ARMREB, Directorate of Armaments, DRDO, India
Dr. Kishore Kumar, Scientist G, Associate Director, TDG, GTRE, Bangalore
Dr. Krishna Prasad Balike, Senior Analyst, Pratt and Whitney, Canada
Dr. Kunigal Shivakumar, Director, North Carolina A&T State University, USA
Dr. Mahesh Hosur, Research Professor, Tuskegee University, USA
Dr. R. Mariappan, Associate Professor, Vel Tech Dr. RR and Dr. SR Technical University, India
Dr. P. Mathiyalagan, Dean SoM, Vel Tech Dr. RR and Dr. SR Technical University, India
Dr.-Ing. Michael Magin, Institute for Composite Materials Ltd, Germany
Dr. Muthukumaran Packirisamy, Professor, CONCORDIA University, Canada
Dr. K. Natarajan, Professor, Vel Tech Dr. RR and Dr. SR Technical University, India
Dr. Pankaj Priyadarshi, Scientist/Engineer 'SG',VSSC, ISRO, India
Dr. Parminder Singh, Additional Director, Directorate of Aeronautics, DRDO, India
Dr. Prahlada, Vice Chancellor, Defence institute of Advanced Technology, Pune
Dr. Prakash Pattnaik, Director, National Research Council, Canada
Dr. J. K. Prasad, Professor, Birla Institute of Technology, India
Dr. M. Y. S. Prasad, Distinguished Scientist/Director, SDSC, SHAR, India
Prof. Rajkumar S. Pant, Aerospace Engineering Department, IIT Bombay, India
Dr. Rama Bhat, Professor, CONCORDIA University, Canada
Dr. Ranganathan, Professor, Vel Tech Dr. RR and Dr. SR Technical University, India
Dr. P. N. Rao, Professor, University of Northern Iowa, USA
Dr. V. K. Saraswat, Former SA, DRDO, India
Dr. S. Senthil Kumar, Vel Tech RR and SR Technical University, India
Shri. Shrikant R. Marathe, Director, ARAI, India

Dr. Simona Badilescu, Senior Research Associate, CONCORDIA University, Canada
Prof. C. S. Sivakumar, Professor, Vel Tech Dr. RR and Dr. SR Technical University, India
Dr. P. Sivakumar, Director, CVRDE, India
Mr. G. Siva Kumar, Assistant Professor, Vel Tech Dr. RR and Dr. SR Technical University, India
Dr. P. Srinivasa Murthy, Scientist 'F', ADE, India
Shri. A. N. Subash, VSSC, ISRO, India
Dr. Tapan K. Sengupta, Professor, IIT Kanpur, India
Dr. V. Venkateswara Rao, Scientist 'G', Advanced Systems Laboratory, India
Dr. Vinayak Kabadi, Professor, North Carolina A&T State University, USA
Dr. K. C. Vora, Head—Academy, ARAI, India
Dr. Yang Jinglei, Professor, Nanyang Technological University, Singapore

About the Editors

Dr. Ram P. Bajpai, Ph.D. (IIT-Delhi), D.Sc. (Hokkaido University, Japan) is presently the Vice-Chancellor of Vel Tech Technical University, Chennai. He has also been the Director General of MM Group of Institution, MM University, Mullana and Vice-Chancellor of MM University Solan till November 30, 2011. At the time, he was also holding charge of three universities as Vice-Chancellor, namely (1) Kurukshetra University, Kurukshetra; (2) Guru Jambheshwar University of Science and Technology, Hisar and (3) Deen Bandhu Chhotu Ram University of Science and Technology, Murthal simultaneously. Before joining GJU, Hisar, Dr. Bajpai was Director/Distinguished Scientist/Scientist at Central Scientific Instrument Organization, Chandigarh, one of the pioneering Research laboratories under CSIR, Government of India, New Delhi. He has excellent knowledge of Japanese and German languages, having a large number of publications in national and international journals of repute and a good number of patents. Dr. Bajpai has guided a large number of Ph.D. and M.S./M. Tech/B. Tech students in their thesis work for the award of degrees in various areas of electronics/microelectronics, materials science, nanoelectronics, optoelectronics, etc. Dr. Bajpai was UNDP Project Co-ordinator/ Director for the Microelectronics Instrumentation Development programme started in 1984 at CSIO, Chandigarh. He has filed and registered a large number of patents. Dr. Ram P. Bajpai has been Scientist/Visiting Professor in several universities at Japan, Germany, Switzerland, France, and USA.

Dr. U. Chandrasekhar is the Director of Engineering Staff College of India (ESCI), an autonomous organ of The Institution of Engineers (India). Before taking charge as Director—ESCI, he was an Additional Director at a Ministry of Defence R&D organisation called Gas Turbine Research Establishment. For the past 26 years, he has been involved in design, analysis, prototyping, rapid manufacturing, and testing of aero gas turbine engines. He set up the first-ever rapid prototyping laboratory in the country. For his research efforts, he received commendation medal from the Scientific Advisor to the Defence Minister. He received a B.E. in Mechanical Engineering from NIT, Suratkal, M.Tech in Design Stream from IIT, Madras and Ph.D. from VTU. For his academic excellence at IIT-Madras, he received an award from former President of India Dr. A. P. J. Abdul Kalam. He was trained in Germany, UK, and Belgium in RP and sensor technologies. He is currently leading a critical technology development project of high-temperature thin film sensors in collaboration with NRC, Canada. He currently serves in the council of the Institution of Engineers and National Design and Research Forum. He was also chosen to represent India in the Young Leaders Convention of World Federation of Engineering Organisations at Geneva. He has been invited by several national and international professional bodies as the keynote speaker on advanced prototyping and sensor technologies.

Prof. Avinash R. Arankalle, is an M.Tech from IIT, Mumbai with a Masters' Diploma in Business Administration (MDBA) from Pune. He possesses +35 years of experience in the automotive industry and in particular in selection of materials, testing and characterization, and failure analysis of automotive components. After working in Kirloskar Oil Engines, Paranjpe Engineering and Foundry Company, and Bajaj Auto Ltd., he retired as a senior executive from the Automotive Research Association of India, Pune. With experience and knowledge in Automotive Materials, now, he is associated with Vel Tech Dr. RR and Dr. SR Technical University in promoting various engineering post-graduate programs, teaching specialized topics on engineering materials, guiding students in their theses, etc. The first of its kind, he designed along with his team in ARAI, Pune, FRP leaf spring and propeller shaft for LCV for which he possesses an Indian patent. He was a member in the sub-committee on Automotive Materials and End-of-life of Vehicles of Core Group of Automotive R&D (CAR), various BIS committees (main and sub groups—PCD1, PCD 3 and PCD 4), as well as AISC sub-committee on elastomers while being in ARAI, Pune and was associated in preparing standards on anti-freeze engine coolant IS-5759, Pull-push type control cables, plastic fuel tank, flammability test, elastomeric materials and components, fuels and lubricants, brake fluid, etc. He has set up a state-of-the-art test facility at ARAI, Pune, iCAT, Mannesarand ARAI-FID, Chakan for material testing and characterization. For over 30 students of engineering colleges, he provided guidance in their thesis work for completion of the graduate and postgraduate degrees. He has published over 40 papers at national and international forums which include papers on fuel quality, significance of diesel fuel flash point, analysis of toxic constituents in fuels that are carcinogenic in nature, advances in ceramic materials for auto engines, automotive plastics and composites, design of FRP leaf spring and propeller shaft, etc. Being one of the founder members of ASM International Pune Chapter and a former Chairman of Education Committee, he conducted various educational courses for the industry which includes the famous one on "Metallurgy for Non Metallurgists." On behalf of ARAI, Pune, ARAI-FID, Chakan and Arkey Conferences, Pune he has conducted over 40 conferences at national and international levels at the levels of Chairman, co-chairman, and member.

Contents

Part I Key Note Address

Stirling Engine Technology and Its Application on Solar Power Generation 3
Chin-Hsiang Cheng and Hang-Suin Yang

Non-Destructive Techniques and Systems for Process and Quality Control in the Automotive and Supplier Industry 11
Eckhardt Schneider and Christian Boller

Future Trends in the Development of Vehicle Bodies Regarding Lightweight and Cost 13
Frank Nehuis, Sebastian Kleemann, Patricia Egede, Thomas Vietor and Christoph Herrmann

Flying with Cracks, Damage Tolerance and NDT: Potential Contributions of the Fraunhofer Institute for Nondestructive Testing IZFP to Aeronautical Safety 23
Christian Boller and Eckhardt Schneider

A Synchronous Wound Excitation Transverse Flux Machine with Solid Rotor 25
Francois Bernot, Alix Bernot and Jean-Claude Vannier

R&T Effort Led by MBDA in the Field of Aero-propulsion 41
Francois Falempin

Technological Trends in the Application of Advanced Materials in Automobiles 45
P. K. Mallick

Recent Developments of Mechanical and Fatigue Analyses of Fiber-Reinforced Structures for Aerospace Applications 47
Michael Magin

Design and Development of Lighter-Than-Air Systems 63
Rajkumar S. Pant

Computational Fluid Dynamics: A Design Tool for Aircraft Industries 65
P. Srinivasa Murthy

Part II Contributed Papers

An Approach to Improve Aviation Quality Management Using Total Quality Management Principles 69
B. Immanuel Lazur, L. Jagadeesh, B. Karthikeyan and M. Shanmugaraja

Design and Field Trials of a Payload Recovery Device for Tethered Aerostats 79
Nitin Sharma, Agnit Mukhopadhyay, Vishaal Sharma, Mayank Milind and Rajkumar S. Pant

Design of an Effective Subsonic Wing Cross-Section Using Viscous-Inviscid Interactive Method 85
L. Prabhu and J. Srinivas

Computational Study of Various Longitudinal Fin Profiles 91
Pankaj Dhaka, Prashant Kumar and Varun Goel

Dynamic Modeling and Simulation of Flapping Wings UAV 105
Balasubramanian Esakki, Vasantharaj Rajagopal and Rusendar Babu Srihari

Off Line Performance Measures of Two Wheeler Engine: In Automatic Objective Method 117
G. Mugunthan, E. Balasubramanian, R. Madhavan and Jebin Sing

Optimization of Variable Speed Turbo Coupling in Boiler Feed Pump 129
R. Pandithurai and I. Ambrose Edward

Cyclic Life Estimation of Turbine Rotor Blade Fitted in a Twin Spool Turbojet Engine 135
N. Sreenivasaraja

CFD Analysis of Swirl Enhancement in a Direct Injection Diesel Engine with Vortex Generator in Inlet Manifold 145
G. Sivakumar and S. Senthil Kumar

Numerical Study of Air-Intake Performance of a Scramjet with Strut Blockage at Various Angle of Attacks 153
S. Sivaraj and S. Senthil Kumar

Numerical Study of Aerodynamic Characteristics of Triangular Flapping Wing for MAV 163
S. Senthil Kumar and Pranesh Rajan

Numerical Analysis of Suppression of Laminar Bubble at Low Reynolds Number Using Different Protrusions 171
S. Senthil Kumar, R. P. Vijay Ganesh, S. Karthikeyan and J. Karthik

Numerical Simulation of Rayleigh-Bernard Convection in Enclosures Filled with Nanofluid 179
S. Senthil Kumar and S. Karthikeyan

A Conceptual Design of Versatile Furtive Craft 189
P. Gowtham, C. Gnanavel, V. Manoj Kumar and R. Naveen

Reduction of the Passage Between the Flame Tubes of the Combustor for a Millimeter Size Gas Turbine Engine in the Art of Micromachine Technology 197
Syed Alay Hashim, Nancy Manish, Deepam Mishra and Krishna Ahuja

Influence of Temperature on Mechanical Characterization of Basalt/Epoxy Owen Fabric Composites 207
J. Alexander and B. S. M. Augustine

Designing and Controlling the Performance of FMS: Application of Principle Component Analysis 215
M. Bragadish, N. Venkatachalapathi and A. Ramakrishna Rao

Improved Methods for Thermal and Compression Testing of Carbon Foams 225
S. Gowthaman, J. M. Babu, P. Mathiyalagan and K. Shivakumar

Integration of eLCAr Guidelines into Vehicle Design 235
Patricia Egede, Frank Nehuis, Christoph Herrmann and Thomas Vietor

Design of Magnetic Wheel Rotor 243
S. Yuvaraj and M. Anantha Kumar

The Effect of Injection Pressure on Engine Performance While the Engine is Running on Electrolytically Generated Hydrogen Oxygen Mixture 253
J. M. Babu, B. Harish Babu, M. Rajesh Kumar, S. Gowthaman, R. Mariappan and P. Mathiyalagan

Design and Fabrication of Air Preheater for Diesel Engine 261
Vikas Sharma, J. M. Babu, R. Naresh, S. Gowthaman and R. Mariappan

Studies on Tropical Conditions in Hot Sunny Days in India and Their Effects on Interior Temperature Rise in Truck Cabin 269
Suraj Mani Chaurasiya, Swanand Kirpekar and Avinash R. Arankalle

Fabrication of Bamboo Fiber Reinforced Polylactide Biocomposites 277
Wen-Bin Young, Bo-Yuan Tian and Yu-Chi Tsau

Emission Control by Selective Exhaust Gas Recirculation Scavenging System in Two-Stroke Engine 285
V. Ramakrishnan, K. Purushothaman and S. Mohammed Ghouse

Design of Airfoil Using Backpropagation Training with Combined Approach 295
K. Thinakaran and R. Rajasekar

Improvement of Performance and Emission Characteristics of Diesel Engines Using TBC and Emerging Techniques: A Review 307
G. Sivakumar, R. Nanda Kumaran, M. G. Varun Kumar and S. Senthil Kumar

Author Index 319

Part I
Key Note Address

Stirling Engine Technology and Its Application on Solar Power Generation

Chin-Hsiang Cheng and Hang-Suin Yang

Abstract In this study, a beta-type 500-W Stirling engine is developed and tested, and a nonideal adiabatic model is built and applied to predict performance of the engine. Engine torque, engine speed, and shaft power output are measured under various operating conditions. Furthermore, as the charged pressure and the heating temperature are set at eight bars and 800 °C and a No. 50 wire mesh is used in the regenerator, the shaft power of the engine can reach 556 W at 1665 rpm with 1.21-kW input heat transfer rate (45.95 % thermal efficiency). The experimental data are compared with the numerical predictions to verify the theoretical model. It is found that the experimental data of the shaft power output closely agree with the numerical predictions. This implies that the theoretical model is valid and helpful to the engine design.

Keywords Stirling engine · Solar power · Experiment · Numerical model

1 Introduction

Stirling engines are referred to as external combustion engines, and hence, they can be operated with a variety of heat sources and have been applied for energy conversion in a number of engineering applications. The ideal regenerative Stirling cycle consists of two isothermal processes and two isochoric processes. In theory, the thermal efficiency of an ideal regenerative Stirling cycle is equal to the Carnot cycle efficiency under the same thermal reservoirs temperatures. Due to the above advantages, Stirling engines have been used in concentrating solar power (CSP) systems that adopt mirrors or lenses to concentrate a large area of solar energy onto a

C.-H. Cheng (✉) · H.-S. Yang
Department of Aeronautics and Astronautics, National Cheng Kung University, No.1, Ta-Shieh Road, Tainan 70101, Taiwan, Republic of China
e-mail: chcheng@mail.ncku.edu.tw

R. P. Bajpai et al. (eds.), *Innovative Design, Analysis and Development Practices in Aerospace and Automotive Engineering*, Lecture Notes in Mechanical Engineering, DOI: 10.1007/978-81-322-1871-5_1,

Fig. 1 20 kW-class concentrating solar power (CSP) systems. (Developed by Power Engines and Clean Energy Laboratory [PEACE Lab.], National Cheng Kung University)

small area. Figure 1 shows the CSP system developed by Department of Aeronautics and Astronautics, National Cheng Kung University. Electrical power is produced when the concentrated sunlight is converted to thermal energy to drive the Stirling engine and connected to an electrical power generator [1]. Since Stirling engine is suitable for various heat source, in addition to CSP systems, it is also suitable for variable external heat sources, including waste heat, geothermal energy, and combustion heat from fossil fuels or hydrogen provided that heating temperature of any external heat source is sufficiently high. Therefore, the Stirling engine technology receives increasing attention from energy-related researchers in recent years.

The first patent of the Stirling engine was filed by Robert Stirling in 1817 who originally intended to invent a hot-air closed-cycle prime mover to serve as an alternative to the explosively dangerous steam engine. In 1871, Schmidt proposed an isothermal second-order analysis of the ideal Stirling engine cycle. In order to modify this model, Finkelstein [2] presented ideal adiabatic models, in which the thermodynamic processes taking place in the expansion and the compression spaces are assumed to be adiabatic. However, the heater and the cooler are still assumed to be isothermal. In a real engine, the thermal conditions are not isothermal or adiabatic. Recently, Cheng et al. [3] developed a nonideal adiabatic model for a beta-type Stirling engine with rhombic-drive mechanism by introducing effectiveness of mechanism [4], effectiveness of regenerator, imperfect heat transfer, and pressure drop. Moreover, Cheng and Yang [5] establish a nondimensional thermodynamic model to optimize the geometry of rhombic-drive mechanism, and give an optimal design for maximum shaft work.

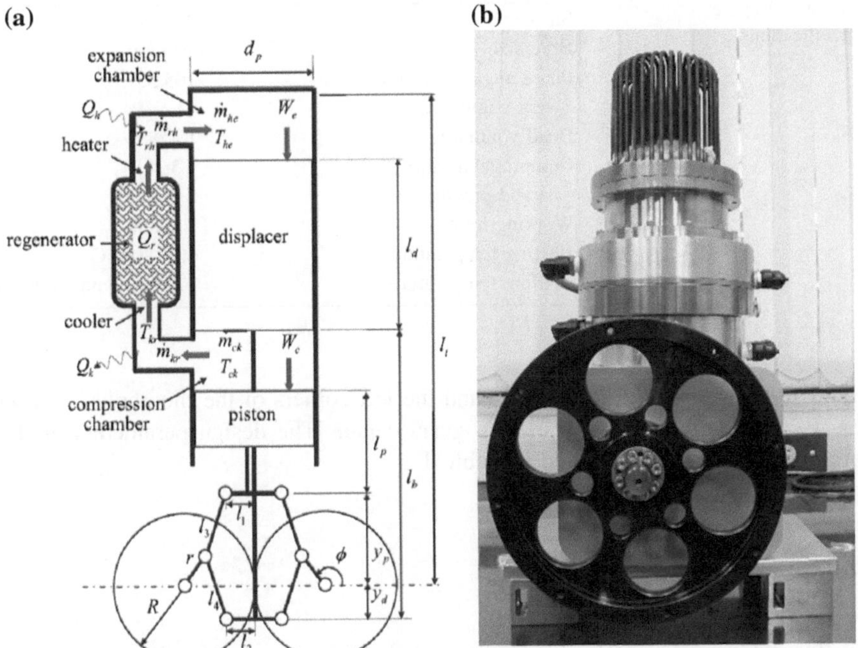

Fig. 2 Schematic diagram and photo of prototype engine. **a** Schematic diagram, **b** Photographic copy

In general, Stirling engines can be divided into three major categories, namely, alpha-, beta-, and gamma-types. And the beta-type engine has higher power density than the other two types, as shown by Cheng and Yang [6]. Therefore, in this study, the beta-type Stirling engine with rhombic-drive mechanism is developed. Experiments are also conducted to test the performance of the engine under various operating conditions. In parallel, a theoretical model is built and applied to predict performance of the engine. Eventually, the obtained experimental data are compared with the numerical predictions to verify the theoretical model.

2 Engine Design

The schematic of the beta-type Stirling engine with rhombic drive mechanism and photo of the prototype engine are shown in Fig. 2. The rhombic drive, which is frequently used with the beta-type Stirling engines, utilizes a jointed rhomboid to convert linear motion of a reciprocating piston to a rotational of flywheel. In the rhombic drive mechanism, one rigid rod is installed connecting the piston to the top corner of the jointed rhomboid, and another rigid rod connecting the displacer to the bottom corner of the rhomboid. In addition, two symmetric gears of equal

Table 1 Engine specifications

Stroke	35 mm
Bore size	70 mm
Phase angle of volume	105.537°
Swept volume ratio	1
Dead volume ratio	1.478
Compression ratio	1.364
Charged pressure	8 atm
Working fluid	Helium
Heating temperature	600–800 °C
Cooling temperature	27 °C (liquid cooling)

diameter are connected to the right and the left corners of the rhomboid fixed on the gears at an offset distance from gears center. The design parameters of the prototype engine are tabulated in Table 1.

3 Theoretical Model

In this study, a nonideal adiabatic model that involves the friction loss due to mechanism, power loss due to pressure drop, and the heat loss due to imperfect regeneration is proposed. The assumptions with the model are:

(1) The pressure is uniform throughout the interior space of the engine.
(2) The expansion chamber and the compression chamber are insulated.
(3) The heat is absorbed in the heater and rejected in the cooler.
(4) The temperature distribution in the regenerator is varied linearly from the heater's temperature to the cooler's temperature.
(5) The working gas is regarded as an ideal gas.

The detail process of nonideal adiabatic model is described in [3]. With the help of nonideal adiabatic model, indicated power and power loss due to friction loss can be calculated. However, mechanism loss has to be considered to obtain shaft power. The correlations of mechanism loss under different engine speed and heating temperature can be obtained by comparing the data which calculated by numerical model and measured by experiments. Generally, mechanism loss can be calculated by following formula.

$$\dot{W}_{m,loss} = A\omega^3 + B\omega^2 + C\omega + D\ (\mathrm{W}) \tag{1}$$

where A, B, C, and D are function of heating temperature, and ω is engine speed.

Fig. 3 Experimental apparatus and measuring devices

4 Experimental Apparatus

The experimental apparatus for the engine performance test is shown in Fig. 3. An electric infrared heater of 1.21 kW (11 A/110 V) is used as the external heat source, and a thermostat is used to control the heating temperature. A K-type thermocouple is placed in the center of the electric infrared heater to monitor the heating temperature and the temperature data are recorded by a data logger. In experiments, the engine load is applied on the crank shaft of the engine by using a hysteresis brake dynamometer. The dynamometer provides brake torque loading independent of the engine speed. The rotation speed of the engine and the brake torque can be measured by a rotation meter and a torque sensor, respectively. The data are also recorded by the data logger. The shaft power is then determined in terms of the brake torque and the engine rotation speed.

5 Results and Discussion

In this section, effects of the mesh number of the regenerator and the heating temperature are measured experimentally and predicted theoretically. Figure 4 conveys the comparison between experimental and numerical data in the shaft power as a function of engine speed and mesh number. The symbols are

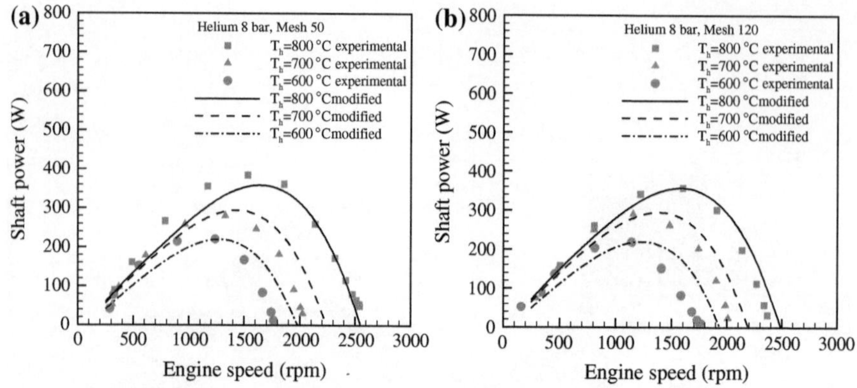

Fig. 4 Experimental data and simulation results with different mesh number of regenerator. **a** 8 atm helium, No.50 **b** 8 atm helium, No.120

Fig. 5 The experimental data with 8 atm helium, No.50 mesh, and enhanced cooling

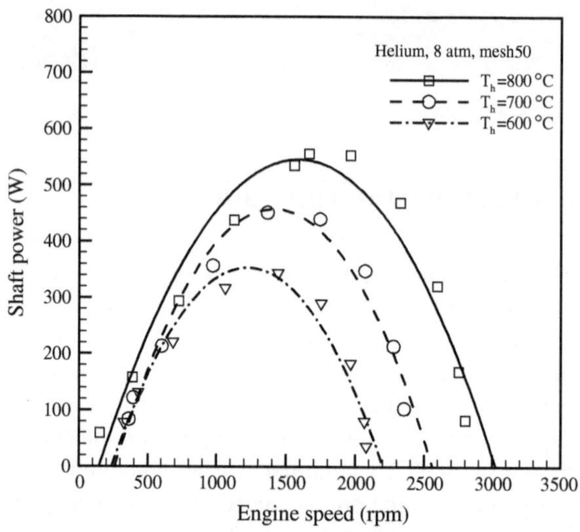

experimental data and the lines are obtained from simulation result. The results reveal that the maximum shaft power output decreases as the mesh number is increased from No. 50 to No. 120. The corresponding maximum shaft power is 384 W at 1,530 rpm and 356 W at 1,595 rpm respectively. And the results also show that the simulation results in accordance with experiment data well, and the error is lower than 17%.

Furthermore, for the case of Fig. 4a, the maximum shaft power can reach about 556 W at 1,665 rpm as the cooling system is enhanced as Fig. 5 shows. With 1.21-kW input heat transfer rate, thermal efficiency can reach 45.95 %.

6 Conclusions

In this study, a domestic scale 550-W Stirling engine with rhombic-drive mechanism is developed and tested. The shaft power of the engine can reach 556 W at 1,665 rpm with 1.21-kW input heat transfer rate, which means a thermal efficiency of 45.95 %. The theoretical model is valid and helpful to the engine design, and the engine can be applied on CSP system.

Acknowledgments Financial supports from the National Science Council, Taiwan, under grant NSC 102-2221-E-006-088, and from the Southern Taiwan Science Park Administration (STSPA), Taiwan, under contract 101CP03, are greatly appreciated.

References

1. Mancini T, Heller P, Butler B, Osborn B, Schiel W, Goldberg V, Buck R, Diver R, Andraka C, Moreno J (2003) Dish-stirling systems: an overview of development and status. J Sol Energy Eng, Trans ASME 125(2):135–151
2. Finkelstein T (1975) Computer analysis of stirling engines. Adv Cryog Eng 20:269–282
3. Cheng CH, Yang HS, Keong L (2013) Theoretical and experimental study of a 300-W beta-type stirling engine. Energy 59:590–599
4. Senft JR (2000) Extended mechanical efficiency theorems for engines and heat pumps. Int J Energy Res 24(8):679–693
5. Cheng CH, Yang HS (2014) Optimization of rhombic drive mechanism used in beta-type Stirling engine based on dimensionless analysis. Energy 64:970–978.
6. Cheng CH, Yang HS (2012) Optimization of geometrical parameters for Stirling engines based on theoretical analysis. Appl Energy 92:395–405

Non-Destructive Techniques and Systems for Process and Quality Control in the Automotive and Supplier Industry

Eckhardt Schneider and Christian Boller

Abstract The objective of the applied research and development at Fraunhofer IZFP is the improvement of product quality—together with the reduction of quality cost—and proof of technical safety of devices, machines, and plants.

Keywords Damage tolerant design · Material properties · Defect detection, sizing · Metallic components · Civil infrastructure

The contribution gives an overview of the Fraunhofer IZFP-developments of nondestructive techniques and systems to:

- characterize the mechanical properties of steel and Al-sheets for the automotive body,
- evaluate the hardness and hardening depth of gear parts and engine components,
- evaluate stress states in sheets and plates as well as in and around welds,
- evaluate the clamp load during fastening processes,
- evaluate the thickness of the clinched joint,
- detect and describe defects in body parts on the surface as well as in the volume of components of the engine and the power train.

Ultrasonic, eddy current, micromagnetic, magnetic flux, thermography, and X-ray methods are optimized to meet the particular requests of the client. Manual and automated systems are developed and introduced into the daily application in manufacturing lines as well as in quality assurance laboratories.

E. Schneider (✉) · C. Boller
Fraunhofer IZFP Saarbrücken, Saarbrücken, Germany
e-mail: Eckhardt.Schneider@izfp.fraunhofer.de

R. P. Bajpai et al. (eds.), *Innovative Design, Analysis and Development Practices in Aerospace and Automotive Engineering*, Lecture Notes in Mechanical Engineering, DOI: 10.1007/978-81-322-1871-5_2, © Springer India 2014

Future Trends in the Development of Vehicle Bodies Regarding Lightweight and Cost

Frank Nehuis, Sebastian Kleemann, Patricia Egede, Thomas Vietor and Christoph Herrmann

Abstract Lightweight design has become relevant to a lot of industry sectors, due to the wish to improve the CO_2 balance of their products and to save resources. To reduce the weight of a vehicle there are different approaches to lightweight design, e.g. material substitution or lightweight construction types. These existing approaches are reaching their limits with conventional materials. For this reason lightweight materials like fibre-reinforced plastics are becoming more and more popular. However, these materials are very cost intensive. Therefore, these materials are generally used, in sports cars or in premium vehicles. In order to solve the conflict between vehicle costs and weight reduction, new lightweight approaches have to be developed for mass production. The opportunities of new materials can only be provided for a large group of customers through new approaches. One such approach is to combine fibre-reinforced plastics with the existing metallic materials. This is called multi-material or hybrid design. In this approach, the advantages of the two types of material are used for a specific application. Only then the lightweight potential can be used effective and at reasonable cost. For a vehicle, the multi-material design can be used in particular for designing the body in white. Here, almost every component can be considered by the new lightweight approach. Hybrid materials, for example, can increase the stiffness and reduce the weight of a rocker at the same time. For this the load cases of the rocker need to be analysed for which the load frequencies and amplitudes are needed. Also the life expectancy has to be determined in order to design a component with thin-walled sections as well as a minimal safety factor. Afterwards the material combination of the hybrid material is determined with respect to the load cases. Generally the fibre-reinforced plastics are placed in areas with tensile stress. Vice versa metallic materials are placed in areas with compressive stress. For designing the components ideally direct load should exist. In this case, the maximum potential for lightweight design can be used. The new lightweight

F. Nehuis · S. Kleemann · P. Egede · T. Vietor · C. Herrmann (✉)
Technische Universität Braunschweig, Braunschweig, Germany
e-mail: C.Herrmann@tu-bs.de

R. P. Bajpai et al. (eds.), *Innovative Design, Analysis and Development Practices in Aerospace and Automotive Engineering*, Lecture Notes in Mechanical Engineering,
DOI: 10.1007/978-81-322-1871-5_3,

approach will be developed within the Open Hybrid LabFactory together with research institutes and industrial companies. For this purpose, we gratefully thank the Federal Ministry of Education and Research for funding the research project.

Keywords Vehicle body design · Muli-material design · Lightweight design

1 Introduction

The amount of customer demands regarding vehicle comfort, safety or driving dynamic is rising continuously. In the past, these demands led to increasing vehicle weight. However, this trend was stopped by developing new vehicle concepts using lightweight materials (light metals, high-strength steel or fibre-reinforced plastics), thus keeping the vehicle weight at a constant or slightly decreasing level, especially in vehicles with conventional drivetrains. This lightweight trend must be continued taking into account the automotive standards in quality, ecology and economy in the future [1, 2].

In the recent years, lightweight design has become an important lever in the automotive industry, e.g. to reduce the CO_2-emissions or to decline the use of raw materials. And in future lightweight design will remain a key driver to reduce the CO_2-emissions. Current studies indicate that the increase of the vehicle weight will continue with the application of electric or hybrid drivetrains. Because of that the use of lightweight materials will still be required in the vehicle body development [3].

The vehicle body design will play a large role in reducing the emissions and in meeting the strict EU legislation, which demand a reduction of the average CO_2-emissions of the vehicle fleet from the automotive industry. In this context, huge potential is identified in redesigning vehicles with lightweight materials in the mass segments. In mass segments even a moderate reduction of the vehicle weight causes a huge effect on the total CO_2-emissions. But conventional design and materials are reaching their limits. For this reason, the use of lightweight materials, such as fibre-reinforced plastics, in mass segments must be pursued [4, 5].

The combination of fibre-reinforced plastics with steel or steel–aluminium structures is currently not available for mass production. This is due to the high cost of the fibre materials. In addition, the process time in producing fibre-reinforced components is high. Thus the production of mixed structures with fibre-reinforced components seems to be more suitable for mass production. Using multi-material design will also have an effect on the overall vehicle. In addition, the production of lightweight materials increases the consumption of energy and needs more resources during the production. Therefore, e.g. the use of carbon fibre-reinforced plastics amortises ecologically with a mileage of 85,000–130,000 km per year, which is only reached in a few use cases [6]. In this context, it is necessary to design innovative vehicle concepts and components in addition to new manufacturing technologies. Also, the entire product life cycle must be examined regarding energy and resource consumption. Taking into account these factors lightweight concepts can be successful.

2 Lightweight in Modular Vehicle Bodies

In modern production processes for assembling vehicle bodies, modularisation concepts are being pursued. Due to modularisation it is possible to realise cost-effective vehicles despite the increasing number of vehicle variants. A key factor for the success of a modular vehicle body concept is to achieve as much flexibility as possible with respect to the total vehicle design. The flexibility gives the designer the possibility to develop different vehicle concepts and to fulfil the individual customer requirements [7, 8]. The number of vehicle variants can be achieved by dividing the vehicle body into standardised modules. For this purpose, the vehicle body will be divided into basic modules as well as assembly modules. In this context, a module includes parts of the supporting structure [9]. If a module entirely consists out of structural elements, it is called a basic module. If the module contains additional elements as well and if it is not assembled in the body shop, it is called assembly module. Hence, an assembly module always includes a basic module [10]. To assemble the structure the vehicle body is divided into modules at specific interfaces. In Fig. 1 typical modules of a vehicle body are shown.

The different modules for assembling the variants of the vehicle body normally differ significantly in their production batches. Thus, different batches of each module have to be produced. Using the example of a roadster this will be clarified: the basic module of the vehicle body can normally be used for several vehicle variants. Because of this it can be produced in a huge batch cost-effective. But the other vehicle body modules of a roadster normally cannot be used in other vehicle concepts. Therefore, these components are produced only in small batches. So they are relatively expensive. However, to apply new materials in the vehicle body, the aspects of modularisation and costs have to be considered.

To implement new lightweight technologies and strategies into the design of the vehicle body, existing and cost-effective production processes should be used. One possibility is to use the existing manufacturing processes, e.g. for producing steel bodies or steel–aluminium mixed structures. Both manufacturing processes are often used in practice and are optimised over many years. In addition, the material-specific joining technologies were developed continuously. However, the weight reduction potential of these manufacturing processes is limited in comparison to the concepts based on fibre-reinforced plastics [11].

To produce a cost-effective vehicle body with a high lightweight potential, the batch size influences the design, the material choice and the manufacturing process. For the basic modules of the vehicle body shell construction are recommended normally. For small batches, e.g. modules can be manufactured in

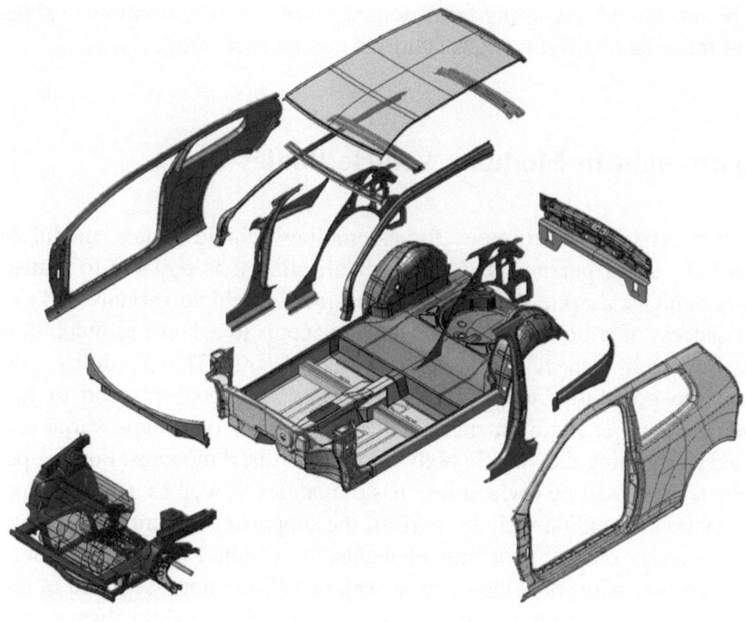

Fig. 1 Typical modules of a vehicle body, which can be used to generate different vehicle variants (in this figure a small car as an example), [10]

profile-intensive design. Here the investment in tools to manufacture the profiles is low compared to shell constructions, which requires large presses, e.g. for deep-drawing. For this reason the basic modules of a vehicle, which are produced in small series, should be produced in shell design and the assembly modules as profiles. Figure 2 gives a view on the material choice according to the produced units per year.

In order to choose the material for different modules one goal is to enhance the axle load distribution and to lower the vehicle's centre of gravity. The final modification will be done by adapting the component's material quality or by using reinforcements in specific areas. The interfaces between the modules must enable a change in material as well as a compensation of tolerances. As an example the interface between the A-pillar and the ground module will be considered: in this area several vehicle body modules are put together. These modules can consist of different materials. One possibility to combine the modules is to design specific node elements. Node elements help to solve the problems, which result from the material combination. A technology for joining different materials is, e.g. adhesive bonding. Adhesive bonding can compensate different tolerances. At the same time, this bonding type can transmit high forces. Furthermore, the use of structural foam allows a certain tolerance compensation and a combination of different structure elements and materials.

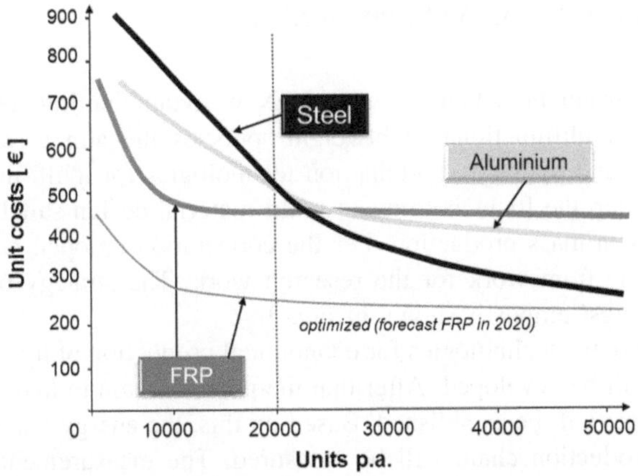

Fig. 2 Choice of different materials for designing the vehicle body according to the manufactured units per year [10]

3 New Materials and Production Processes

Load path-optimised components and structures using fibre-reinforced plastics can lead to weight reduction in the vehicle body. At the same time, rigidity and strength of the components can be maintained or even increased. The extensive use of fibre-reinforced plastics can be seen in the aerospace industry. Due to the high material costs, the energy consumption in manufacturing and processing, incompatibilities with existing manufacturing processes and low level of automation, these materials are currently rarely used in the automotive industry. Despite intensive research, they are only used for sports or luxury cars. Besides, the energy and resource flows of the production processes have to be modelled to evaluate the economic and ecological influences. The possibilities on the production side must be clarified and possible measures for improvement must be derived next to the design perspective. Conventional lightweight strategies like geometry optimisation or material substitution are already maxed out, so there is a need for new lightweight strategies to continue the weight reduction. One new strategy is multi-material systems [4]. They are characterised by the interaction of various materials within a component, which are adapted to function and load. In the automotive industry, existing approaches are normally focused only on subsystems or modules [5], pursue new vehicle concepts (e.g. BMW iConcept), which require new process chains and manufacturing processes, or substitute one material by another (e.g. SuperLight Car). But multi-material systems will have an influence on the components and the total vehicle design. To integrate multi-material systems into already existing process chains, the component's requirements regarding material properties as well as joining and manufacturing technologies are affected. To understand the relations between these requirements is a great challenge in multi-material design [12].

4 Future Trends for Vehicle Bodies

The research group ForschungsCampus—'New manufacturing technologies for economic and multifunctional lightweight' pursues the aim to develop design methodologies and innovative production technologies for multi-material design. Within this group the focus is to make multi-material design suitable for flexible vehicle bodies in mass production. For the cooperation an open research factory will provide the framework for the research work. The strategy of the research group ForschungsCampus is shown in Fig. 3.

In a first step, the technologies for economical production of hybrid lightweight components will be developed. After that the process chain to manufacture multi-material systems will be established. Based on this, the energy and material flows of the new production chain will be measured. The measurements will help to improve the production process regarding energy demands and material consumption. The measurements will also enable to calculate exact life cycle data. Based on this data the research group can evaluate the economic and ecological impact of multi-material systems. The capability of the process chain can also be estimated.

Furthermore the research group will develop component-specific design concepts taking into account the new production technologies and the life cycle analysis. The goal in the design is to provide general design rules for the combination of multi-material systems with conventional materials. These rules can be used afterwards to design a multi-material component with lightweight potential in every specific scenario. In addition to the design rules to develop multi-material components, the research group will guide current and future industry projects to transfer the technology into practice [13].

Furthermore, a demonstrator will enable the group to validate the research results starting from the product requirements till production taking into account the entire product life cycle. The research group will provide guidance to integrate the results from lightweight material and production research for designing ecological and economical sustainable vehicle bodies with a focus on vehicles, which are built in mass production.

The combination of established and innovative materials to hybrid materials or composites is a key issue for transferring fibre-reinforced plastics into mass production. To establish the new possibilities in the automotive industry, methods and tools must be provided. These methods and tools will give recommendations for an optimal material and process combination. Within the research group, these are the five main objectives:

1. **Identification of mass production use cases, which are relevant for vehicle bodies in lightweight design**:
   Identification of user profiles and application scenarios for different vehicle bodies, where the use of lightweight design can be taken into account. Determination of the optimal use of lightweight materials from an economic and ecological point of view.

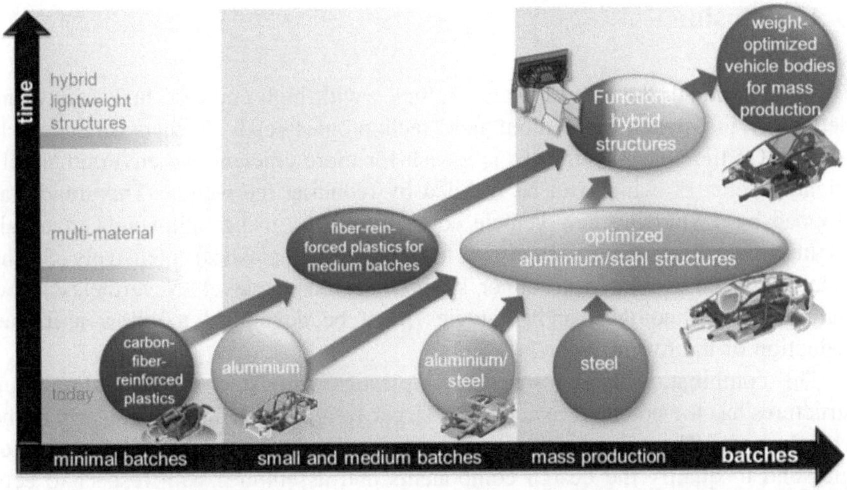

Fig. 3 Strategy to enable new lightweight potential in the automotive industry within the research group ForschungsCampus

2. **Investigations on the interactions and influences of lightweight concepts on the vehicle bodies' requirements**:
   Identification of restrictions and potentials of multi-material systems to reduce the vehicle body weight. Recommendations for a use-case specific application of multi-material systems.
3. **Establish a model to give an answer on the economic and ecological impacts of multi-material design throughout the life cycle (Life Cycle Prediction model)**:
   Life cycle-oriented analysis and evaluation of hybrid lightweight materials depending on the application scenario, the design parameters, the selected materials and the different production scenarios.
4. **Development of vehicle body concepts with multifunctional modules**:
   Components and assemblies will be developed as demonstrators to show the potential of multi-material design. The design rules will be used to design the demonstrators. Those will show the potential with regards to the technological, ecological and economic aspects.
5. **Analysis and evaluation of the energy and resource efficiency of the new process chains:**
   The continuous measurement and software-based modelling of energy and resource flows of the different process chains will give the possibility to calculate the ecological and the economic impact. In addition to the design, the production-related areas of action will also be identified. The analysis will give measures to improve the process chains.

5 Conclusion

On the one hand, the trend towards vehicles with high comfort, high safety and better driving dynamics without new technologies leads to increasing vehicle weight. On the other hand, there is a wish for more efficient and environmentally friendly vehicles, which can be fulfilled by reducing the weight. This trend was stopped by developing new vehicle concepts and by using lightweight materials (light metals, high-strength steel and fibre-reinforced plastics) intensively. So the vehicle weight is at a constant or slightly decreasing level. Nevertheless, new lightweight technologies and strategies must be developed to allow a further reduction of the overall vehicle weight.

The combination of fibre-reinforced plastics with steel or steel–aluminium structures has the potential to achieve a larger weight reduction, particularly in the vehicle body. Therefore, it is necessary to develop new manufacturing technologies and to qualify the design components and assemblies with respect to ecological and economical aspects. The research group ForschungsCampus pursues the aim to develop specific design methodologies and innovative production technologies to qualify multi-material design for mass production.

Acknowledgment The research group will develop the new production technologies and design rules within the ForschungsCampus—'New manufacturing technologies for economic and multifunctional lightweight'. In this group, they will elaborate innovative multi-material components and assemblies for the vehicle body. The results of this project will provide basic knowledge in production processes and product designs for lightweight vehicles.

The authors gratefully thank the German Federal Ministry of Education and Research (BMBF) for supporting the Project ForschungsCampus and also the companies involved in this project.

References

1. Wallentowitz H, Freialdenhoven A, Olschewski I (2008) Strategien in der Automobilindustrie—Technologietrends und Marktentwicklungen, 3. Vieweg+Teubner Verlag, Wiesbaden. ISBN 9783834807250
2. Nehuis F, Sánchez Ruelas J, Stechert C, Vietor T (2013) Robust conception of vehicles considering region-specific requirements. In: Proceedings of the 19th international conference on engineering design (ICED'13), Seoul
3. Nationale Plattform Elektromobilität (NPE) (2011) Zweiter Bericht der Nationalen Plattform Elektromobilität. Gemeinsame Geschäftsstelle Elektromobilität der Bundesregierung (GGEMO), Berlin
4. Gies S, Lesemann M, Sahr C, Thoennes M (2008) Research needs and future trends in lightweight design. In: Proceedings of the international conference innovative developments für lightweight vehicle structures, Wolfsburg, Germany
5. Eckstein L, Schmitt F, Hartmann B (2010) Leichtbau bei Elektrofahrzeugen. ATZ Automobiltechnische Zeitung 112:S.788–S.795
6. Duflou J, de Moor J, Verpoest I, Dewulf W (2009) Environmental impact analysis of composite use in car manufacturing. CIRP Ann Manufact Technol 58:9–12

7. Ponn J, Lindemann U (2011) Konzeptentwicklung und Gestaltung technischer Produkte: Optimierte Produkte—systematisch von Anforderungen zu Konzepten. Springer Verlag, Berlin. ISBN 9783540685623
8. Nehuis F, Ibe M, Stechert C, Vietor T, Rausch A (2013) Clustering regional-specific requirements as a methodology to define the modules of a car concept. In: Chakrabarti A (ed) Proceedings of the 22nd CIRP design conference, Bangalore, India, Bd. Sustainable Product Development, pp 239–249. ISBN 9781447145066
9. Kroker J (2005) Schnittstellensystematik für modulare Fahrzeugkarosserien. PhD-Thesis, Technische Universität Braunschweig, Braunschweig
10. Franke H, Crull S, Koschorrek F, Krusche T (2006) Faszination Karosseriebau—Innovationen für zukünftige Karosseriekonzepte. Gesamtzentrum für Verkehr, Braunschweig, pp 13–31. ISBN 3937655034
11. Song Y, Youn J, Gutowski T (2009) Life cycle energy analysis of fiber-reinforced composites. Compos A 40:1257–1265
12. Prinz A, Nehuis F, Vietor T, Stechert C (2012) The effects of regional specific requirements on the development of vehicle concepts. In: Lienkamp M (ed) Proceedings of the conference on future automotive technology (Munich) Bd. Focus Electro Mobility. ISBN 9783658011406
13. Nehuis F, Stechert C, Vietor T (2011) Use of design methodology to accelerate the development and market introduction of new lightweight steel profiles. In: Culley BJ, McAloone TC, Howard TJ, Dong, ASJ, Hicks HSJ (eds) Proceedings of the 18th international conference on engineering design (ICED'11), Kopenhagen, Bd. 10. Design methods and tools, pp 321–330. ISBN 9781904670308

Flying with Cracks, Damage Tolerance and NDT: Potential Contributions of the Fraunhofer Institute for Nondestructive Testing IZFP to Aeronautical Safety

Christian Boller and Eckhardt Schneider

Abstract The application of the damage tolerance design concept together with the continuously improved modelling of light weight structures and the use of updated non-destructive and slightly invasive testing techniques contribute to the high safety standard of aeroplanes. The objective of the applied research and development of the Fraunhofer Institute for Non-Destructive Testing IZFP is the improvement of established testing techniques and the development of new approaches and systems as well as the development of quality and safety assurance concepts based on NDT.

Keywords Clamp load · Hardness, case depth · Mechanical properties · Surface and bulk defects · Defect sizing

Concerning testing of metallic components Fraunhofer IZFP developments include ultrasonic and micromagnetic techniques to evaluate the residual stress states of Al-and steel plates and sheets to be machined for particular parts of the structure. High frequency eddy current and ultrasonic phased array technology are developed to evaluate layer thickness, to characterize surface treated superalloy components and to detect and reconstruct defects. The combination of low frequency eddy current technique with GMR sensors (giant magnetic resistors) yields high defect detectability even in some mm of depth. Electro-magnetic ultrasonic transducers (EMATs) generate guided waves in order to detect corrosion between the deeper layers of GLARE components. A nano-agent based anticorrosion paint allowing corrosion detection in its incubation stage is under test.

The development of NDT techniques to inspect black CFRP parts came into the foreground during the recent years. Piezoelectric sensors for guided waves are developed to be incorporated into the fibre reinforced structure contributing to the Structural Health Monitoring (SHM) concept. An eddy current system is

C. Boller · E. Schneider (✉)
Fraunhofer IZFP, Saarbrücken, Germany
e-mail: Eckhardt.Schneider@izfp.fraunhofer.de

R. P. Bajpai et al. (eds.), *Innovative Design, Analysis and Development Practices in Aerospace and Automotive Engineering*, Lecture Notes in Mechanical Engineering,
DOI: 10.1007/978-81-322-1871-5_4,

developed to test multilayer carbon fibre tissue. Computed X-ray laminography is adapted to inspect large reinforced composite parts with small thickness. A modified Sampling Phased Array ultrasonic technique proved to have high potential to detect and reconstruct defects in Carbon Fibre Reinforced Plastic (CFRP) taking advantage from the modelling of the elastic anisotropy and from a tomography algorithm. The reliable and long lasting bonding between different materials is a challenge. The potential of the optically excited pulse-video thermography and the newly developed multispectral pulse thermography are under investigation together with Nuclear Magnetic Resonance (NMR) technology.

The presentation will briefly address the basics of the techniques considered before describing the applications including rewarding benefits in further detail.

A Synchronous Wound Excitation Transverse Flux Machine with Solid Rotor

Francois Bernot, Alix Bernot and Jean-Claude Vannier

Abstract This research study introduces a new synchronous transverse flux electrical machine, with iron solid rotor, stator in SMC iron powder, and wound excitation attached to the stator. The three-dimensional magnetic flux circulation fully exploits the soft magnetic composites isotropic properties. The steel rotor allows functioning at high speeds and in high ambient temperatures. Following the presentation of the machine, a simple analytical model confirms the concept. A finite element analysis allows a refined model and tests on a prototype motor validate the concept in the end.

Keywords Transverse flux machine · Soft magnetic composites · Wound excitation synchronous machine · Reluctance analytical model · High temperature electrical machine

1 Introduction

Soft magnetic composites are magnetic materials, mainly composed of iron powder, that present several advantages to realize magnetic circuits for electrical machines. The eddy current losses are reduced, and the material is isotropic, allowing a three-dimensional flux circulation. The manufacturing process, through compression and sintering, allows making complex shapes at low cost. However, SMC also have several drawbacks that prevent them from being used more widely: their magnetic permeability and saturation induction are limited, and the hysteresis losses are high, leading to higher iron losses below 2–3 kHz.

F. Bernot · A. Bernot · J.-C. Vannier (✉)
Supélec, plateau du Moulon, 91192 Gif sur Yvette, France
e-mail: Jean-Claude.Vannier@supelec.fr

R. P. Bajpai et al. (eds.), *Innovative Design, Analysis and Development Practices in Aerospace and Automotive Engineering*, Lecture Notes in Mechanical Engineering,
DOI: 10.1007/978-81-322-1871-5_5,

Replacing laminating iron sheets by iron powder-based materials generally leads to a performance degradation of 20–30 %. At low frequencies, the use of the iron powder potential will truly be done by exploring new architectures with three-dimensional magnetic flux circulation. With this in mind, the claw-pole machine is often studied, generally associated with a permanent magnet rotor.

The study starts with a claw-pole machine structure associated with a wound excitation. The aim was to propose a wound excitation synchronous machine with the excitation attached to the stator inducing the flux in the rotor. Thus, the rotor is turned into a unique solid iron piece with very low manufacturing cost and reinforced properties. It can particularly easily rotate at very high speed. All the heat sources are at stator, so the facilitated cooling combined with the absence of magnets will allow this motor to function at high ambient temperatures. The synchronous machine topology was chosen to ensure good torque and power per mass ratios.

2 Machine Presentation

The base element of the transverse flux studied machine is a single-phase claw-pole machine stator. It is made of two SMC half stators, each of which carry the claws in the opposite direction. A homopolar coil winded around the machine rotation axis is inserted within the two claws (Fig. 1).

When a current flows through the winding of this statoric element, a magnetic flux will circulate from one tooth to its neighbor embracing the winding. Each tooth can then be represented as a north or south pole, all the teeth of the same SMC piece having the same polarity. The current flowing through this statoric element will be alternative, synchronous with the rotor movement (Fig. 2).

In the classical machine, the statoric element is associated with a permanent magnet rotor with as many magnets as teeth. The main flux path goes from one tooth to the rotor, advances through the other tooth, and comes back to the original tooth embracing the copper winding (Fig. 3).

Two claw-pole machine stators are aligned in the studied transverse flux machine, separated by a magnetic short-circuit ring, which allows the flux circulation from one stator to the other. The excitation winding is put up under the magnetic short-circuit ring; it is a homopolar simple linear coil winded around the rotation axis of the machine (Fig. 4).

The rotor is a unique solid iron piece that can be extruded. It is a cylinder on which are regularly mounted teeth that have the same width as the stator teeth. There is one rotor tooth for two stator teeth. The shaft is not represented on the picture below; rotor and shaft can be the same part. The rotor does not require very high magnetic properties, so it can be made out of stainless steel (Figs. 5 and 6).

When a DC current flows through the excitation winding, the magnetic flux will circulate in the rotor teeth, climb through a claw-pole statoric element, cross the magnetic short-circuit ring, and come back in the rotor at the opposite side through

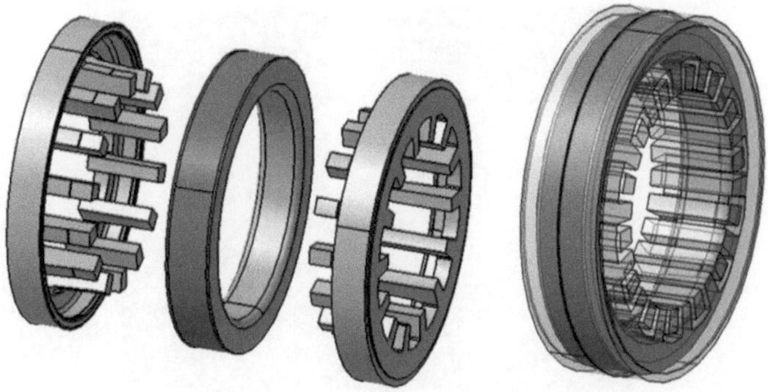

Fig. 1 Claw-pole machine stator, base element of the studied transverse flux machine

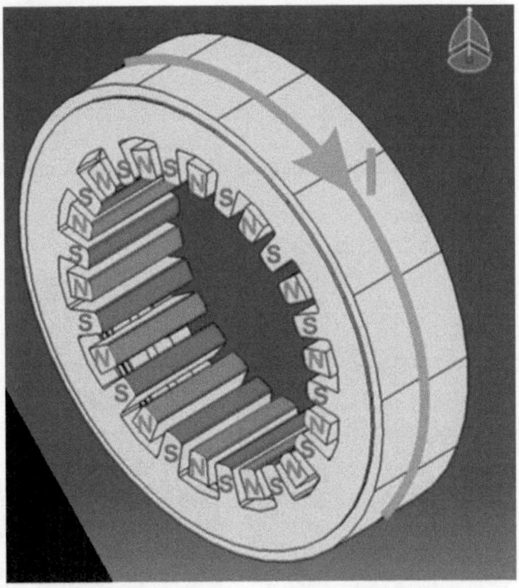

Fig. 2 Magnetization generated by the current through a claw-pole statoric element

the other claw-pole statoric element. When the rotor rotates, the magnetic flux will always flow in the same direction, and the rotor teeth will always keep the same polarity: north pole on the one side and south pole on the other side. The two claw-pole statoric elements windings are connected in series, an opposite current flowing through them. This current is synchronous with the rotor movement, magnetizing the stator teeth synchronously with the rotor teeth, and making up a single-phase wound excitation synchronous machine. The interaction between the flux generated by the excitation winding and flux generated by the claw-pole statoric element winding generates the electromagnetic torque. The current flowing through the excitation winding is DC current, so the rotor teeth magnetization will never change (Fig. 7).

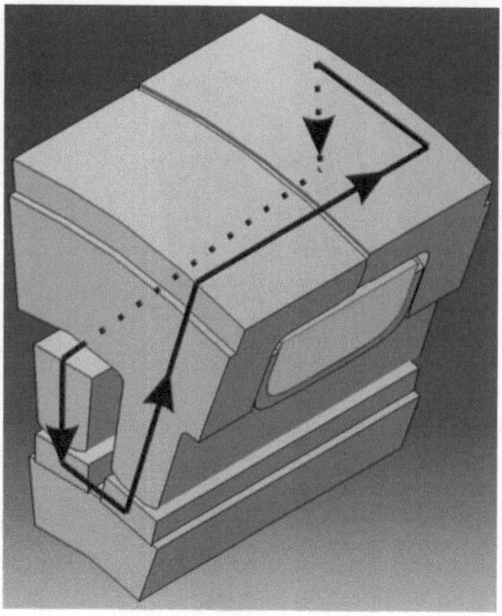

Fig. 3 Permanent magnet claw-pole machine (detail of one pair of poles)

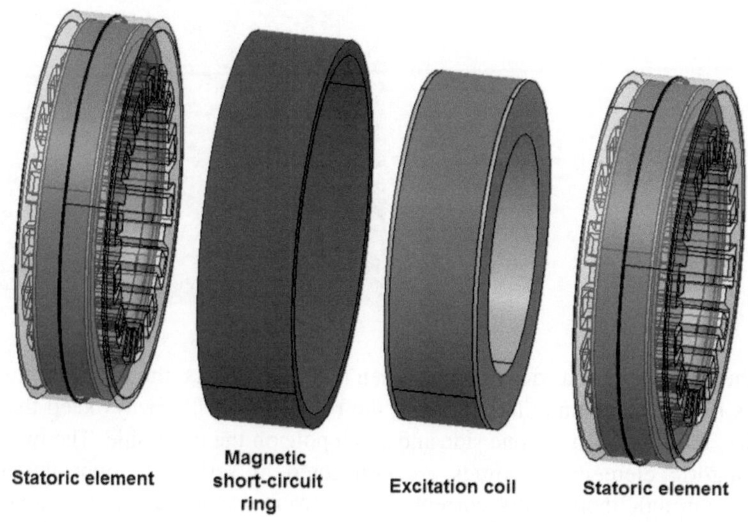

Fig. 4 Complete assembly of the studied transverse flux machine stator

3 Simplified Analytical Model

To make a simplified analysis of the machine, we consider that the iron (both SMC stator and stainless steel rotor) has an infinite magnetic permeability. The study is reduced to one pair of poles, studying one rotor tooth and two stator teeth per

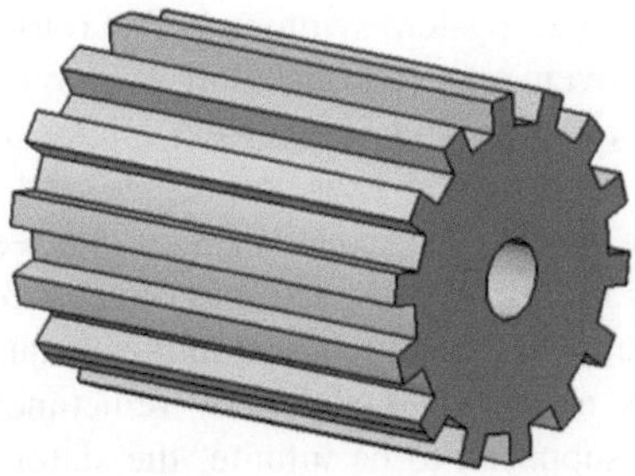

Fig. 5 Rotor of the studied transverse flux machine

Fig. 6 Complete assembly of the studied transverse flux machine

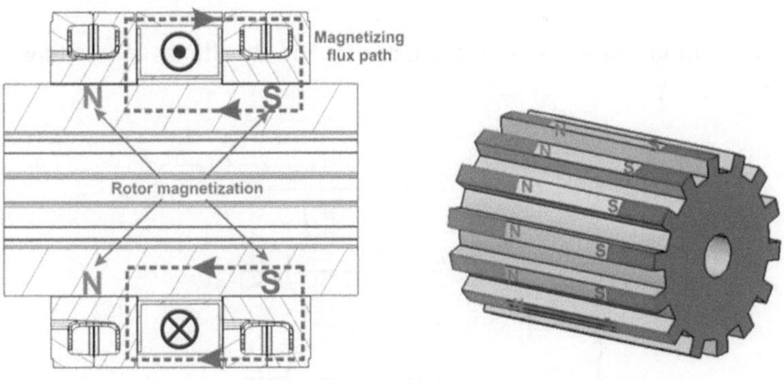

Fig. 7 Flux path and rotor magnetization in presence of an excitation current

statoric element, thanks to the system symmetry. The rotor curve is also neglected, assimilating the rotor movement to a translation. We consider that the flux comes out of the stator teeth radially, and we neglect all the flux pathes that are not beneath a stator tooth. The flux crosses the airgap and enters the rotor, either through a rotor tooth or through the space between two teeth. The airgap between each statoric element is in this way divided into three or four areas, depending on the rotor position; we suppose that the induction is constant in each of these areas.

These magnetic flux areas are modeled by reluctances. Since the magnetic permeability of iron is supposed to be infinite, the stator and the rotor lead perfectly the magnetic flux without any magnetic potential drop. The coil currents are magnetomotive forces, which are inserted in the reluctance diagram. The current flowing in the claw-pole statoric elements generate magnetomotive forces named nI1 and nI2 (with nI1 + nI2 = 0 because the coils are connected in series), and the current flowing in the excitation winding generates a magnetomotive force named nIf (Figs. 8 and 9).

To simplify the calculation, the pairs of reluctance in parallel are replaced by a single equivalent reluctance, leading to a unified magnetic circuit. The final system has four unknown parameters, the magnetic fluxes. Gauss's law for magnetism around the rotor is expressed (the sum of the fluxes is zero) and Ampère's law is expressed on the three simple loops, leading to four independent equations. The loops chosen for Ampère's law are pictured with dashed green lines below (Fig. 10).

The four equations governing the system are:

$$\begin{cases} \varphi_1 + \varphi_2 + \varphi_3 + \varphi_4 = 0 \\ R_{\text{éq1}} * \varphi_1 + nI_1 - R_{\text{éq2}} * \varphi_2 = 0 \\ R_{\text{éq2}} * \varphi_2 + nI_f - R_{\text{éq1}} * \varphi_3 = 0 \\ R_{\text{éq1}} * \varphi_3 + nI_2 - R_{\text{éq2}} * \varphi_4 = 0 \end{cases}$$

The resolution leads us to the expression of the fluxes depending on the currents:

$$\begin{cases} \varphi_1 = \dfrac{-nI_1\left(\frac{1}{R_{\text{éq1}}}+\frac{2}{R_{\text{éq2}}}\right)-nI_f\left(\frac{1}{R_{\text{éq1}}}+\frac{1}{R_{\text{éq2}}}\right)-\frac{nI_2}{R_{\text{éq2}}}}{2\left(1+\frac{R_{\text{éq1}}}{R_{\text{éq2}}}\right)} \\ \varphi_2 = \dfrac{\frac{nI_1}{R_{\text{éq1}}}-nI_f\left(\frac{1}{R_{\text{éq1}}}+\frac{1}{R_{\text{éq2}}}\right)-\frac{nI_2}{R_{\text{éq2}}}}{2\left(1+\frac{R_{\text{éq2}}}{R_{\text{éq1}}}\right)} \\ \varphi_3 = \dfrac{\frac{nI_1}{R_{\text{éq1}}}+nI_f\left(\frac{1}{R_{\text{éq1}}}+\frac{1}{R_{\text{éq2}}}\right)-\frac{nI_2}{R_{\text{éq2}}}}{2\left(1+\frac{R_{\text{éq1}}}{R_{\text{éq2}}}\right)} \\ \varphi_4 = \dfrac{\frac{nI_1}{R_{\text{éq1}}}+nI_f\left(\frac{1}{R_{\text{éq1}}}+\frac{1}{R_{\text{éq2}}}\right)+nI_2\left(\frac{2}{R_{\text{éq1}}}+\frac{1}{R_{\text{éq2}}}\right)}{2\left(1+\frac{R_{\text{éq2}}}{R_{\text{éq1}}}\right)} \end{cases}$$

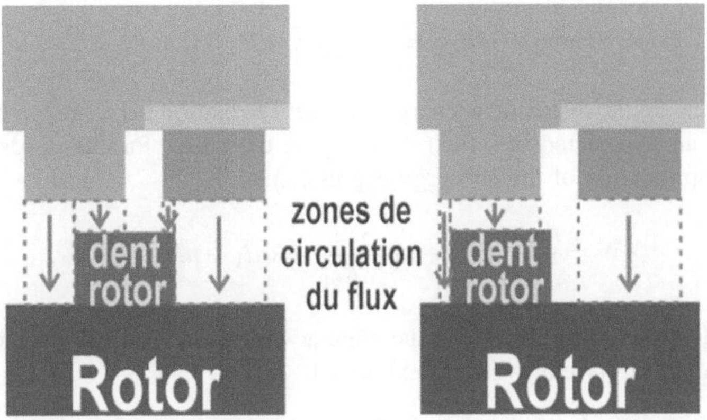

Fig. 8 Magnetic flux circulation areas

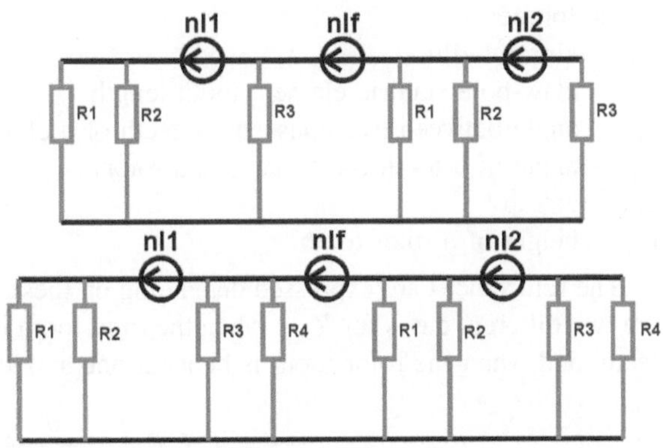

Fig. 9 Equivalent magnetic circuit of the studied transverse flux machine Rotor tooth beneath one (*left*) or two (*right*) stator teeth

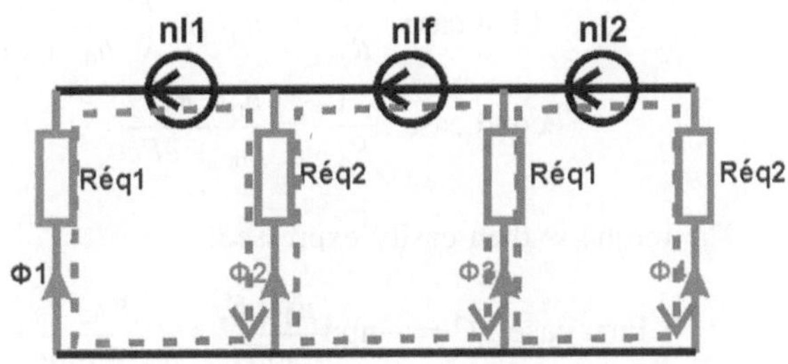

Fig. 10 Equivalent magnetic circuit of the studied transverse flux machine

To calculate the torque, the overall magnetic coenergy of the machine is expressed. npp is the number of pair of poles, introduced because the diagram represents only one pair of poles, a fraction of the total machine. The expression of the overall magnetic coenergy is:

$$W = \text{npp} * \frac{1}{2}\left(R_{\text{éq1}}\varphi_1^2 + R_{\text{éq2}}\varphi_2^2 + R_{\text{éq1}}\varphi_3^2 + R_{\text{éq2}}\varphi_4^2\right)$$

To calculate the magnetic coenergy, we use the relation nI1 + nI2 = 0 (the two windings are connected in series). Developing the above formula leads to a dramatic simplification of the coenergy expression:

$$W = \frac{\text{npp}}{4}\left(\left(nI_1 + nI_f\right)^2 \frac{1}{R_{\text{éq1}}} + \left(nI_1 - nI_f\right)^2 \frac{1}{R_{\text{éq2}}}\right)$$

The electromagnetic torque is the angular differentiate of the magnetic coenergy at constant current, which is expressed:

$$C = \frac{\text{npp}}{4}\left(\left(nI_1 + nI_f\right)^2 * \frac{\partial}{\partial\theta}\left(\frac{1}{R_{\text{éq1}}}\right) + \left(nI_1 - nI_f\right)^2 \frac{\partial}{\partial\theta}\left(\frac{1}{R_{\text{éq2}}}\right)\right)$$

The geometric parameters of the machine are named to develop the expression of the torque:

Re airgap radius
Lg claw-pole statoric element tooth length
α_{id} angle between two consecutive teeth of a claw-pole statoric element
α_{d} angle of a tooth (both stator and rotor)
eFe airgap
h_{dr} height of a rotor tooth.

The reluctances are expressed depending on these geometric parameters. There are two different cases for $R_{\text{éq2}}$: when the rotor tooth is beneath two different stator teeth, and when the rotor tooth is beneath one single stator tooth.

$$\frac{1}{R_{\text{éq1}}} = \mu_0 L_g R_e \left(\frac{\theta}{h_{\text{dr}} + eFe} + \frac{\alpha_d - \theta}{eFe}\right)$$

$$\text{First case}: \frac{1}{R_{\text{éq2}}} = \mu_0 L_g R_e \left(\frac{\alpha_d + \alpha_{\text{id}} - \theta}{h_{\text{dr}} + eFe} + \frac{\theta - \alpha_{\text{id}}}{eFe}\right)$$

$$\text{Second case}: \frac{1}{R_{\text{éq2}}} = \frac{\mu_0 L_g R_e \alpha_d}{h_{\text{dr}} + eFe}$$

The torque is then easily expressed:

$$\text{First case}: C = -\text{npp}\frac{\mu_0 L_g R_e}{eFe} * \frac{h_{\text{dr}}}{h_{\text{dr}} + eFe} * nI_1 * nI_f$$

$$\text{Second case}: C = -\text{npp}\frac{\mu_0 L_g R_e}{eFe} * \frac{h_{\text{dr}}}{h_{\text{dr}} + eFe}\left(\frac{nI_1 * nI_f}{2} + nI_f^2 + \frac{nI_1^2}{4}\right)$$

The first case is purely the couple of a perfect synchronous machine: $\frac{\mu_0 * nI_f}{2 * eFe}$ is the induction in the airgap generated by the excitation winding (retrieved with

Ampère's law), which multiplied by $L_g * R_e$ (airgap surface) is proportional to the overall excitation flux. The torque is proportional to the number of pair of poles, and the currents are independent from the number of pair of poles because the windings are homopolar (collinear with the rotation axis), so they are common to all the machine poles. This torque gain is limited by the maximum frequency in the magnetic materials. $\frac{h_{dr}}{h_{dr}+eFe}$ is a geometric parameter between 0 and 1 which expresses that without rotor teeth (smooth rotor) there is no torque. Its value is around 1 when the airgap is small compared to the rotor teeth's height.

The torque is proportional to $\frac{1}{eFe}$ because the iron has an infinite magnetic permeability and does not saturate, so a very small airgap leads to very high induction in this model. Introducing an actual B(H) curve for both the rotor steel and the stator iron will lead to a natural torque limitation.

In the second case, the term in nI_f^2 is the cogging torque due to the excitation current, which average is zero one full period. The term in $\frac{nI_1^2}{4}$ is the cogging torque due to the winding in the claw-pole statoric elements (armature current). The term in $\frac{nI_1 * nI_f}{2}$ is the synchronous couple, half the other one because only one stator tooth out of two is used to generate torque. The volume of the excitation winding also does not generate torque, so the overall torque of the studied transverse flux machine will be around 40 % of the torque of a standard permanent magnet claw-pole machine in the same volume. The improvement of the structure of the machine by introducing two rotor teeth without any magnetic short circuit could lead up to a 100 % torque increase.

4 Finite Element Modeling

The machine is simulated with the finite element methods, with the software Flux3D. The magnetic materials are modeled by their first magnetization curves that were experimentally measured. The chosen material for the rotor is X20Cr13 1.4021 stainless steel, and the chosen material for the claw-pole statoric elements and the magnetic short-circuit ring is Somaloy 1000 3P from Höganäs (Fig. 11).

The reference statoric element has an 80 mm outer diameter and 20 mm length, because they are the dimensions of the standard SMC blanks supplied by Höganäs. A prototype motor has been manufactured later and tested with these dimensions to validate the performances. The airgap is 0.25 mm. The teeth are 3.6 mm wide, and the claw-pole statoric element interteeth space is 1.8 mm wide with 14 pair of poles. The teeth height is 4.5 mm at the edge. The motor outer frame and the magnetic short-circuit ring have a 3 mm height. The claw-pole statoric element has a 10 mm × 6 mm cross section, with a 1 mm insulation distance with the SMC parts. The two claw-pole statoric element coils have 120 turns of copper wire of 0.3 mm diameter, and the excitation coil has 250 turns of copper wire of 0.5 mm diameter. The magnetic short-circuit ring is 10 mm long.

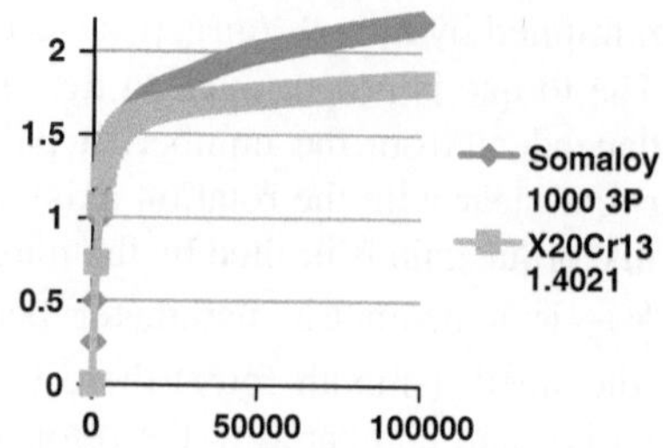

Fig. 11 Magnetization curves of the magnetic materials

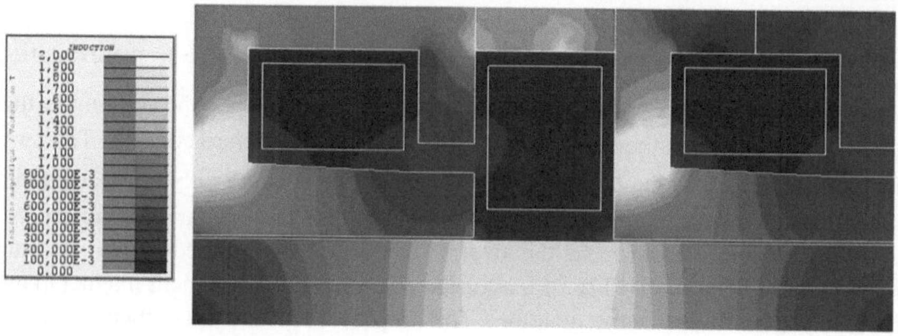

Fig. 12 Induction for nIf = 500 A.turns

The simulation was done on one pole pair (two stator teeth per statoric element, one rotor tooth), exploiting the symmetry of the motor. The induction and torque were calculated in one electrical turn. The simulation is done for a sine current in phase with the stator rotation (nI1 = 0 A when the stator teeth are aligned with the rotor teeth) (Fig. 12).

The major problem of the architecture is the saturation of the root of the stator teeth. The inductions are very high (up to 1.9 T) while the SMC has very low magnetic permeability at such inductions and is almost saturated. An improvement of the surface for magnetic flux passage in the root of the stator tooth would lead to a dramatic increase of the output torque and is one of the next steps to improve the studied transverse flux machine.

The simulation results are interesting and sometimes counter-intuitive. At lower excitation currents (If = 1 A), an armature current increase will lower the torque. The torque period is half the mechanical period, which is due to the motor symmetry. The cogging torque is very high (100 %), because the motor is single phased. The torque is not proportional to the current: when the armature current doubles, the torque increase is 30 % at If = 2 A and 65 % at If = 3 A. With an armature I1 = 1.5 A, an excitation current also leads to a torque decrease. These results may be due to the high saturation and the leakage fluxes (Fig. 13).

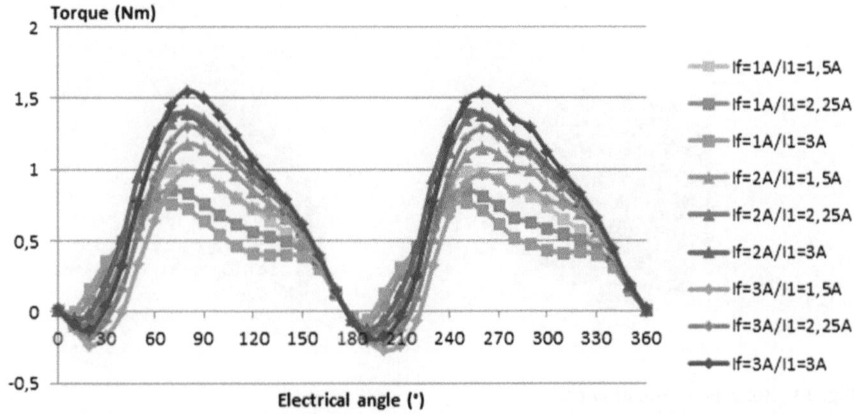

Fig. 13 Finite element model torque

Table 1 Finite element model average torque

Average torque (Nm)	If = 1 A	If = 2 A	If = 3 A
I1 = 1.5 A	0.508	0.538	0.437
I1 = 2.25 A	0.461	0.664	0.588
I1 = 3 A	0.404	0.705	0.714

Table 2 Simplified analytical model average torque

Average torque (Nm)	If = 1 A	If = 2 A	If = 3 A
I1 = 1.5 A	0.669	1.339	2.008
I1 = 2.25 A	1.004	2.008	3.012
I1 = 3 A	1.339	2.677	4.016

The simplified analytical model proposed in part **3** leads to the following average torque:

The simplified analytical model is very optimistic for high currents (up to 500 % above finite element model) because it does not consider magnetic saturation. For lower currents, the simplified model seems to give a good rough estimate of the torque (Tables 1 and 2).

The strange torque/current behavior is partly due to the very high magnetic saturation of the SMC stator. An optimization of the magnetic circuit will increase the performances in the dramatic way, and the motor behavior should be closer to that of a classical linear wound excitation synchronous motor with the torque proportional to both excitation current and armature current, as in the simplified analytical model.

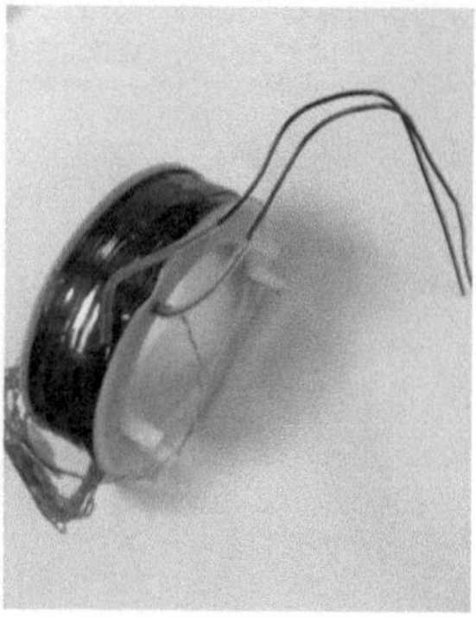

Fig. 14 Rotor and excitation coil

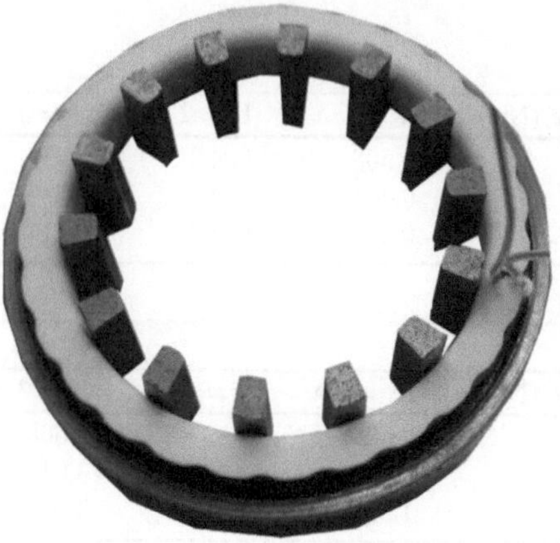

Fig. 15 Half claw-pole statoric element with its winding before assembly (*left*), Claw-pole statoric element assembled and resined (*right*)

5 Test of the Prototype Machine

A prototype machine has been made with the dimensions of the machine simulated with finite elements method. The SMC claw-pole statoric elements were machined in Höganäs Somaloy prototyping material. It has been turned and wire cut to obtain the final shape. The magnetic short-circuit ring was also wire-cut with Somaloy prototyping material, and the rotor was machined. Each coil has been wound on a mandrel and inserted manually. The claw-pole statoric elements were filled with resin under vacuum. The claw-pole statoric elements have been aligned with small pins integrated on the excitation coil mandrel. All the stator elements have finally been inserted in an outer aluminum frame with a press (Figs. 14, 15 and 16).

Fig. 16 Motor mounted on the test bench (position sensor on the *left*)

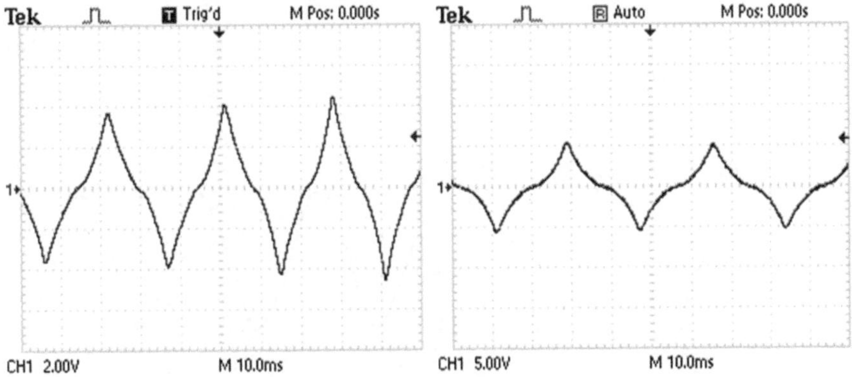

Fig. 17 Electromotive force, If = 1.5 A (*left*) and If = 3 A (*right*)

The machine is first rotated with an external system, while the excitation winding is powered with regulated DC current. The voltage between the claw-pole statoric elements coils connected in series is measured.

The electromotive force is not a sine wave; its shape will depend mainly on the teeth shape. The teeth shape can later be optimized to improve the electromotive force harmonics. The electromotive force is 0.18 V/Hz with an excitation current of 3 A and 0.11 V/Hz with an excitation current of 1.5 A. This is a 63 % voltage increase when the excitation current doubles (Fig. 17).

The machine is then powered by a three-phase IGBT inverter which has a 250 V DC bus. A position sensor is mounted on the shaft. The machine is powered with a single-phase sine voltage synchronous with the rotor position. The angle between the rotor position and the voltage is adjusted experimentally to achieve the best torque/current ratio. The excitation winding is powered with a regulated DC current. The torque measurements have been done at 150 rpm. The machine

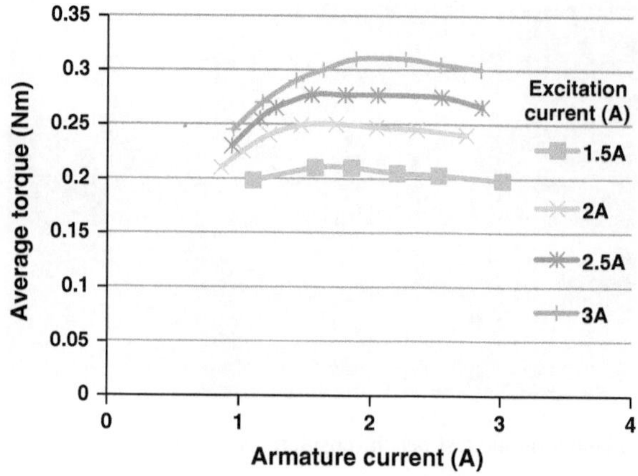

Fig. 18 Measured torque

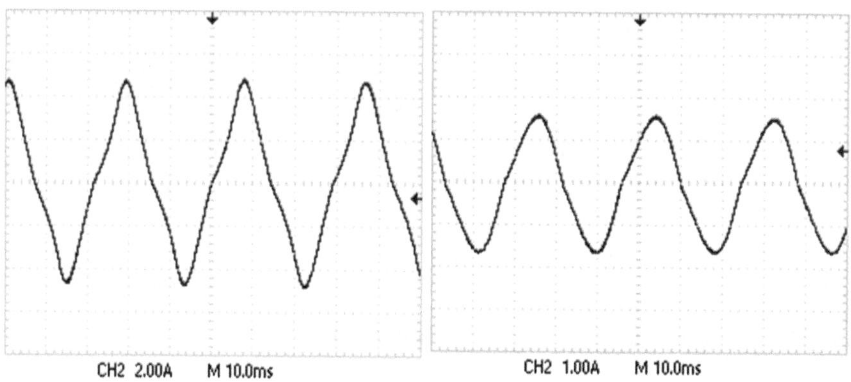

Fig. 19 Armature Current; the excitation current is 1.5 A (*left*) and 3 A (*right*)

cannot start alone (it is a single-phase machine), so it had to be rotated at a sufficient speed first to be able to rotate by itself.

The torque measured decreases when the armature current increases too much, we retrieve here the behavior simulated with the finite element model. This is probably linked with the magnetic material high saturation. An increase in excitation current leads to a smaller torque gain. The results are very different from the finite element model, with a difference above 50 % (Fig. 18).

The current shape has also been measured. It is not sinusoidal and it is more regular when the excitation current increases (Fig. 19).

6 Conclusion and Perspectives

The studied transverse flux machine has several advantages. The manufacturing cost of the whole machine is less, especially with a low winding cost and a small number of parts. Its major interest is the simplicity of the rotor, which is a solid stainless steel piece. It allows to reach very high speeds and to function at high ambient temperatures (300 °C). All the heat sources are at the stator, so cooling is facilitated. The machine also has the advantages of the wound excitation, in terms of safety and ease of control. The performances measured are small, especially the torque/volume ratio, but should rise to levels comparable with other wound excitation synchronous machines after optimization. The mockup has validated the concept of this synchronous wound excitation transverse flux machine.

On the other hand, several drawbacks must be mentioned. The efficiency is limited by higher copper losses due to the excitation winding and higher iron losses due to high number of pair of poles. The torque ripple is also high; since the machine is naturally a single-phase machine, associating several single-phase machines will lead to a multiphase machine with less torque ripple and the possibility to start in every position.

Several trails of improvement have been drawn. A geometric optimization must be performed, especially on the claw-pole statoric elements to reduce the saturation at the root of the teeth. The architecture can also be improved to propose a system with the same number of rotor teeth as stator teeth to increase the torque. The teeth shapes (both stator and rotor), especially at the airgap side are to be optimized to improve the voltage and current harmonics.

R&T Effort Led by MBDA in the Field of Aero-propulsion

Francois Falempin

Abstract MBDA is leading an important research and technology effort on aero-propulsion covering a large set of propulsion technologies from high-speed dual-mode ramjet to detonation wave propulsion and hybrid electrical/chemical propulsion. MBDA is also working on the key issue of thermal management and developed innovative approach to control external flow thanks to near surface electrical discharges.

Keywords Aeropropulsion · High-speed airbreathing propulsion · Detonation · Hybrid propulsion · Flow control · Electrical discharge

Thanks to permanent Research and Technology (R&T) effort and innovation over the past decades, MBDA is today one of the leaders in the field of missile systems. Competition being more and more aggressive to hold that position, MBDA pursues a very intensive R&T effort covering a very large panel of disciplines and technologies. As part of that, R&T activities related to aero-propulsion and powered airframe are very intensive and address different possible options for powering (propulsion and on-board energy) future missiles:

- High-speed air-breathing propulsion,
- Detonation wave propulsion,
- Plasma flow control,
- Hybrid air-breathing propulsion,
- Energy recovery and storage.

In the field of high-speed air-breathing propulsion, the main effort is to extend the flight domain of ramjet-powered systems by mastering the dual-mode ramjet technology. Based on past efforts within the framework of several R&T programs,

F. Falempin (✉)
MBDA, 1 Avenue Réaumur, Le Plessis-Robinson 92350, France
e-mail: francois.falempin@mbda-systems.com

R. P. Bajpai et al. (eds.), *Innovative Design, Analysis and Development Practices in Aerospace and Automotive Engineering*, Lecture Notes in Mechanical Engineering,
DOI: 10.1007/978-81-322-1871-5_6,

which will be briefly mentioned in this chapter, the effort today is focused on two key issues:

On one hand, technologies related to the dual-mode ramjet combustion chamber are developed to ensure its performance and its thermal and mechanical strength:

- Variable geometry needed to optimize the performance,
- Fuel used as coolant for combustion chamber structure,
- Fuel-cooled structure itself.

On the other hand, and before any operational application (civilian or military), it is mandatory to demonstrate our ability to predict the aero-propulsive balance (generalized thrust-minus-drag balance) of a hypersonic vehicle, providing sufficient margins prior to start a costly full-scale development program. Considering this mandatory step, MBDA France and ONERA are leading a specific flight test program, called LEA, aiming at:

- Defining a methodology for the development of a hypersonic vehicle using ground tests and numerical simulation,
- Developing the tools (experimental or numerical) required for this purpose,
- Applying this methodology to the development of a simplified and scientific experimental vehicle,
- Validating this methodology through a series of flight tests to be performed in Russia.

Beside these effort on technology development, some system studies are performed, e.g., within the European LAPCAT 2, ATLLAS 2, and HIKARI programs aiming at preliminary designing a high-speed civilian transport airplane able to perform very long range and acquiring some associated necessary knowledge.

In the field of Detonation Wave Propulsion, MBDA is developing some specific and innovative concepts for both Pulsed Detonation Engine (PDE) and Continuous Detonation Wave Engine (CDWE). A dedicated test bench is under preparation at MBDA ramjet test facility allowing to test a complete PDE system able to power a small UAV and providing specific mission capabilities which is studied in cooperation with Singaporean DSO. In parallel, a large-scale demo is under development for the application of CDWE concept to hydrocarbon fuel in air-breathing mode. Such development could lead to large improvement of performances for both turbojet and ramjet systems. Based on previous works performed with Lavrentiev Institute in Novosibirsk (Siberian Branch of Russian Academy of Science), the program is now performed with a contribution from the Institute Von Karman in Belgium.

For high-speed cruising systems, the thermal management is a key point and some dedicated activities are led by MBDA to determine the best options in terms of technology components and system design to optimize the performance of the thermal management system on-board future products. As a part of that, a specific research line is dedicated to flow control using near surface electrical discharge. After basic testing performed at Joint Institute for High Temperature in Moscow, a large-scale supersonic air intake has been tested at the Institute of Theoretical and

Applied Mechanics in Novosibirsk demonstrating the possibility to control the shock-on-lip Mach number by providing electrical discharge on the compression ramp. Other applications of near surface electrical discharge such as, i.e., the control of the boundary layer transition on a forebody.

Another direction of research is aiming at developing a hybrid air-breathing system taking benefit of recent and huge progress made in the domain of electrical motors and electrical energy storage systems. The first demonstration of a complete hybrid engine will be performed with MBDA test facility in 2014 resulting from a specific effort dedicated to the development of innovative solutions for electrical motors led in cooperation with French SME FranceCol. By another way, some cooperation is led with different suppliers to develop the needed hybrid energy storage as, for example, works made with Prollion in France.

The presentation will describe all these research works and will give a brief status on each of them.

Technological Trends in the Application of Advanced Materials in Automobiles

P. K. Mallick

Keywords Advanced steel · Light alloys · Composites · Weight reduction · Fuel economy

This keynote presentation will address the technological trends in the application of advanced materials in automobiles. It will start with a brief discussion on the vehicle fuel economy requirements and methods of meeting these requirements. It will review the current materials scenario and the application of advanced steels, light alloys, and carbon fiber composites. Challenges in meeting vehicle weight reduction using advanced materials are discussed. The effect of weight reduction on fuel economy and environment is presented. The application of advanced materials in Indian automobiles is discussed.

P. K. Mallick (✉)
Center for Lightweighting Automotive Materials and Processing, University of Michigan-Dearborn, Dearborn, MI 48128, USA
e-mail: pkm@umich.edu

R. P. Bajpai et al. (eds.), *Innovative Design, Analysis and Development Practices in Aerospace and Automotive Engineering*, Lecture Notes in Mechanical Engineering,
DOI: 10.1007/978-81-322-1871-5_7,

Recent Developments of Mechanical and Fatigue Analyses of Fiber-Reinforced Structures for Aerospace Applications

Michael Magin

Abstract Recent generations of passenger aircraft show an increased use of fiber-reinforced composite materials for structural components due to their outstanding weight-specific strength and stiffness and fatigue resistance. These materials allow higher load capacity, high functional integration, and weight reduction. Notwithstanding their high performance in the areas mentioned, fatigue, fracture, and impact resistance as well as out-of-plane properties require particular attention in the design process. A significant number of aerospace applications need fiber-reinforced materials in the form of thin-walled structures, allowing the use of relatively simple analysis techniques for the majority of load cases. Thick-walled structures and fatigue, impact, and out-of-plane load cases necessitate advanced models, which are still subject of intensive research. A newly developed fatigue analysis code, based on the Critical Element concept [1], uses nonlinear material laws and failure models to predict damage evolution, stress state, and failure of carbon-fiber-reinforced structural components under cyclical fatigue loading. Experimental studies were used to determine the required material laws and failure models and a finite element analysis enabled the validation of the procedure for mechanical components with complex stress states in the vicinity of a cut-out. Computational tomography and X-ray analyses accompanying cyclical tests confirmed the validity of failure and life prediction of the code.

Keywords Fatigue · Finite element analysis · Composites · Critical element model

M. Magin (✉)
Institut für Verbundwerkstoffe, Erwin-Schrödinger-Str. Geb. 58 67663
Kaiserslautern, Germany
e-mail: michael.magin@ivw.uni-kl.de

R. P. Bajpai et al. (eds.), *Innovative Design, Analysis and Development Practices in Aerospace and Automotive Engineering*, Lecture Notes in Mechanical Engineering,
DOI: 10.1007/978-81-322-1871-5_8,

1 Introduction

Recent generations of passenger aircraft show an increased use of fiber-reinforced composite materials for structural components due to their outstanding weight-specific strength and stiffness. These materials allow higher load capacity, high functional integration, and weight reduction. Fiber-reinforced composites own outstanding fatigue resistance compared to metals, but analytical procedures for their design are remarkably fewer than those existing for metal fatigue.

Notwithstanding their high performance in the areas mentioned, fatigue, fracture, and impact resistance as well as out-of-plane properties require particular attention in the design process. For a quasi-static design, physically based fracture criteria already exist, whereas analysis models to quantify the complex fatigue failure process are still subject of intensive research.

These fatigue analysis models will be needed when higher requirements are made for the design of composite structures to increase the overall performance of aircraft structures.

2 Analysis Models for Mechanical and Fatigue Analysis of Fiber-Reinforced Structures

2.1 Quasi-Static Mechanical Analysis and Failure Criteria

The quasi-static mechanical analysis is based on the material law, e.g., *Young's* modulus, and the geometry to compute deformation, stress, and strain.

A proven nonlinear material law is the Ramberg–Osgood equation to describe the stress–strain relation in an unidirectional lamina in *i*-*j*-material orientation, the indices indicating fiber parallel or perpendicular direction, E is *Young's* or the shear tangent modulus (Index t) in the quasi-static (Index 0) test, and ξ and n fitting parameters for experimental results.

$$\varepsilon_{ij}\left(\sigma_{ij}\right) = \frac{\sigma_{ij}}{E_{t,0,ij}} + \xi_{ij}\left(\frac{\sigma_{ij}}{R_{ij}}\right)^{n_{ij}} \tag{1}$$

Fracture criteria are used to determine the limiting load in terms of strength of the material. For fiber-reinforced composites structures, two classes of failure criteria exist [2]: generalized or global criteria and the differentiating criteria. Failure criteria of the first group, for example the Tsai/Wu criterion, compute the failure load of the laminate without giving additional information about the underlying stress component. They do not allow identifying the "weak" part of the structure, breaking of fibers (fiber failure, FF) necessitating a reinforcement of on-axis plies, or the development of matrix cracks in off-axis laminae (interfiber failure, IFF), hence requiring additional reinforcement of these plies. Differentiating criteria, on

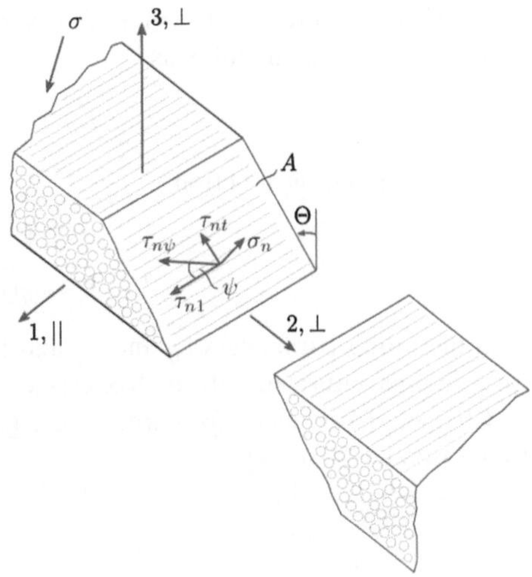

Fig. 1 Fracture plane *A* and governing stresses in a single ply of a composite laminate

the other hand, allow the distinction between FF and IFF. Puck's failure criterion, a differentiating criterion with interaction, computes the interaction between stress components $\sigma_\perp$ perpendicular (Index $\perp$) to the fiber direction and shear stress components $\sigma_{\perp||}$ or $\sigma_{\perp\perp}$ (Index || indicating fiber parallel direction). The criterion iteratively analyzes the stress components on possible fracture planes, or action planes (Index *A*), perpendicular to the fiber direction but inclined by the angle Θ to the axis of the thickness direction.

The interfiber failure effort $f_{\text{IFF}}(\Theta)$ is a function of the strength of the action plane $R^A_{\perp\psi}$, strength of the material with respect to shear loading $R_{\perp||}$, or fiber-perpendicular loading in tension (Index t) or compression (Index c), $R^t_\perp$ respectively $R^c_\perp$, the normal stress σ_n, the angle ψ of the resultant of the action plane normal shear stress τ_{nt} and shear stress in 1-direction τ_{n1}, Fig. 1, and a fitting factor $p^t_{\perp\psi}$ or $p^c_{\perp\psi}$ describing the strength in the transition zone from fiber-perpendicular stress to shear stress. For fiber-perpendicular tension stress, the effort is defined as

$$f_{\text{IFF}}(\Theta) = \sqrt{\left[\left(\frac{1}{R^t_\perp} - \frac{p^t_{\perp\psi}}{R^A_{\perp\psi}}\right)\sigma_n\right]^2 + \left(\frac{\tau_{nt}}{R^A_{\perp\perp}}\right)^2 + \left(\frac{\tau_{n1}}{R_{\perp||}}\right)^2} + \frac{p^t_{\perp\psi}}{R^A_{\perp\psi}}\sigma_n \tag{2}$$

and for fiber-perpendicular compressive stress as

$$f_{\text{IFF}}(\Theta) = \sqrt{\left(\frac{\tau_{nt}}{R^A_{\perp\perp}}\right)^2 + \left(\frac{\tau_{n1}}{R_{\perp||}}\right)^2 + \left(\frac{p^c_{\perp\psi}}{R^A_{\perp\psi}}\sigma_n\right)^2} + \frac{p^c_{\perp\psi}}{R^A_{\perp\psi}}\sigma_n \tag{3}$$

Interfiber failure results in stiffness decrease which can be computed for *Young's* and shear modulus as

$$E_{ij} = \eta_{ij} E_{0,ij} \tag{4}$$

using a degradation function

$$\eta_{ij} = \frac{1 - \eta_{R,ij}}{1 - c_{ij}(f_{IFF} - 1)^{\chi_{ij}}} \tag{5}$$

with the fitting parameters c_{ij} and χ_{ij} and the maximum stiffness degradation $\eta_{R,ij}$ for a lamina saturated with matrix cracks.

Characteristic values for carbon- and glass-fiber-reinforced composites can be found in VDI 2014 [2].

2.2 Fatigue Analysis of Fiber-Reinforced Structures

Basis of a significant part of existing finite element fatigue analysis programs are analytical fatigue models describing the cycle-dependent degradation of relevant material parameters due to cyclical fatigue loading. Degrieck and van Paepegem introduced in a review article [3] a classification of existing models in three categories: fatigue life, phenomenological, and progressive damage models.

Fatigue life models are among the first models developed, originally used for iron and steel, e.g., Wöhler's method to measure bending cycles to failure with regard to the governing loading [4], and the subsequently derived forms of stress fatigue life cycle diagrams, also used for fiber-reinforced composites (FRP). These fatigue life models lack consideration of the fatigue-induced damage mechanisms.

Phenomenological models, in contrast to fatigue life models, take the damage development with respect to the fatigue life into account. Examples are residual strength, stiffness degradation, or crack density models.

Progressive damage models sometimes use comparable representations as the previously mentioned models, but interactively include different damage mechanisms and represent the interaction of the individual damage mechanisms in the complex failure process of the FRP laminate. The critical element model [1], developed by Reifsnider and Stinchcomb, was included in a finite element-based program to predict fatigue failure of carbon-fiber-reinforced composite (CFRP) with complex stress states at the vicinity of a drill hole, commonly found in structural applications of CFRP materials [5].

2.2.1 Fatigue Life Models

S-N-curves to predict cycles to failure are only applicable to uniform loading, e.g., sinusoidal cyclic fatigue loading. For FRP applications, a modified form, e.g., Sendeckyi's formula [6], having a smooth transition from quasi-static strength to the low-cycle fatigue regime, are recommended [2].

To assess the differences in fatigue life due to the influence of reigning tension or compression loading regimes, *Goodman* diagrams and variations thereof are commonly employed, their applicability determined to a large extent by the quality of the underlying experiments and the fitting procedures chosen [7]. Although some developed fatigue life models are the bases for enhanced models and are used for the design of FRP structures, e.g., in the Germanischer Lloyd "Guideline for the Certification of Wind Turbines" [8] or the *VDI* guidelines for the analysis of FRP structures [2], the missing consideration of damage mechanisms requires further research to establish simple but reliable models which allow a validation using a reduced number of specimens within a limited testing procedure.

2.2.2 Residual Strength Models

Residual strength models consider the loading history and presume a monotone decrease of the quasi-static strength of a structure after cyclic fatigue loading of the part. Hahn and Kim [9] proposed the existence of a "unique relationship" between static strength and fatigue life, expressed as "strength-life equal rank assumption" by Chou and Croman [10]. If this assumption is valid, the residual strength can be expressed using a deterministic mathematical function of the number of cycles n. Thereby, the residual strength function possesses an inherent failure criterion, predicting the fatigue-induced failure when the acting stress in a certain volume of the part reaches or exceeds the momentarily governing residual stress [11]. This implies that a certain level of residual stress is reached when a certain damage state is present in the material, independent of the loading history. Although these models possess an important inherent failure criterion, they show several disadvantages [12–14], as they are dependent on the statistical distribution of strength in the material examined, and some materials show over a large number of fatigue cycles only small decrease in strength, but a sudden drop at the, individual, failure point of the specimen tested, the so-called sudden-death phenomenon. Broutman and Sahu [15] developed a commonly used formula for the residual stress as

$$R_R(n) = R - (R - \sigma_{\max})\left(\frac{n}{N}\right) \tag{6}$$

with R_R the residual strength, the quasi-static strength R, the maximum stress $\sigma_{\max}$ at the cycle n, and the total number of cycles to failure N.

2.2.3 Stiffness Degradation Models

Contrary to residual strength, requiring the destruction of a specimen or structure to determine the required value, stiffness degradation can be measured in a nondestructive way, using a dynamic modulus measured during a cyclic fatigue test [16–18] but possesses no inherent failure criterion.

The dynamic modulus can be expressed as a function of upper σ_u and lower stress $\sigma_l$and upper $\varepsilon_u$and lower $\varepsilon_l$strain during a sinusoidal fatigue loading cycle:

$$E_{\text{dyn}} = \frac{\sigma_u - \sigma_l}{\varepsilon_u - \varepsilon_l} \tag{7}$$

Hwang and Han [19] introduced a fatigue modulus to describe damage progression. Ogin [20] introduced a description of the ratio of the degraded stiffness E to the stiffness of a virgin specimen E_0 as

$$\frac{E}{E_0} = 1 - \left(A \left(\frac{\sigma_{\max}}{E_0} \right)^B n^C \right) \tag{8}$$

with the fitting parameters A, B, and C.

A reduction of the dynamic modulus to a certain level, e.g., 20–50 % of the quasi-static modulus, can for some applications be regarded as fatigue failure.

2.2.4 Damage Accumulation Models

Damage accumulation models calculate from the damage d_i induced during each individual fatigue loading cycle i, each possibly having a distinct frequency, stress amplitude, and mean stress, the total damage D induced during fatigue loading of n cycles. A commonly used model is the Palmgren-Miner-Rule [21]

$$D = \sum_i d_i = \sum_i \frac{n}{N_i} \tag{9}$$

Failure of the structure is predicted when the damage reaches $D = 1$.

This rule does not permit the consideration of sequence effects, e.g., high-to-low or low-to-high stress amplitudes during a fatigue experiment. For the analysis of FRP structures the VDI 2014 [2] uses *Mattheij's* proposal, to limit the maximum damage to $D = 0.1$.

2.2.5 Critical Element Concept

The critical element concept is at the moment one of the most sophisticated models for fatigue analysis of composite materials. It describes the change of the stress state and cycle-dependent material properties of a representative volume of a structure under cyclical load. The representative volume is divided into a sub-volume, the critical element, which is representative for the structural integrity of the laminate, and the subcritical element or elements. The critical element usually corresponds to the 0° layers, oriented in the direction of the dominant loading, and its failure is determined using a residual strength criterion. The critical elements stress state does not correspond to the external loading, since developing damage induces a stiffness degradation in the subcritical element, thereby affecting a stress transfer from subcritical to critical element. Overviews and applications of the critical element concept can be found for example in [3] and [22–24].

3 Development of a Critical Element Concept-Based Fatigue Analysis Using the Finite Element Method

The newly developed fatigue analysis code [5], based on the Critical Element concept, uses nonlinear material laws and failure models to predict damage evolution, stress state, and failure of carbon-fiber-reinforced structural components under cyclical loading.

3.1 *Material Laws, Stiffness Degradation, and Residual Strength*

Nonlinear material laws are implemented using a quasi-inverse of the Ramberg–Osgood equation to describe the stress–strain relation as a function $\sigma_{ij}(\varepsilon_{ij})$ due to the requirements of the FE code. Parameters κ, m, ε_0, and $E_{t,0,ij}$ are required to fit the original Ramberg–Osgood description of the experimental results using a least-square procedure

$$\sigma_{ij}(\varepsilon_{ij}) = \varepsilon_{ij}E_{t,0,ij} - \kappa_{ij}\left(\frac{\varepsilon_{ij}}{\varepsilon_0}\right)^{m_{ij}} \tag{10}$$

Its implementation in the FE code ANSYS uses a USERMAT function, i.e., an interface of the ANSYS code to user-developed external computer programs, written in the FORTRAN language to achieve high numerical reliability and minimized computing times.

The material model is called by the FE code during each computing step k for each integration point, Fig. 2, of the finite element, providing stress, strain, and further state variables.

Stiffness degradation is implemented using the function of Eq. (5) for the quasi-static case and Ogin's model, Eq. (8), for the fatigue-induced stiffness reduction.

Residual strength is implemented using transposed nonlinear residual strength equation, cf. Eq. (6), as proposed in [25], to obtain an equivalent cycle number n_e necessary for the FE algorithm. A nonlinear form of the *S-N*-curve [26]

$$\sigma(N) = \frac{R}{1 + \left(a(\log N)^b\right)^c} \tag{11}$$

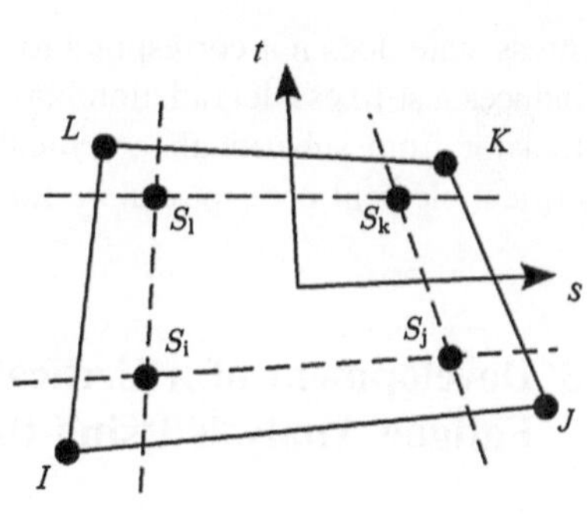

Fig. 2 Finite element with nodes *I*; *J*; *K*; *L*, and integration points $S_{i;j;k;l}$ in the *s*-*t*-element coordinate system

with the fitting parameters *a*, *b*, and *c* as well as the *Miner* rule, Eq. (9) is used to determine the equivalent cycle number n_e.

Figure 3 shows a flowchart of the FE fatigue analysis code. In the preprocessor, analysis variables, e.g., loading, stress ratio and control parameters, geometry and laminate lay-up as well as material parameters as stress–strain relations, residual strength, stiffness degradation, and *S-N*-curve parameters are provided. The preprocessor generates the model geometry and the FE mesh and sets boundary conditions (applied force and boundary conditions). The solver attributes material models to elements, calculates residual strength and damage development, and uses a nonlinear analysis with the USERMAT routine to compute resulting stress. Element stresses are transferred to the postprocessor, where the externally computed FF and IFF effort are evaluated and the element failure due to FF is computed. If subsequent failure of elements reach the boundary of the model geometry (x_{limit} or y_{limit}), it is assumed that the developed "cracks," in the form of IFF and resulting FF, have destroyed the complete material.

4 Experimental Examinations

Experimental studies were conducted [5] on specimens produced using the resin transfer molding process of an FRP material certified for aviation use; the fiber type was Hexcel HexTowAS7-J-12 K and the resin type was Hexcel Hex-Flow RTM-6.

Specimens used to determine the required material laws and failure models had unidirectional or cross-ply laminate lay-up. Validation specimens had a quasi-isotropic laminate lay-up and a circular cut-out. Computational tomography, X-ray analyses, and digital image correlation accompanied the cyclical tests to confirm the validity of failure and life prediction of the code.

Figure 4 shows penetrant-enhanced X-ray images of a 25 mm wide and a 110 mm long specimen with 10 mm free length between end tabs, approximately 2.5 mm thick, quasi-isotropic lay-up $[0/\text{-}45/90/45]_S$ at different stages of fatigue-induced failure under sinusoidal loading ($R = -1$, $\sigma_{x,\infty} = \pm 198$ MPa, 5 Hz.

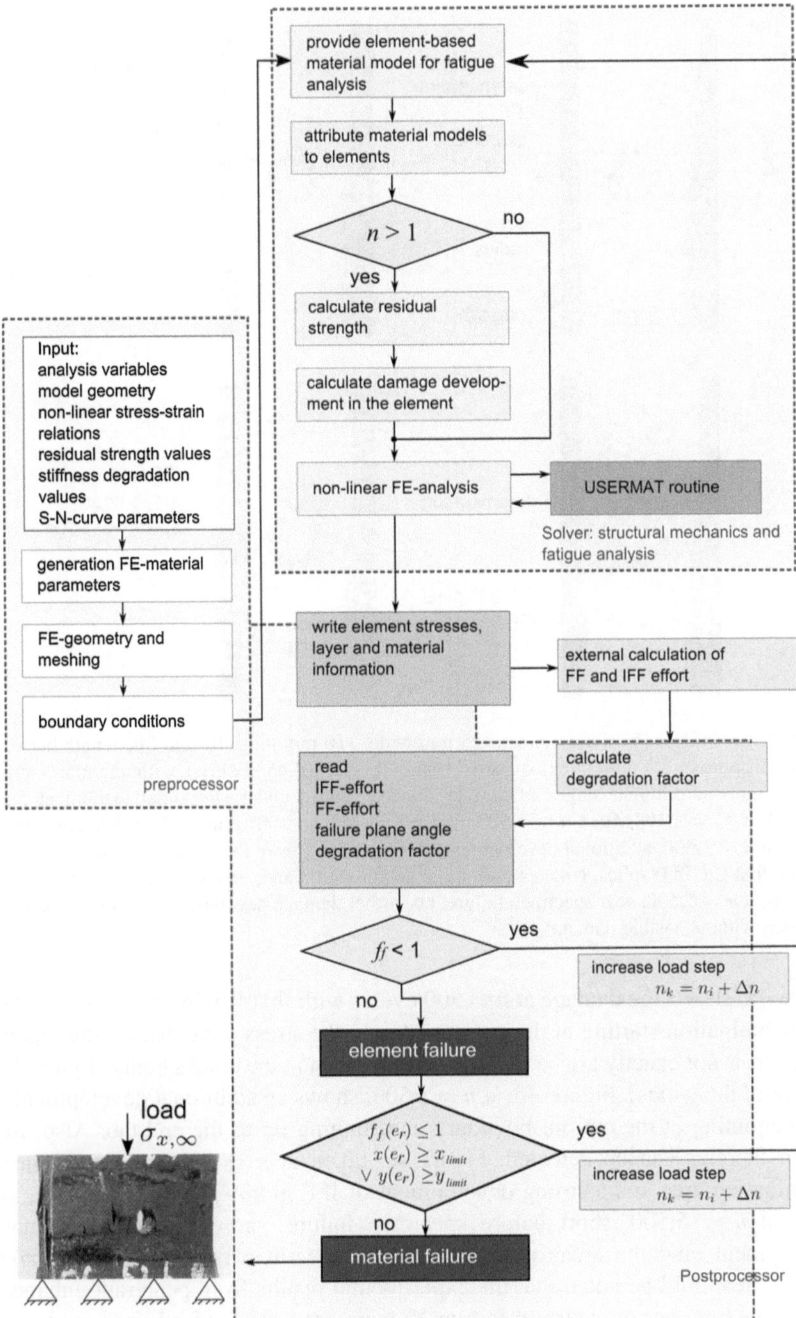

Fig. 3 Flowchart of the FE fatigue analysis code

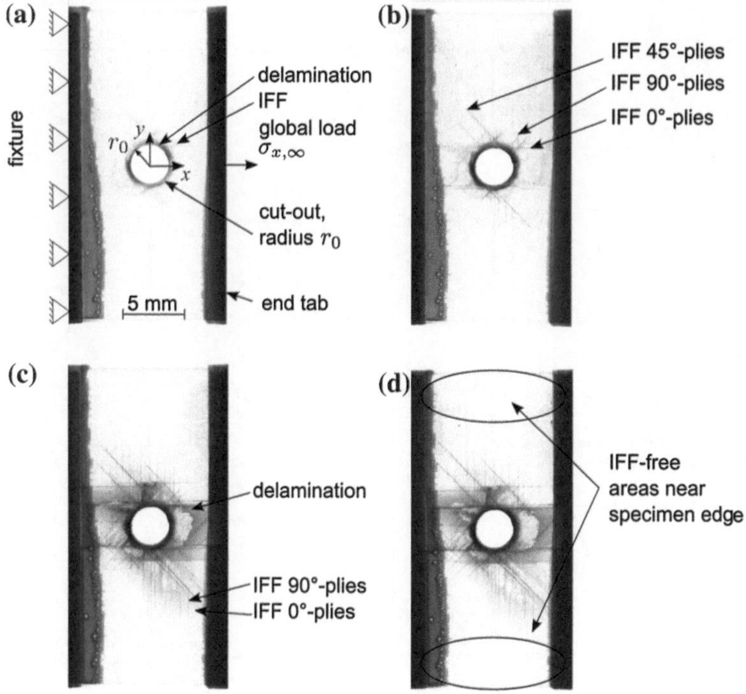

Fig. 4 X-ray images [5] of a specimen (25 mm wide, 110 mm long, 10 mm free length between end tabs, approx. 2.5 mm thick, quasi-isotropic lay-up $[0/\text{-}45/90/45]_S$) with circular cut-out ($r = 1.6$ mm) at different stages of fatigue-induced failure under sinusoidal loading at 5 Hz ($R = -1$, $\sigma_{x,\infty} = \pm 198$ MPa), **a** $n = 500$, first IFF in 45°- and 90°-plies and delamination at the cut-out, **b** $n = 1{,}500$, additional development of IFF, starting from the cut-out boundary up to the end-tab, first IFF in 0°-plies, **c** $n = 4{,}500$, large delamination area, strong development of IFF in 90°-plies, **d** $n = 5{,}500$, near specimen failure, no further damage development, areas near edge of specimen without visible damage

Figure 4a) shows the damage at $n = 500$ cycles with first IFF in 45°- and 90°-plies and delamination starting at the cut-out (due to the stress combination the highest IFF effort is not exactly at $y = 0$, which can be seen at the cracks being slightly left or right of the y-axis). Figure 4b, at $n = 1{,}500$, shows an additional development of IFF, beginning at the cut-out boundary and running up to the end-tab. Also, first IFF in 0°-plies can be detected. Figure 4c, taken at $n = 4{,}500$, shows a large delamination area and a strong development of IFF in 90°-plies. Finally, Fig. 4d, taken at $n = 5{,}500$, short before specimen failure, shows no further damage development, and the areas near the edge of specimen remain without visible damage. It should be noted that the experimental results from penetrant-enhanced specimens (number of cycles to failure N) were not used for further computation, since the repeated installation and removal of the specimen in the testing machine may have induced damage earlier, i.e., at lower cycles n, compared to specimens that were not analyzed using X-ray.

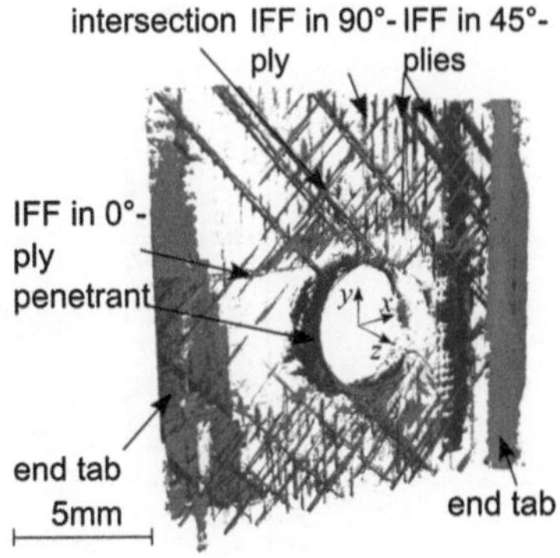

Fig. 5 Computed tomography image [5] (resolution 5 μm) of a quasi-isotropic specimen with circular cut-out ($r = 1.6$ mm) after cyclic loading ($R = -1$, $\sigma_{x,\infty} = \pm 138$MPa, 5 Hz) until $n = 10^6$ cycles showing the damage distribution to individual plies in the laminate

Computed tomography images of the quasi-isotropic specimen allowed the attribution of damage to individual plies of the laminate. Figure 5 shows such an image of a quasi-isotropic specimen (similar to the one shown in Fig. 4) after cyclic loading ($R = -1$, $\sigma_{x,\infty} = \pm 138$ MPa, 5 Hz) until $n = 10^6$ cycles. The computed tomography showed that IFF reached beyond the end tab into the areas of the specimens required for gripping in the testing machine. Also, it could be seen that IFF cracks in 90°-plies expanded throughout the complete thickness of the respective ply, whereas IFF cracks in 45°-plies were only visible near the ply boundaries. IFF in 90°-plies were concentrated approximately 0.1 mm from intersections of cracks in 45°-plies. Few IFF in 0°-plies were visible.

5 Example of Numerical Analyses

The validation specimens, shown in Figs. 4 and 5, were modeled [5] in the FE-program ANSYS, using 2,688 SHELL281 elements with eight layers generating a quasi-isotropic $[0/\text{-}45/90/45]_S$ lay-up, Fig. 6. The load was set from $\sigma_{x,\infty} = \pm 138$ MPa to $\sigma_{x,\infty} = \pm 275$ MPa and the resulting cycles to failure N computed. Using the lower load of $\sigma_{x,\infty} = \pm 138$ MPa, the program did not detect material failure up to 1.1×10^6 cycles, so that this result was regarded as a run-out. The SHELL281 elements showed sufficiently good computation speed, but lack the possibility to compute stresses in the thickness direction, necessary for the analysis of delamination.

The FE fatigue analyses, Fig. 7, showed a conservative prediction of cycles to failure, FE values being aligned with the 50 % probability of survival *S-N*-curve. Differences are possibly due to the chosen criterion of first damage at model boundary (x_{limit} or y_{limit}) indicating destruction of the complete material, whereas the computed tomography images showed that first damage extended into the area of the end tabs at high cycles of failure.

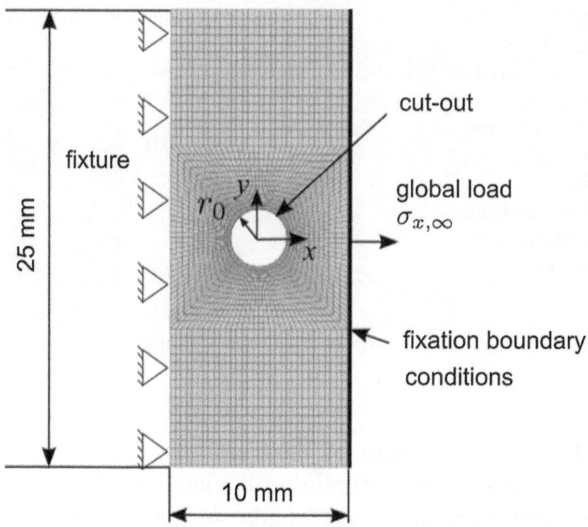

Fig. 6 Finite element model [5] in the FE-program ANSYS, using 2,688 SHELL281 elements with 8 layers, quasi-isotropic $[0/\text{-}45/90/45]_S$ lay-up

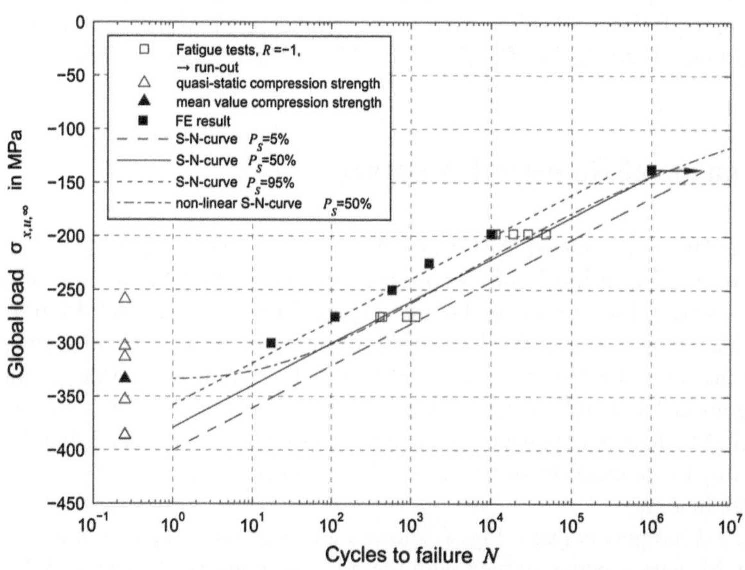

Fig. 7 Comparison of predicted cycles to failure of quasi-isotropic $[0/\text{-}45/90/45]_S$ specimens with cut-out (radius 1.6 mm) using a nonlinear material law compared to experimental results with sinusoidal load of $\sigma_{x,\infty} = \pm 138\,\text{MPa}$, $\sigma_{x,\infty} = \pm 198\,\text{MPa}$ or $\sigma_{x,\infty} = \pm 275\,\text{MPa}$, linear and nonlinear *S-N*-curves with different probability of survival P_S [5]

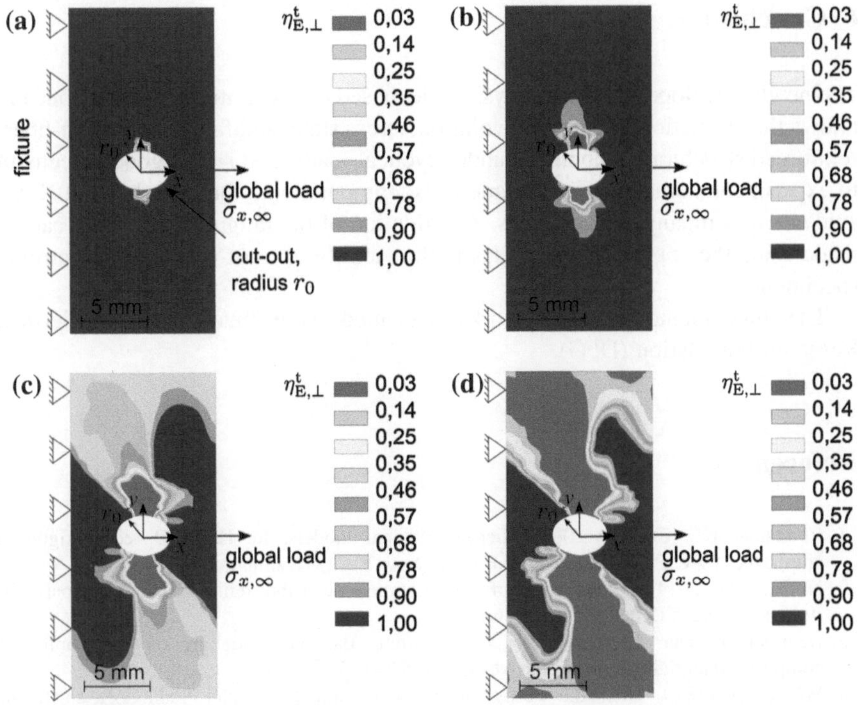

Fig. 8 Computed stiffness degradation factor $\eta^t_{E,\perp}$ for the fiber-perpendicular *Young's* modulus $E_\perp$ in a −45°-ply of the quasi-isotropic [0/-45/90/45]$_S$ specimen with cut-out (radius 1.6 mm) at $\sigma_{x,\infty} = \pm 198\,\text{MPa}$ and $R = -1$, **a** at the beginning ($n = 500$) point-symmetric degradation near the cut-out at $x = 0$, **b** at $n = 1{,}500$ larger degraded areas from the cut-out in positive and negative y-direction, **c** at $n = 4{,}500$ large areas of higher degradation inclined to the y-axis, **d** nearly complete degradation at $n = 9{,}700$ [5]

Figure 8 shows the computed stiffness degradation factor $\eta^t_{E,\perp}$ for the fiber-perpendicular *Young's* modulus $E_\perp$ in a −45°-ply of the quasi-isotropic [0/-45/90/45]$_S$ specimen with cut-out (radius 1.6 mm) at $\sigma_{x,\infty} = \pm 198\,\text{MPa}$ and $R = -1$. The predicted degradation shows good agreement to the matrix cracks visible in the penetrant-enhanced X-ray images, Fig. 4. For lower cycles n at $n = 500$ (Fig. 8a) and at $n = 1{,}500$ (Fig. 8b), the damage prediction of point-symmetric degradation near the cut-out at $x = 0$ and larger degraded areas from the cut-out in positive and negative y-direction is nearly congruent to the X-ray images. For higher n (Fig. 8c) at $n = 4{,}500$ and (Fig. 8d), the predictions show large areas of higher degradation inclined to the y-axis and nearly complete degradation at $n = 9{,}700$ reaching into the area of the specimen edge, a clear difference to the X-ray images, where the specimens showed little visually detectable crack development in this region.

6 Conclusion

The newly developed fatigue analysis code, based on the Critical Element concept, allows the prediction of damage evolution, stress state, and failure of carbon-fiber-reinforced structural components under cyclical loading. It shows good agreement to experimental studies using fatigue tests, computational tomography, and X-ray analyses accompanying these tests. A validation of the fatigue analysis cut can be made using the complex stress states in the vicinity of a cut-out in quasi-isotropic specimens.

The fundamental research works presented were funded by the German Research Foundation (DFG).

References

1. Reifsnider KL, Case S (2003) Micromechanical models. In: Harris B (ed) Fatigue in composite. CRC Press, BocaRaton, Boston, New York, Washington
2. N. N.: VDI 2014: Development of FRP components (fibre-reinforcedplastics), part III: Analysis. Verein Deutscher Ingenieure, 2006
3. Degrieck J, Van Paepegem W (2001) Fatigue damage modelling of fibre-reinforced composite materials: review. App Mech Rev 54(4):279–300
4. Wöhler A (1870) Über die Festigkeits-Versuche mit Eisen und Stahl. Zeitschrift für Bauwesen 1–3:73–106
5. Magin M (2012) Schadensfortschrittsentwicklung durch zyklische Belastung und deren numerische Modellierung unter Berücksichtigung nichtlinearer Werkstoffgesetze bei endloskohlenstofffaserverstärkten Polymerwerkstoffen. Technical University Kaiserslautern, PhD-Thesis
6. Sendeckyj G (1981) Fitting models to composite materials fatigue data. In: Chamis C (ed) ASTM STP 734—Test methods and design allowables for fibrous composites. American Society for the Testing of Materials, West Conshohocken, PA, USA, pp 245–260
7. Vassilopoulos AP, Manshadi BD, Keller T (2010) Influence of the constant life diagram formulation on the fatigue life prediction of composite materials. Int J Fatigue 32:659–669
8. Guideline for the Certification of Wind Turbines. Germanischer Lloyd, Edition 2010 (1st July 2010)
9. Hahn H, Kim RY (1975) Proof testing of composite materials. J Compos Mater 9:297–311
10. Chou PC, Croman R (1978) Residual strength in fatigue based on the strength-life equal rank assumption. J Compos Mater 12:177–194
11. Schaff JR, Davidson BD (1997) Life prediction methodology for composites structures. Part I—Constant amplitude and two-stress level fatigue. J Compos Mater 31
12. Fong JT (1982) Whatisfatiguedamage?, p 40–62. In: Reifsnider KL (ed) ASTM STP 775—Damage in composite materials. American Society for the Testing of Materials, West Conshohocken, PA, USA
13. Talreja R (1991) Statistical conderations. In: Reifsnider KL (ed) Fatigue of composite materials. Elsevier, New York, Oxford, Amsterdam, Tokyo, pp 485–501
14. Sendeckyj GP (1991) Life prediction for resin-matrix composite materials. In: Reifsnider KL (ed) Fatigue of composite materials. Elsevier, New York, pp 431–483
15. Broutman LJ, Sahu S (1972) A new theory to predict cumulative fatigue damage in fibreglass reinforced plastics. In: Corten HT (Hrsg.) ASTM STP 497—Composite materials: testing and design, p 170–188. American Society for the Testing of Materials

16. Lee L, Fu K, Yang J (1996) Prediction of fatigue damage and life for composite laminates under service loading spectra. Compos Sci Technol 56:635–648
17. O'Brien TK, Reifsnider KL (1981) Fatigue damage evaluation through stiffness measurements in Boron-Epoxy laminates. J Compos Mater 15:55–69
18. Poursartip A, Ashby MF, Beaumont PWR (1986) The fatigue damage mechanism of a carbon fibre composite laminate: I - development of the model. Compos Sci Technol 25:192–218
19. Hwang W, Han KS (1986) Fatigue of composites—fatigue modulus concept and life prediction. J Compos Mater 20:154–165
20. Ogin SL, Smith PA, Beaumont PWR (1985) Matrix cracking and stiffness reduction during the fatigue of a (0/90)s GFRP laminate. Compos Sci Technol 22:23–31
21. Palmgren A, Die Lebensdauer von Kugellagern (1928) Zeitschrift des Vereins Deutscher Ingenieure, p. 339–341
22. Diao X, Ye L, Mai Y (1995) A statistical model of residual strength and fatigue life of composite laminates. Compos Sci Technol 54:329–336
23. Subramanian S, Reifsnider K, Stinchcomb W (1995) A cumulative damage model to predict the fatigue life of composite laminates including the effect of a fibre- matrix interphase. Int J Fatigue 17:343–351
24. Halverson H, Curtin W, Reifsnider K (1997) Fatigue life of individual composite specimes based on intrinsic fatigue behaviour. Int J Fatigue 19:369–377
25. Loverich JS (1997) Life predictionof composite armor in an unbonded flexible pipe. Virginia Polytechnic Institute and State University, Diplomarthesis
26. Bach C (2008) Beitrag zur Modellierung des Schwingermüdungsverhaltens und zur rechnerischen Lebensdaueranalyse von endlos kohlenstofffaserverstärkten Vinylester-Matrixsystemen. Technical University Kaiserslautern, PhD-Thesis

Design and Development of Lighter-Than-Air Systems

Rajkumar S. Pant

Keywords Lighter-Than-Air systems · LTA systems · Blimps

Micro Airship @ IIT Bombay

Mini Airship @ Pune

Macro Airship @ Manali

Lighter-Than-Air (LTA) systems are aerial platforms that overcome gravity primarily by the buoyancy force acting on them due to a lighter-than-air gas (e.g., Helium) enclosed in their envelopes. The two main types of LTA systems are Tethered Aerostats and Airships; both of which are ideally suited for missions involving prolonged aerial observation of any activity (e.g., aerial coverage of sporting events or security patrolling) and product promotion. Aerostats can be used as platforms for several scientific and commercial applications such as remote sensing, measurement of local wind profiles, and solar irradiation.

Design development and flight testing of prototypes of several LTA systems have been carried out at the Lighter-Than-Air systems Laboratory of IIT Bombay for more than a decade. Several outdoor remotely controlled airships and tethered aerostat systems have been designed, fabricated, and field tested. The laboratory has a facility for in-house fabrication of the LTA envelopes using an RF sealing machine. More than 40 students from 20 different engineering colleges in India

R. S. Pant (✉)
Lighter-Than-Air Systems Laboratory, Aerospace Engineering Department, Indian Institute of Technology Bombay, Powai, Mumbai, India
e-mail: rkpant@aero.iitb.ac.in

R. P. Bajpai et al. (eds.), *Innovative Design, Analysis and Development Practices in Aerospace and Automotive Engineering*, Lecture Notes in Mechanical Engineering, DOI: 10.1007/978-81-322-1871-5_9, © Springer India 2014

and abroad have successfully completed their UG/PG final year projects and/or internships in this laboratory, during periods ranging from 2 months to 1 year. International tie-ups with Texas A&M University, USA, and Instituto Tecnológico de Aeronáutica, Brazil have also been accomplished. Several R&D projects with a total financial outlay of more than '1,000,000 have been carried out /are in progress in this laboratory. Some idea of the past projects can be obtained from http://www.aero.iitb.ac.in/~airships.

This talk will provide an overview of Lighter-Than-Air systems and highlight the key design challenges. Two case studies will be presented; the first one related to a relocatable wireless communication system for providing data and voice communications during disaster relief operations in remote areas. The second case study will showcase an outdoor remotely airship developed as a technology demonstrator for snow cover evaluation over lower Himalayas. Details of an ongoing research project related to development of prototype of an autonomous outdoor airship will be provided.

In the end, a few suggestions toward the possibility of some R&D projects and research work to be carried out in association with the faculty and students of Vel Tech University will be presented.

Computational Fluid Dynamics: A Design Tool for Aircraft Industries

P. Srinivasa Murthy

Abstract Design of Aircraft is generally based on three stages which are conceptual design, preliminary/detailed design and prototype testing. Some of the problems which are anticipated in flight can be solved during design stage itself by making use of modern technology, that is Computational Fluid Dynamics (CFD). This chapter gives an overview of what has been achieved in aircraft design using CFD and what are the future challenges which are required to be solved by upgrading CFD technology.

Keywords Flow separation · Unsteady flow · Flow transition · Laminar separation bubble · Unmanned Aerial Vehicle Design

Computation of aerodynamic characteristics of propeller in the presence of Aircraft flow field is essential in the integrated design of propeller aircraft combinations. Since propeller operates at various altitudes and Mach numbers its efficiency in producing the required thrust to achieve the mission objective in terms of aircraft performance is to be computed before wind tunnel and flight tests can be undertaken. There are a number of theories such as blade momentum theory, actuator disk theory, and empirical estimates based on ESDU data sheets available to predict the performance of the propeller. But these are approximate and do not consider the flow interference effects between the propellers and the aircraft. Computational fluid dynamics (CFD) is a design tool, which gives the complete solution including flow interference effect between propeller slipstream and aircraft flow field.

To start with a simple configuration, Wing Engine Propeller (WEP) combination has been investigated with three-bladed two propellers in tractor mode rotating clockwise direction. It has been found that both the propellers develop equal thrust. But when Fuselage Wing Engine Propeller (FWEP) combination has been analyzed both the propeller develop asymmetric thrust. Flow asymmetry

P. Srinivasa Murthy (✉)
Aeronautical Development Establishment (ADE), Bangalore, India
e-mail: dr_ps_murthy@yahoo.co.in

R. P. Bajpai et al. (eds.), *Innovative Design, Analysis and Development Practices in Aerospace and Automotive Engineering*, Lecture Notes in Mechanical Engineering,
DOI: 10.1007/978-81-322-1871-5_10,

developed due to induced cross flow effect of the fuselage. Design modification has been done to overcome induced cross flow effect of the fuselage. CFD analysis of the aircraft configuration with propeller running is very important during the design stage to identify the critical phases of flight wherein the basic stability is required to be augmented. At the same time, it helps to identify the critical engine failure of twin engine tractor propellers at all phases of flight and take suitable remedial measure for safe control of the aircraft.

This talk also covers a brief overview of future aircraft design with CFD as an analysis tool in aircraft industry in designing unmanned combat Aircraft, manned fighter aircraft, mini and micro aerial vehicles of various types such as fixed wing MAV, rotary wing MAV, and flapping wing MAV which are the future challenge for the aircraft designers. The aerodynamic challenges are many with interaction of flow among various discipline such as structures, controls with smart actuators, smart materials, and MEMS systems. The flow complexity at low Reynolds number is very high involving flow separation, transition, reattachment, and flow unsteadiness due to kinematics of flapping motion and flow-induced motion due to unsteady flow separation. To handle such complex flow physics, it is required to solve the full nonlinear fluid dynamic equation, which requires solution by numerical methods involving high performance computing systems.

Part II
Contributed Papers

An Approach to Improve Aviation Quality Management Using Total Quality Management Principles

B. Immanuel Lazur, L. Jagadeesh, B. Karthikeyan and M. Shanmugaraja

Abstract In today's competitive economic environment, aviation industry realizes more and more the importance of ensuring a positive customer experience. Airline owners and stakeholders have over the years shifted their focus from "larger and faster" equipment to awareness in addressing customer wants and needs. Airports too, have come to realize the quality of the passenger experience correlates directly, along with aerodrome security and emergency response capabilities, in garnering new airlines and expanding revenue. For any aviation organization, the primary goal is to improve financially, physically, and organizationally at the least possible cost. This paper proposes the concept of Total Quality Management (TQM) for both airports and airlines as a "systems approach" to manage the aviation quality. The ultimate focus of this paper is to enhance the overall departmental effectiveness and overall employee efficiency in providing quality customer-oriented aviation. To prepare for aviation quality management, the paper presents an overview study of elements, function, and methodologies of TQM. An attempt is made to show how TQM can be taken from theory and operationalized at aviation processes. Exclusive attention is paid on aircraft maintenance and safety management in which the influence of TQM principles is reported in this paper. The possible limitations for this attempt and further research avenues are addressed at the end.

Keywords Aviation · Customer experience · Airline · Airport · Aerodrome · Security · Total quality management · Quality · Aircraft maintenance · Safety management

B. Immanuel Lazur · L. Jagadeesh · B. Karthikeyan · M. Shanmugaraja (✉)
Department of Aeronautical Engineering, Park College of Technology, Karumathampatti, Coimbatore, India
e-mail: shanmugarajam@gmail.com

R. P. Bajpai et al. (eds.), *Innovative Design, Analysis and Development Practices in Aerospace and Automotive Engineering*, Lecture Notes in Mechanical Engineering,
DOI: 10.1007/978-81-322-1871-5_11,

1 Introduction

Worldwide, the aerospace industry is a multibillion dollar industry. Annual increase in air traffic is estimated at 7.5 % between 1995 and 2000, and 7 % between 2000 and 2010. The Asian aviation market is projected to be the highest growth sector in commercial aviation, accounting for 28 % of the world's passengers by the year 2010. Although air operators around the world have succeeded in offering a quality product that is highly safe and usually affordable (meeting another customer requirement: low fares), the air operators have not been rewarded for the quality of their services. The airline industry is notorious for never having paid returns to its shareholders in the aggregate. The problem of the profitability of the industry needs to be urgently targeted. For efficiency and profitability, airlines can benefit from an advanced form of quality management, total quality management (TQM). This tool goes well beyond satisfying the customer or offering quality products. TQM is a management approach in which all members of an organization participate in improving processes, products, services, and the culture in which they work. Airlines can benefit from TQM because it is widely agreed that the industry needs cost reduction and control, without losing the focus on product safety. TQM emphasizes, among other things, eradicating defects and waste from operations, reducing development cycle times, reducing product and service costs, and challenging quantified goals and benchmarking.

The objective of this is to familiarize the readers and aviation professionals with TQM principles and tools for implementing them successfully in the field of aviation to enhance the aviation quality. An aircraft has an economic life of about 25 years. To remain in serviceable condition, regular checks and repairs are conducted. Some of the repair and overhaul work involves parts and components that can be removed from the plane, namely, engine turbine blades and landing gear. The frequency of such repair activities depends on the manufacturer's specification. As maintenance costs form an integral part of the total operating costs, airlines are constantly sourcing for cost-effective and quality service methodologies. To this end, application of TQM principles will prove useful.

2 Literature Review

In this paper work, the research precedence's for introducing TQM for aviation quality management is studied using the data sources like "inderscience," "emerald in sight," "inter science," "Elsiever," and "science direct." "European Journal for Operational Research," "International Journal of Automotive Technology," "International Journal of Productivity and Performance," "International Journal of Operation and Management," and "Journal of Quality in Maintenance Engineering" are the few sources identified in this literature survey as much contributing publications to the objective of the paper. With the general search in these journals, numerous literatures are found. From that some are shortlisted and further reviewed.

2.1 Need for Aviation Quality Management

The deregulation of the airline industry in most part of the world marked the beginning of a new realm of competition in the industry. The deregulation ensured that airlines set fares and service levels based on the market situation. In trying to gain competitive advantage, airlines try to outshine their competitors by providing quality services that meet or exceed the expectation of customers. Thus, customer satisfaction in the airline industry is never ending as they face numerous challenges and competition daily. This makes Quality management critical to the airlines as they strive to continuously improve their services to meet customers' expectation. Service quality is necessary in the airline industry as it is a major determinant of competitiveness. Airlines paying strict attention to service quality will be differentiated from others and will in the course of doing this gain competitive advantage. Although it has been argued that price is a major determinant of airline choice by customers and most airline would rather compete on it than on service quality. However, not managing quality will mean no added and assuring value to the airlines. The use of a strategic approach to quality management by airlines will therefore improve their competitiveness. This approach ensures that airlines remain customer focused. Committed leadership to the strategy ensures that the airline offerings are continually updated so as to meet or exceeds customers' requirements. Few researchers have noted that the main cause of poor performance by service firms is that they do not know what is expected from them by customers. Airlines need to consistently monitor the perception of customers to their services so as to know where the gap lies between the offered service and expectations of customers. On the other hand, some academic practitioners have noted that acceptable level of customer satisfaction is attained when passenger's expectation have been met or exceeded. To ensure customer satisfaction, everybody in the organization including suppliers will need to have the customers at heart and work toward meeting their needs and expectations. For the airlines to be innovative in its offerings, it requires a flexible structure which permits cooperation between different functions. It is noted that intensive cooperation, empowerment, and open communication facilitate innovation. While several researches have been carried out on service quality in the airline industry, no research has been carried on the effects of TQM implementation in the airline industry.

2.2 Overview of TQM

In today's competitive environment, the existence of a company depends upon its business and thus its customers. Satisfying the customer's expectation must be the company's primary objective. Customers expect quality because they want a product or service that is reliable, meet their requirements, value for money, on-time delivery and of course' they do not want any problem. "Total quality management is a comprehensive and integrated way of managing any organization in order to:

- Meet the needs of the customer consistently;
- Achieve continuous improvements in every aspect of the organization's activities;
- Quality is determined by the customer; use it to your advantage or ignore it at your peril;
- TQM is a journey toward the company's vision;
- Measures must be used to reflect progress improvement;
- Goal clarity is absolutely essential at all levels of the organization;
- Measure your service or product the same way as the customer does.

TQM is a process aiming at meeting the customer requirements through the quality of all the tasks carried out in the company. It is a set of principles, procedures, guidelines, and methods organized as a comprehensive strategy with the goal of mobilizing the entire company in order to achieve the customer's specifications and greatest satisfaction at the lowest cost. The management approach should bring the increase in productivity, cost control and/or reduction, customer service, and product quality together in a strong combination.

2.3 Need for Quality in Aviation Maintenance

The aviation industry in itself has various divisions, all of which play a vital role in its function and working. These division ranges from design to product and from product to sale. These will help in the healthy functioning of the industry. From the various divisions of the aviation industry, aircraft maintenance division is focussed in this paper. Aircraft maintenance checks are carried out periodically after specified time or usage. These checks are categorized into A, B, C, and D checks with the light checks referred to as A and B checks and the heavy maintenance checks regarded as C and D checks. The aviation maintenance industry is primarily responsible for the retaining or restoring of aircraft parts in or to a state in which they can perform their required design function(s). This includes the combination of all technical and corresponding administrative, managerial, supervisory, and oversight activities. Inclusive of the described periodic checks, maintenance-type activities are principally the servicing, repair, modification, overhaul, inspection, and determination of condition of the aircraft. The main role of the aviation maintenance sector can thus be summarized as the arm of the aviation industry that is essentially responsible for the provision of a fully serviceable aircraft when required by the operator at affordable and reasonable cost with optimum quality.

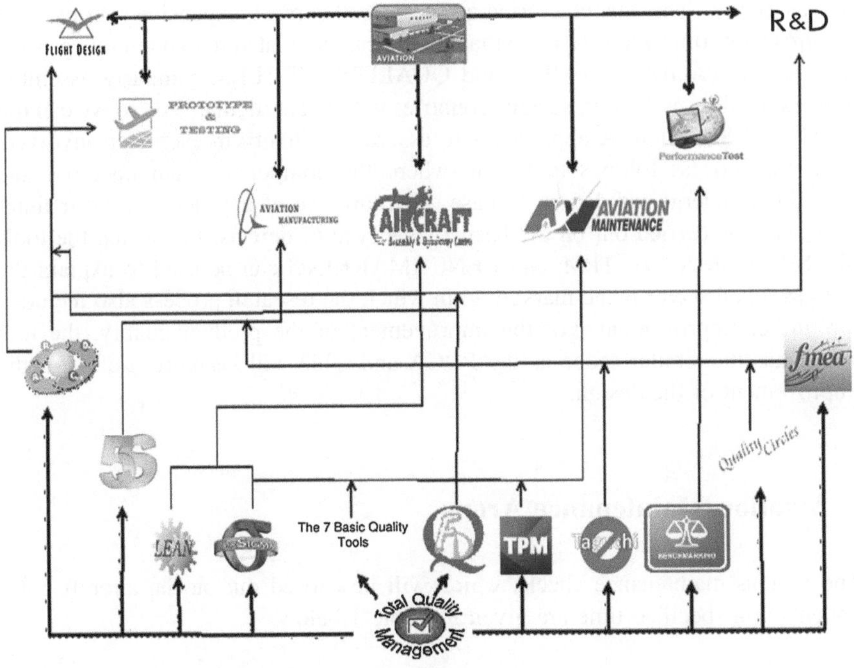

Fig. 1 Proposed model to incorporate TQM in aviation industry

3 The Proposed Methodology

In this research attempt to introduce TQM principles and tools in aviation industry, an interface model is developed by illustrating the implementation of various tools in major divisions of the aviation industry as shown in Fig. 1.

The aviation industry in itself has various divisions; all play a vital role in its function and working. These divisions include design to product and from product to sale. The interface model proposed in this paper is developed based on the scope of each divisions and the scope of the TQM tools. One tool can be related to many divisions because of its wider scope. For example, Six Sigma may be used in maintenance division and assembly division. Similarly each division will have more than one tool to use in it. The design process starts with the aircraft's intended purpose. The purpose may be to fit a specific requirement, e.g., as in the historical case of a British Air Ministry specification, or fill a perceived gap in the market. Some of the TQM tools have the scope which will fulfill the requirements of the design process such as Plan—Do—Check—Act (PDCA) and Quality Function Deployment (QFD). Next to the design sector is prototype and testing. The manufacturer will decide whether to carry out small scale or large-scale testing. It can be explained using the PDCA cycle which has a similar scope. One of the major and important sectors in the aircraft industries is the manufacturing and assembly. The manufacturing process is the most expensive process and requires more time and safety. So many tools can be related to explain this process such as 5S, TAGUCHI, FMEA, and QUALITY CIRCLES. Similarly assembly process requires parts from various countries within a particular time so we can use LEAN and SIX SIGMA approaches to assure the quality in each step involved. Then the process follows to testing where the components manufactured and assembled undergo validation. In case of occurrence of any defects, the maintenance will be carried out on the basis of the type of defects. For which the tools like SIX SIGMA, Q7, TPM, and BENCHMARKING can be used to explain the process. Final sector is the marketing, in which the research process also included for further improvement. For the improvement of the product quality, the tool which has the similar scope as the PDCA and QFD will be more useful for the improvement of the design.

4 Aviation Maintenance Arena

The various maintenance checks which will be carried out on the aircraft components in a specified time are given in Table 1 below:

4.1 Six Sigma in Aviation Maintenance

Six Sigma originated at Motorola in the 1980s and enjoyed an early start in aerospace companies such as General Electric and United Technologies. Since that time, Lockheed Martin, Raytheon, and many other aerospace companies have adopted both Six Sigma and Lean methodologies to both drastically reduce defect rates and cycle times. While DPMO (defects per million opportunities) is a primary measure in Six Sigma, to focus solely on the reduction in defect rate may limit the understanding of the power of Six Sigma. The data-driven tollgate methodology in Six Sigma requires a disciplined approach far beyond reducing DPMO. Six Sigma has five phases with each phase requiring a tollgate review resulting in one of three outcomes: continue to the next phase, stop the project, or continue study in current phase. These five phases are: Define Measure, Analyze, Improve, and Control. This DMAIC methodology is designed specifically for improvement of existing processes. The methodology consists of five phases: (a) Define business opportunities, (b) Measure performance, (c) Analyze opportunity, (d) Improve performance, and (e) Control performance. The practice of Si Sigma DMAIC approach in aviation maintenance is attempted in this paper as described below;

Table 1 Aviation maintenance scenario

Maintenance	A check	B check	C check	D check
Flight hours	500–800 h	4–6 months	15–21 months	5 year
Man hours	20–50 h	150 h	6,000 h	50,000 h
Components	Oil, control surface, wheel break etc.	Tires, landing gear, wear and tear components etc.	Whole aircraft components	Painting, fuselage metal skin inspection and complete disassembly take place
Duration	One night	1–3 days	1–2 weeks	2 months

Problems human errors, time delay, lower quality of service, rework time, inadequate man power etc.

4.1.1 I Phase of DMAIC: Define

The first step is to define the problem and its major concerned area. In this step, goal has to be set to complete the whole process within a specified time. For example, experience revealed that there are urgent needs to reduce the repair turn time of the turbine blade section. As the sales contribution from the blade section was significant, improving the turn time will increase the firm's profit margin and productivity. The aim was to reduce the repair turn time to less than 37 days.

4.1.2 II Phase of DMAIC: Measure

Once the base line for the problem is identified, the data collection should be carried out. To measure adequately the effectiveness of any programme, it is necessary to collect data before and after implementation. It is necessary to obtain accurate data about the current situation and compare the results obtained after implementation to determine the level of improvement. In this study, repair turn time and rework data should be collected from the company database or any rework record. After implementing the new method, data were collected at different periods in time and check sheets can be used to analyze the post-implementation data.

4.1.3 III Phase of DMAIC: Analyze

Once the problem and root cause are found, the next step is to analyze whether the root cause identified is creating the problem or not. This can be done by analyzing the data collected. A CE diagram shown in Fig. 2 can be used to summarize the problem of a long turn time and its causes.

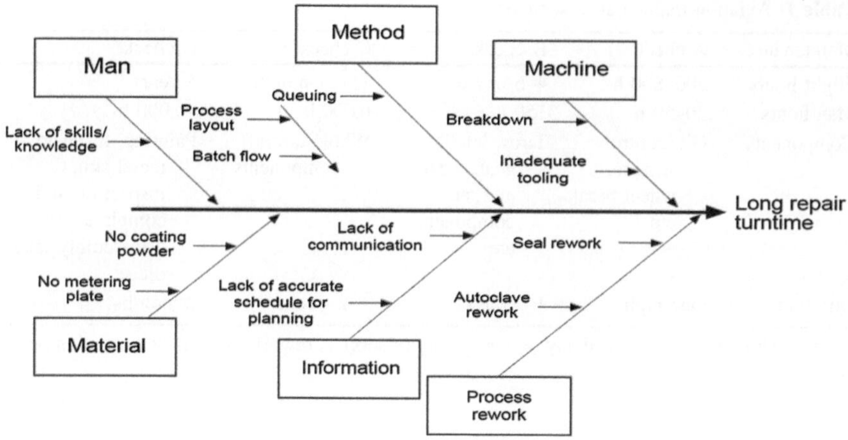

Fig. 2 Fishbone diagram

4.1.4 IV Phase of DMAIC: Improve

The fourth step in this process is to improve the defect found either by repairing or replacing the defected component or system. This can also be done by implementing a new method to overcome the problem. Before implementing a new technique to overcome the problem, it requires a careful planning which leads to successful implementation. In this study, the repair area can be changed from process layout to manufacturing cell layout which will reduce the travel time of the components. One piece flow can be used to repair the components in a sequential order which will avoid the deflection of concentration from the work.

4.1.5 V Phase of DMAIC: Control

In order to maintain the focus, we should narrow down the vital few measurements for the performance. We should maintain the improvement in other areas of the organization which can be done by documenting the procedures followed. Control charts can be used to record the improvements in a periodic time to analyze the improvements. The sharing of project success leads to greater momentum of change within the organization. Six Sigma drives more success.

4.2 Tools Used in Six Sigma DMIAC Approach

The following tools listed in Table 2 are the most commonly used in DMAIC implementations.

Table 2 Six Sigma tools

Process	Tools
Define	SIPOC
	Flow chart
	QFD
Measure	Cause and effect diagram
	Data collection plan
	CTQ
Analysis	Pareto chart
	Cause and effect diagram
	Histogram
	Project charter
Improve	Brainstorming
	PDCA
	Design of experiments
Control	Control chart
	Benchmarking
	Documentation
	Sigma calculation

5 Expected Outcome of the Proposed Model

The effective implementation of TQM in aviation industry to improve the quality of product and service will increase customer satisfaction with the service offerings as indicated in Fig. 3. Quality enhances customer loyalty through satisfaction; this in turn can generate repeat business and lead to the attraction of new customers through positive word of mouth. The word of mouth communication will help in cost reduction. This will provide competitive edge to the company. The improvement in quality will result in increased market share and profitability. The implementation of TQM would ensure that every worker in the organization does his work with quality the first time, thus improving the efficiency of operation and avoiding some cost associated with waste. This in turn will offer more value to customers in terms of price and service quality, thus making them satisfied. Implementation of TQM further ensures that organizations change how they perform activities so as to eliminate inefficiency, improve customer satisfaction, and achieve the best practice. Constant improvement in the effectiveness of operation is essential but not a sufficient factor for the organization to be profitable. TQM helps in improving the quality of products and also reduces the scrap, rework, and the need for buffer stock by establishing a stable production process. TQM will reduce the cost of production and time of production. Continuous improvement which is a feature of TQM is said to reduce the product cycle time thus improving productivity; many other TQM practices such as training, information system management, relationship with suppliers etc., have a positive impact on operational performance. The efficient management handling of these practices will improve efficiency and with no doubt affect the profitability of the firm.

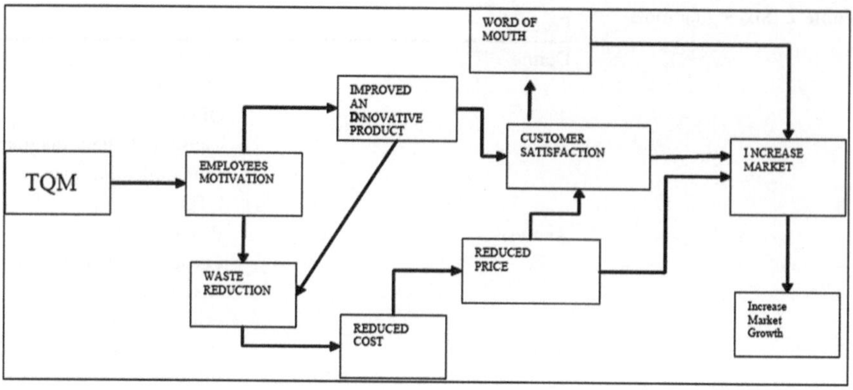

Fig. 3 Effect of adoption of TQM on aviation quality management

6 Conclusion and Future Work

The arguments of this research were focused on the need for aviation industries to adopt the TQM principles, due to the benefit derived from it in terms of customer satisfaction, operational effectiveness, and employee satisfaction. In the airlines industry, the main essence of TQM is to provide services which will make customers satisfied, this in turn will bring about repeated purchase thus increasing the sales and profitability of the organization. A literature survey was carried out to know about the scope of the TQM tools and the scope of the aero divisions both in past history and in present days. It was clear from the literature review, in order to mitigate the continuously increasing competition within the aviation industry, many organizations are turning to TQM philosophy because of the benefits it is perceived to offer. However these perceived benefits have to be clearly understood in the context of the environment where it is to be implemented. One of such tools is the Six Sigma, which is used in maintenance division. Six Sigma is much more than just a number crunching exercise. The visual and measurable outputs of six sigma allowed all the team members, from workers to management, to have a clear understanding of the processes involved, and their importance and contribution in delivering a solid fill, compacted within the specifications. It is recommended that more studies should be carried out, which covers the whole departments of airlines to establish the effectiveness of the implementation of TQM in the aviation Industry. Also, it will be of great benefit to ascertain the true perception of customers to service quality as no prior research has been carried out in that field. This will give a clue to what the customers' desire most in terms of airlines service delivery. However, future work of this research might be continued in following steps;

1. Implementing the suggested interface model in industries for experimental results;
2. Correlate this technique by using management statistical software.

Design and Field Trials of a Payload Recovery Device for Tethered Aerostats

Nitin Sharma, Agnit Mukhopadhyay, Vishaal Sharma, Mayank Milind and Rajkumar S. Pant

Abstract This paper describes the design fabrication and testing of Payload{a} Recovery Device (PRD) that gets activated automatically if tether of an aerostat breaks. The recovery of payload is accomplished in three stages, viz., tether breakage detection, signal transmission, and device activation. The envelope of aerostat is deflated by deploying a mechanism that results in a small circular hole on its top. As the LTA gas in the envelope escapes, the buoyancy of the aerostat decreases and it gradually descends. The size of the hole controls the gas leak rate and hence the velocity of descent. A model for simulating the breakaway characteristics of a spherical aerostat was also developed, which estimates the amount of drift and touchdown velocity as a factor of various parameters such as the hole diameter and deployment altitude, for a single and double chamber envelope. The prototype system was then field tested in an indoor and outdoor aerostat.

N. Sharma (✉)
B.Tech Mechanical Department, NIT Hamirpur, HP India, Hamirpur, Himachal Pradesh, India
e-mail: nitin.sharma151993@gmail.com

A. Mukhopadhyay
B.Tech Aerospace Engineering Department, Punjab Engineering College, Chandigarh, India
e-mail: agnitm@gmail.com

V. Sharma
IIT-Bombay, Mumbai, India
e-mail: aero.vishal10@gmail.com

M. Milind · R. S. Pant
Aerospace Department IIT-Bombay, Mumbai, India
e-mail: immayankmilind@gmail.com

R. S. Pant
e-mail: rkpant@aero.iitb.ac.in

N. Sharma · A. Mukhopadhyay · V. Sharma · M. Milind · R. S. Pant
Lighter-Than-Air Systems Laboratory, Department of Aerospace Engineering, Indian Institute of Technology Bombay, Mumbai, India

R. P. Bajpai et al. (eds.), *Innovative Design, Analysis and Development Practices in Aerospace and Automotive Engineering*, Lecture Notes in Mechanical Engineering, DOI: 10.1007/978-81-322-1871-5_12, © Springer India 2014

Keywords Payload · Deflation

1 Introduction

Lighter-Than-Air (LTA) systems derive major part of their lift from aerostatic buoyant force, which is in contrast with heavier-than-Air aircraft that which use relative motion with ambient air for generating the lift force to overcome gravity. LTA systems can be broadly classified into three main categories, viz., Balloon, Airship, and Aerostat. Aerostats operate in lower atmosphere, and there is a high probability of their interaction with sudden unexpected weather changes, leading to gusts or storms, that can result in breakage of tether. Hence, the tether of an Aerostat may break due to many inadvertent reasons. After breakaway, an Aerostat will continue to ascend up due to the *free lift*. During its ascent, the pressure differential between internal and ambient pressure will build up, since the ambient pressure decreases with altitude. This will eventually cause rupture of envelope, resulting in sudden loss of lift and hence a high descent rate, causing damage to the payload mounted onboard the aerostat when it hits the ground. During this time, the envelope will also drift in the ambient wind direction; and can end up in undesirable locations or interfere with the motion of other aircraft in the airspace.

Hence, it is necessary to install a device onboard an aerostat that can trigger automatically when the tether is broken, and ensure that the payload can be safely recovered. Such a device has to meet the conflicting requirements of minimum drift of the breakaway aerostat from location of deployment and the minimum velocity of impact of the payload at touchdown.

This work is an extension of that was earlier done by Bhat and Pant [1].

2 Design of PRD

A survey of literature was launched to identify the details for existing PRD type devices for tethered aerostats. Though the various types of devices that presently exist differ slightly in their working principles, however, they initiate the same action. Once the signal is sent for deflation{b}, all the mechanisms transmit a signal to a current carrying loop, attached on the envelope (usually at the top). The current starts flowing in the loop and eventually, it heats up the wire so much that it creates a hole in the envelope by melting the fabric, and hence letting the gas to escape out, which brings the aerostat back to the ground. This system is commonly known as a hot wire mechanism.

In our present design, we have also used this hot wire mechanism due to its reliability compared to other methods. The wire is fixed on the top of the envelope in the shape of a trapezium with the base absent attached to a circular wooden support element and is powered by a suitable battery of proper wattage depending upon the LTA Gas used. An electronic circuit box is designed so that the battery supplies power on detecting tether breakage. When the heated Nichrome wire melts a cavity in the envelope, the gas contained in the envelope escapes and hence the buoyancy reduces. So the aerostat gradually descends to the ground, and the payload is recovered.

The PRD circuit is a group of three subsystem, viz., actuation system, processing system, and execution system, which are described in the subsections that follow.

2.1 Actuation System

The mechanism used to detect tether breakage consists of a mechanical spring switch, which is a Push Rod system. One end of the actuation system is attached to the ground through the tether cable, and the other end is connected to all the confluence lines. When the tether exerts tension on the Push Rod, it is pulled downward and the spring gets compressed. And when the tether breaks, as the tension no longer acts, the restoring force of the spring pushes the Push Rod upwards. When this happens, the PRD is activated by pressing the push switch.

2.2 Processing System

As soon as the push switch is pressed, it completes the circuit and a signal is sent to the Arduino UNO board. We have placed Arduino UNO board for processing in case of momentary tension loss in the tether due to transients, or atmospherics gusts. To take care of this, the controller is pre-programmed to wait for a specific amount of time; and triggers the circuit only if the tether tension is not regained during this time. This delay has been kept as programmable so that it can be used in various conditions and the delay can be set accordingly, for instance, chances of momentary tension loss is quite high in windy conditions than in still weather. Once the relay gets switched on after the pre-programmed time elapses, it completes the third and final circuit.

2.3 Execution System

When the relay completes the final circuit, a high discharge Lithium Polymer (LiPo) battery sets out, passing high current through the Nichrome coil placed on the aerostat envelope. This high density current heats the coil to a temperature at which the envelope material starts burning, resulting in a hole, which allows the

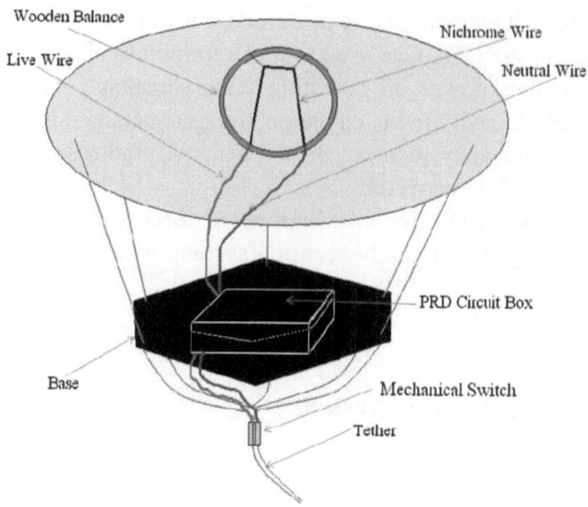

Fig. 1 Schematic of the PRD system

LTA gas to leak out from the aerostat envelope at a desired rate. The controller switches off the circuit automatically after a certain time, ensuring that the envelope is not damaged further.

Figure 1 presents the schematic of the PRD system showing its various components.

3 Field Testing of PRD Prototype

Field testing of the PRD prototype was carried out in the Gymkhana grounds of IIT Bombay in mid-September. The trials were conducted on a spherical aerostat of 2.75 m (9 ft.) diameter filled with helium. The PRD system was mounted on a platform attached at the Confluence Point.

The trials were carried out in the early hours just after sunrise when the ambient wind speed was recorded below 0.5 m/s by a hand-held anemometer. The first trial was conducted with a delay of 8 s, during which the aerostat rose to a height of around 60 feet and gradually came down after the circuit was activated. The small delay of 8 s ensured that the aerostat's trajectory was confined in a small area. Since the wind speed was low, the next trial was conducted with a delay of 24 s. This time the aerostat rose significantly higher than the previous case, but due to low ambient wind speed, the drift was almost negligible and the aerostat landed in the confined region. The PRD worked during both the trials and was successful in bringing down the aerostat without much damage to the payload. During these trials, a GoPro Digital Camera was used as the payload, to record its trajectory. After landing of the deflated aerostat envelope, the camera was found in pristine conditions without any damage (Fig. 2).

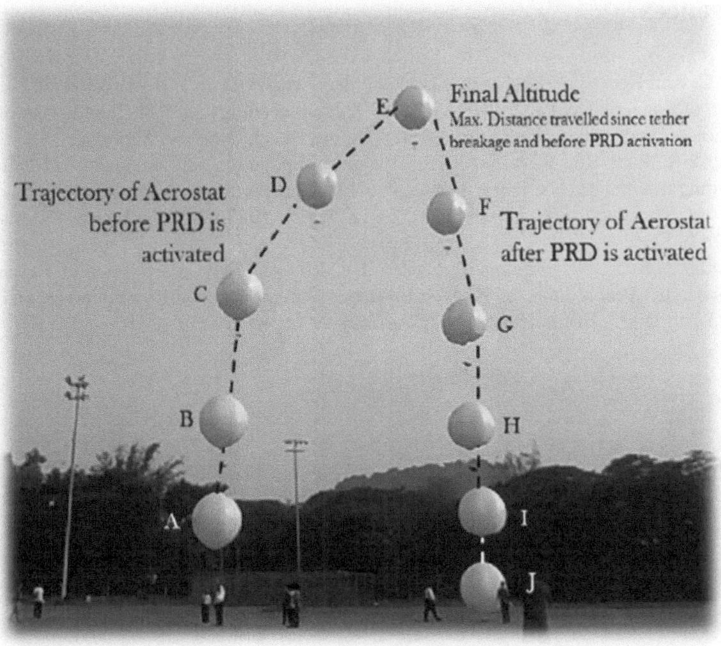

Fig. 2 Trajectory plot of the aerostat explaining various stages of PRD

4 Ongoing Work

An attempt was also made to create a mathematical model so as to predict the position of payload as it descends to ground. A point mass model developed by Krishna et al. [2] and used to verify the ascent of a HASI 2003 (Huygens Atmospheric Structure Instrument), a high altitude scientific balloon for which the experimental data were available. According to Palumbo et al. [3], the drag coefficient is a function of shape of the balloon, and hence they suggested the use of a variable drag co-efficient depending on Reynolds Number, Froude Number, and envelope shape for accurately modeling the equations of motion. The simulations were carried out with a constant drag coefficient as well as variable drag coefficient.

The PRD trials that were conducted had no instrumentation to record trajectory data and hence further trials are scheduled to take place so that the model can be verified. A six-degree of freedom model was also developed for aerostats but lacks sources for validation.

References

1. Bhat C, Pant RS (2011) Design of a payload recovery device in case of accidental breakage of tether of an aerostat. Paper AIAA-2011–7022, proceedings of 19th lighter-than-air (LTA) technology conference, 20–22 Sep 2011, Virginia Beach, Norfolk, Virginia, USA
2. Krishna R, Pant RS, Bodi K (2013) Dynamic simulation of breakaway tethered aerostat including thermal effects. Paper AIAA-2013-1341, proceedings of 20th AIAA lighter-than-air systems technology (LTA) conference, 25–28 March 2013, Daytona Beach, Florida. doi:10.2514/6.2013-1341
3. Palumbo R, Russo M, Filippone E, Corraro F (2007) ACHAB: analysis code for high-altitude balloons. In: Proceedings of AIAA atmospheric flight mechanics conference and exhibit, 20–23 Aug 2007, Hilton Head, South Carolina

Design of an Effective Subsonic Wing Cross-Section Using Viscous-Inviscid Interactive Method

L. Prabhu and J. Srinivas

Abstract In this paper, an approach for optimum airfoil design is proposed for enhancing the lift to drag ratio with respect to the base NACA 2312 airfoil for subsonic incompressible flow regime at four degree angle of attack. An iterative inverse design method is followed to obtain airfoil for desired lift to drag ratio. Differential evolutionary (DE) optimization technique with micro population and elitism is employed and lift to drag ratio has been increased in two ways: with and without constraint on the maximum thickness. An inviscid-viscous interactive method is used as a flow solver in which vortex panel method is used to solve the inviscid flow; Thwaite's method is used to get the laminar boundary layer; the transition is determined by Michel's relation and Head's method is used to get turbulent boundary layer. The approach is implemented with a user-interactive program developed in MATLAB and effectiveness of optimization scheme is validated with forward design tools.

Keywords Airfoil · Aerodynamic characteristics · Lift to drag ratio · Viscous-inviscid interaction · Differential evolution

1 Introduction

Design of modern aircraft relies more and more on the development of techniques reducing the drag forces and improvising the lift. So airfoil selection process and designing still remains as an important part of aircraft design. Airfoil designing is performed by two methods: one is direct and other is inverse approach. In direct method approaches, alternative formulation of the problem is employed so that the

L. Prabhu (✉) · J. Srinivas
Department of Mechanical, NIT, Rourkela 769008 Odhisa, India
e-mail: prabhulakshmananr@gmail.com

R. P. Bajpai et al. (eds.), *Innovative Design, Analysis and Development Practices in Aerospace and Automotive Engineering*, Lecture Notes in Mechanical Engineering, DOI: 10.1007/978-81-322-1871-5_13,

surface coordinates appear as dependent variables. Thus, the unknown boundaries and the flow-field are obtained together. Direct design methods provide the trial solutions close to required profiles. Conversely, inverse design techniques rely on repeated shape modifications such that each iteration consists of flow solution followed by geometry-updating scheme. For example, the method allows the pressure distribution to be directly controlled rather than anticipated from geometry perturbations. Iterative techniques like optimization methods are being increasingly used in solving the geometry/shape design problems. In optimization approach, inverse design problem is often coined as an error minimization task.

There are several methods starting from panel methods to Navier-Stokes flow solvers for analyzing the flow over an arbitrary body. The simple and low computational cost method is viscous-inviscid interaction (VII), which produces result that well matches with experiment. A collection of computational tools for formulation of potential flow over an airfoil and its solution was presented by Lipatov [1]. Peltonen [2] examined viscous-inviscid interactive method for subsonic flow which takes compressibility into account with the Prandtl-Glauret coordinate transformation by obtaining airfoil for target velocity distribution.

Numerical optimization procedures for design and development of high-lift systems configurations were developed by Kim et al. [3]. Fazil and Jayakumar [4] used reverse engineering to get the control points from the coordinates of an airfoil and these control points are used to control the airfoil profile in CAD environment.

Nonconventional optimization tools such as genetic algorithms, differential evolution particle swarm optimization, harmonic search method can be effectively employed in target-based inverse design applications. Parsopoulos [5] presented the work which combines the main concepts of cooperative evolutionary algorithms and micro evolutionary algorithms. Micro differential evolution (μDE) technique based on small number of population is often employed to achieve rapid convergence. In present work, μDE is used to achieve the desired lift to drag ratio of unsymmetrical airfoils. The wing cross section is achieved under certain constraints such as minimum lift coefficient, maximum drag coefficient, and the maximum thickness. The effectiveness of the solution is presented with forward design tool.

2 Mathematical Description

2.1 Parameterization of Airfoil

Design of an airfoil by optimization is of great interest for aviation industry. The process becomes tedious when all coordinates of geometry is considered for an optimization. To overcome the barrier, Bezier parameterization is used to obtain the airfoil geometry by few control points, which are physical points in the plane.

2.2 Panel Method

Viscous-Inviscid interaction method is used as the flow solver. Inviscid flow is solved by panel method [6] to calculate lift coefficient and pressure coefficient. Panel method is a technique for solving incompressible flow over geometry, in which airfoil surface is divided into piecewise straight line segments or panels. Vortex sheets of strength γ are placed on each panel. The density of circulation on each panel varies from one end to the other. The number of panels is m. At an angle of attack α, uniform flow velocity V_∞ and m vortex panels, the velocity potential at the ith control point is

$$\phi(x_i, y_i) = V_\infty(x_i \cos\alpha + y_i \sin\alpha) - \sum_{j=1}^{m} \int_j \frac{\gamma(s_j)}{2\pi} \tan^{-1}\left(\frac{y_i - y_j}{x_i - x_j}\right) \mathrm{d}s_j. \quad (1)$$

The velocity has only a tangential component at the panel surface because of the normal component is zero. Let t_i designate the unit tangential vector on the ith panel, the local dimensionless velocity defined as $(\partial\phi/\partial t_i)/V_\infty$. The pressure coefficient at the ith control point is determined by

$$C_{pi} = 1 - V^2. \quad (2)$$

2.3 Boundary Layer Model

In this part, the computation of an incompressible boundary layer flow over a surface under an arbitrary pressure gradient is presented. To carry out the computation, the stagnation point in the inviscid solution must be identified and the length x in the boundary layer is obtained using the panel discretization.

Laminar boundary layer growth is predicted by Thwaite's method, which relies on the momentum integral equation and it is shown below:

$$\frac{\mathrm{d}\theta}{\mathrm{d}x} + \frac{\theta}{V_e}(2 + H)\frac{\mathrm{d}V_e}{\mathrm{d}x} = \frac{1}{2}C_f. \quad (3)$$

$H = \delta^*/\theta$ which is called as shape factor of the boundary-layer velocity profile, and C_f is the skin friction coefficient.

The turbulent boundary layer is calculated using Head's method. It is based on entertainment velocity.

$$E = \frac{d}{\mathrm{d}x}(U_e \theta H_1) \quad (4)$$

where $H_1 = \frac{\delta - \delta^*}{\theta}$.

Head assumed that the dimensionless entrainment velocity E/V_e depends only on H_1 and that H_1, in turn, is a function of $H = \delta^*/\theta$.

Prediction of the transition point must be accurate for calculating drag. Michel's criteria is based on transition starting at a particular value of Reynolds number based on the distance x from the start of boundary layer. For incompressible flows without heat transfer, Michel [7] examined a variety of data and concluded that, for airfoil-type applications, transition should be expected when

$$\mathrm{Re}_\theta > 1.174\left(1 + \frac{22,400}{\mathrm{Re}_x}\right)\mathrm{Re}_x^{0.46}. \tag{5}$$

2.4 Computation of Drag Coefficient

The Squire-Young formula [8] shown below predicts the profile drag by using the momentum thickness, shape factor, dimensionless velocity at the trailing edge on upper and lower surface.

$$C_d = \left[2\theta_{\mathrm{TE}}(V_i)_{\mathrm{TE}}^{\frac{H_{\mathrm{TE}}+5}{2}}\right]_{\mathrm{UP}} + \left[2\theta_{\mathrm{TE}}(V_i)_{\mathrm{TE}}^{\frac{H_{\mathrm{TE}}+5}{2}}\right]_{\mathrm{LOW}}. \tag{6}$$

3 Present Methodology

In this present study, μDE is employed for optimization so the initial population size is taken as five and they are NACA 0012, NACA 1112, NACA 1412, NACA 2312, and NACA 2314 airfoils. Maximum generation considered is 100. The fitness function can be written as:

$$\text{Minimize } \frac{1}{\left[(Cl/Cd)_{\mathrm{Iter}} - (Cl/Cd)_{\mathrm{Baseline}}\right]} \tag{7}$$

Target $= \left|\frac{Cl}{Cd} \geq 72\right|_{\alpha=4}$ subject to $Cl \geq 0.79$ and $Cd \leq 0.01092$.

For angle of attack 4° with constraint over maximum thickness, the condition is $10 \leq$ max thickness ≤ 11 % of chord.

4 Results and Discussion

For the prediction of aerodynamic characteristics like lift coefficient and drag coefficient of airfoil design, VII MATLAB coding is employed. This code helps in flow analysis over subsonic regime. Flow considered is subsonic incompressible

Table 1 Coefficients of airfoils

Wing cross section	VII		Javafoil	
	C_l	C_d	C_l	C_d
NACA 2312	0.720	0.0104	0.646	0.0072
Optimized	1.008	0.0109	0.975	0.0072
Optimized (constraint over maximum thickness)	1.153	0.0092	1.179	0.0048

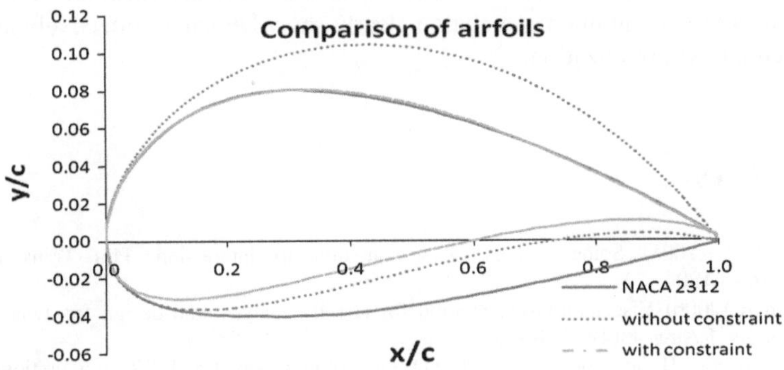

Fig. 1 Comparison of baseline and optimized airfoils

with Mach number 0.1, angle of attack 4°, and Reynolds number 686211. Table 1 shows the drag and lift coefficients of different airfoils. For the case without maximum thickness constraint, the DE successfully increased lift to drag ratio by 34 %. It drastically reduced the drag coefficient of wing cross section, which is the barrier for aircraft performance. The maximum thickness of optimized airfoil is 13 % of chord. Comparison of baseline and optimized airfoil is shown in Fig. 1.

When maximum thickness constraint is considered increment of lift to drag ratio to 125.32 which is 81.62 % compared to baseline is clearly obvious from Table 1. The maximum thickness of optimized airfoil is 10.44 % of chord which is within specified range.

4.1 Validation of VII with Javafoil

Computation of lift and drag coefficient in this work is done by VII which has to be validated, so the optimized airfoils are tested in Javafoil at same working environment, and the results are tabulated in Table 1.

Thickness in airfoil design. From Table 1 it is clearly shows for the maximum lift, the airfoil must be thin so that the lower surface produces maximum lift as well.

5 Conclusions

The VII method was used as a flow solver, and the results are very close when compared with Javafoil, which clearly shows that this method is fit for airfoil design. Optimization for aerodynamic performance is performed using μDE algorithm. Bezier curve was used for parameterization of airfoil geometry, optimization was performed with an in-house built MATLAB code. The drastic increments of the lift to drag ratio value with and without constraint over maximum thickness in minimum generation clearly show DE can be effectively utilized for such airfoil optimization.

References

1. Lipatov I (2005) Some models of viscous-inviscid interaction. Phil Trans R Soc 363:1169–1180
2. Pelton R (2006) Viscous-inviscid method for airfoils analysis and design for aviation and windmills. J Aircr 43(4):1069–1081
3. Kim S, Alonso JJ, Jameson A et al (2002) Design optimization of high-lift configurations using a viscous continuous adjoint method. In: Paper presented at the 40th AIAA aerospace sciences meeting and exhibit, Reno, NV, 14–17 Jan 2002
4. Fazil J, Jayakumar V (2011) Investigation of airfoil profile design using reverse engineering Bezier curve. APRN J Eng Appl Sci 6:43–52
5. Parsopoulos KE (2009) Cooperative micro-differential evolution for high-dimensional problems. In: Rothlauf F (ed) Proceedings of the 11th annual conference on genetic and evolutionary computation
6. Kuethe AM, Chow CH (1998) Foundations of aerodynamics. Wiley, New York
7. Michel R (1951) Etude de la transition sur les profiles d' ailes. ONERA report 1/1578A
8. Squire HB, Young AD (1938) The calculation of the profile drag of Aerofoils. R & M. no. 1838 A.R.C. Technical report, London

Computational Study of Various Longitudinal Fin Profiles

Pankaj Dhaka, Prashant Kumar and Varun Goel

Abstract In this paper, a computational study of longitudinal fins of different profiles has been carried out. For this purpose, various fin profiles have been modeled in CATIA and the further simulations were carried out in ANSYS Fluent. The present research is focused on the performance parameters of fins such as temperature distribution, fin effectiveness, and fin efficiency for different fin geometries. The values obtained from the CFD software have been compared with the theoretical values to validate the results. Also an attempt has been made to compare the weight and cost associated so as to develop a strong basis for the selection of fins.

Keywords Computational · Longitudinal fins · Simulations · CFD · Performance parameters

1 Introduction

Fins are the extended surfaces which are designed specifically to enhance heat transfer from heated surfaces. They find their immense applications in areas like automobiles and aerospace applications, refrigeration and cryogenic applications, electrical and electronic equipments, in gas turbines and boilers, either for cooling the heated surfaces or to keep the operating temperatures optimum for better performance. The demands for the above applications have placed particular

P. Dhaka (✉) · P. Kumar · V. Goel
Department of Mechanical Engineering, National Institute of Technology, Hamirpur, Himachal Pradesh, India
e-mail: pdhakanith@gmail.com

P. Kumar
e-mail: prashant98.nith@gmail.com

R. P. Bajpai et al. (eds.), *Innovative Design, Analysis and Development Practices in Aerospace and Automotive Engineering*, Lecture Notes in Mechanical Engineering,
DOI: 10.1007/978-81-322-1871-5_14,

emphasis on the compactness of the heat exchanger surfaces, fin efficiency, and fin effectiveness, particularly on those surfaces that induce small pressure gradients in the fluids circulated through them.

There are various ways by which the heat transfer rate from these extended surfaces may be increased. These include increasing fluid velocity, reducing fluid temperature, or increasing the surface area across which convection occurs. Generally reducing fluid temperature is impractical and compactness of lays restriction on the surface area to be used. Also increasing fluid velocity demands for extra power requirements to feed blower or pump. But yet another option is to optimize the profile of fin surfaces so as to avail improved performance of fins.

1.1 Literature Review

Various analytical studies have been carried out on the performance parameters of the fins. Fourier [1] and Despretz (1822, 1828a, b) published mathematical analyses of the temperature variation of the metal bars or rods. Harper and Brown [2] provided thorough analytical solutions for the two-dimensional model for both rectangular and wedge-shaped longitudinal fins and the circumferential fin of uniform thickness. Alessandro [3] provided the analytical method for the optimum thermal design of open and closed array configuration. Armstrong [4], in his paper presented the effect of various geometric parameters of a pin fin. Vanfossen [5] carried computations for the heat transfer through short pin fins. Natrajan et al. [6] have also studied the effect of thermal boundary conditions in case of pin fin arrays. Bejan and Morega [7] have presented the methodology for the optimization of geometry for pin fin array. But till date, no one has made a cumulative comparison of different profiles based on the efficiency, effectiveness, and economy of fins which can be useful in the selection of fins for a particular application. The objective of this study is to find out optimum type of fin profile.

2 Methodology

First, theoretical values have been obtained using mathematical formulae for the various fin profiles based upon the certain assumptions [8] which are

- The heat flow in the fin and its temperatures remain constant with time.
- The effect of radiation has been neglected.
- The fin material is homogeneous, its thermal conductivity is the same in all directions, and it remains constant.
- The convective heat transfer coefficient on the faces of the fin is constant and uniform over the entire surface of the fin.

- The temperature of the medium surrounding the fin is uniform.
- The fin thickness is small, compared with its height and length, so that the temperature gradients across the fin thickness and heat transfer from the edges of the fin may be neglected.
- The temperature at the base of the fin is uniform.
- There are no heat sources within the fin itself.
- The heat transferred through the tip of the fin is negligible compared with the heat leaving its lateral surface.
- Heat transfer to or from the fin is proportional to the temperature excess between the fin and the surrounding medium.

2.1 Longitudinal Fin with Rectangular Profile

The efficiency for rectangular fin is given by (Fig. 1)

$$\eta = \frac{\tanh mb}{mb}.$$

Dissipation of heat

$$q_b = k\delta_b L m \theta_b \tanh mb.$$

Temperature difference between any arbitrary point and ambient temperature

$$\theta_a = \frac{\theta_b \cosh mx}{\cosh mb}$$

where

$$m = \left(\frac{2h}{k\delta}\right)^{1/2}$$

and

$$\theta(x = b) = \theta_b.$$

2.2 Longitudinal Fin with Triangular Profile

The efficiency for triangular fin is given by (Fig. 2)

$$\eta = \frac{I_1(2mb)}{mbI_0(2mb)}.$$

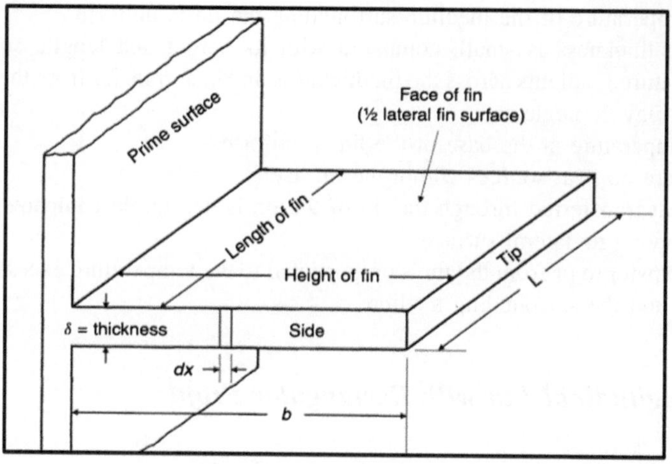

Fig. 1 Geometry of rectangular fin [8]

Fig. 2 Geometry of triangular profile [8]

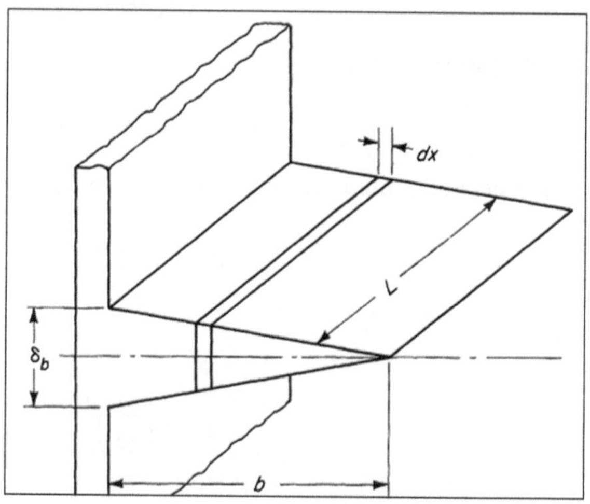

Dissipation of heat by the fin

$$q_b = \frac{2hL\theta_b I_1(2mb)}{mI_0(2mb)}.$$

Temperature difference between any arbitrary point and ambient temperature

$$\theta_a = \frac{\theta_b I_0\left(2m\sqrt{bx}\right)}{I_0(2mb)}.$$

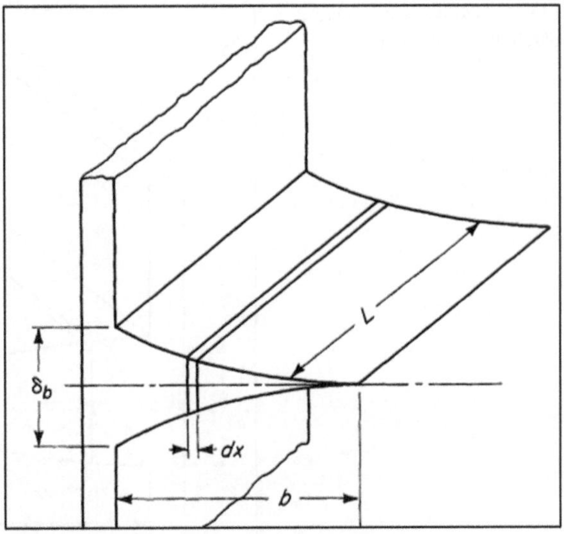

Fig. 3 Geometry of concave parabolic profile [8]

2.3 Longitudinal Fin with Concave Parabolic Profile

The efficiency for concave parabolic profile is given by (Fig. 3)

$$\eta = \frac{2}{1 + \sqrt{1 + (2mb)^2}}.$$

Dissipation of heat by the fin

$$q_b = \frac{k\delta_b L\theta_b}{2b}\left[-1 + \sqrt{1 + (2mb)^2}\right].$$

Temperature difference between any arbitrary point and ambient temperature

$$\theta(x) = \theta_b\left(\frac{x}{a}\right)^\alpha$$

where

$$\alpha = -0.5 + 0.5\ (1 + 4m^2b^2)^{0.5}.$$

2.4 Longitudinal Fin with Convex Parabolic Profile

The efficiency for convex parabolic profile is given by (Fig. 4)

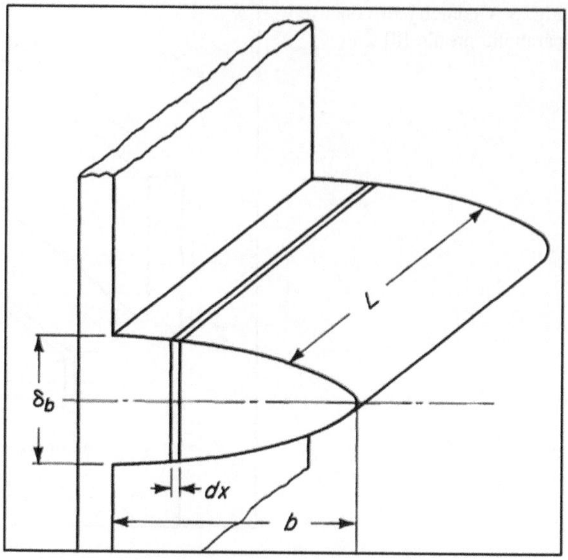

Fig. 4 Geometry of convex parabolic profile [8]

$$\eta = \frac{1}{mb}\frac{I_{2/3}\left(\frac{4}{3}mb\right)}{I_{-1/3}\left(\frac{4}{3}mb\right)}.$$

Dissipation of heat by the fin

$$q_b = k\delta_b L m\theta_b \frac{I_{2/3}\left(\frac{4}{3}mb\right)}{I_{-1/3}\left(\frac{4}{3}mb\right)}.$$

Temperature difference between any arbitrary point and ambient temperature

$$\theta(x) = \theta_b \left(\frac{x}{b}\right)^{\frac{1}{4}} \frac{I_{-1/3}\left(\frac{4}{3}mb^{1/4}x^{3/4}\right)}{I_{-1/3}\left(\frac{4}{3}mb\right)}.$$

3 Computational Analysis

For computational analysis, first of all, the geometric modeling of various fin profiles was done using Catia V5. Further analysis was carried out by importing these geometries in Ansys Fluent. The various input parameters for the computational analysis are:

Table 1 Meshing statistics

Type of fin	Number of nodes	Number of elements
Rectangular	1,104	476
Triangular	1,104	680
Concave parabolic	1,104	680
Convex parabolic	1,104	708

3.1 Solver

Pressure-based solver was used in the analysis because it solves the equations (continuity, momentum and energy) in a segregated or uncoupled manner and requires lesser memory which was the requirement of present analysis segregated or uncoupled manner. *Steady-state* analysis has been carried out and Energy equation has been solved to compute temperature values along the surface of fin and plot the temperature contours (Table 1).

3.2 Geometry Parameters

- Width of the fin base $\delta_b = 0.8\,\text{cm}$
- Fin height b = 10 cm
- Fin length L = 1 cm.

3.3 Material Properties

- Density = 8,030 kg/m^3
- Specific heat = 502.48 j/kg-K
- Thermal conductivity = 30 W/m-K.

3.4 Boundary Conditions

- Temperature of fin base = 363 K
- Ambient temperature = 293 K
- Convective heat transfer coefficient = 40 W/m^2-K.

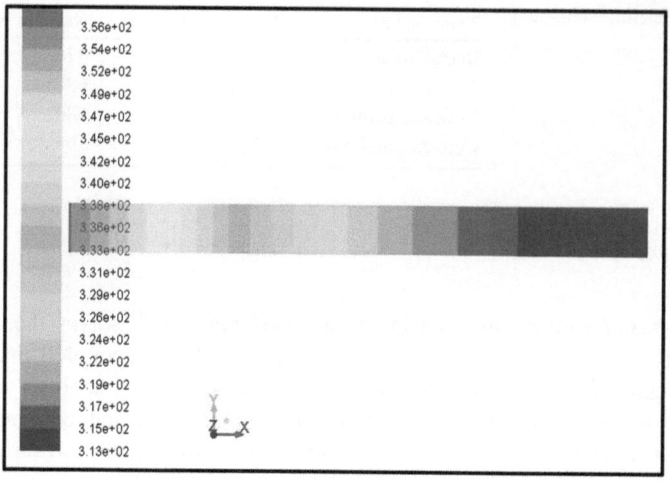

Fig. 5 Temperature contours for rectangular fin

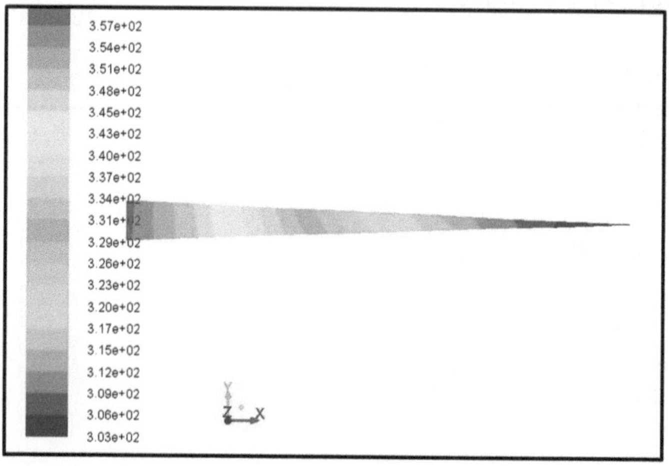

Fig. 6 Temperature distribution along the length of the rectangular fin

4 Results and Discussion

By performing the Ansys analysis, we arrived at following results which are presented in the form of temperature contours and plots of temperature along the fin profiles (Figs. 5, 6, 7, 8, 9, 10, 11, and 12).

Fig. 7 Temperature contours for triangular fin

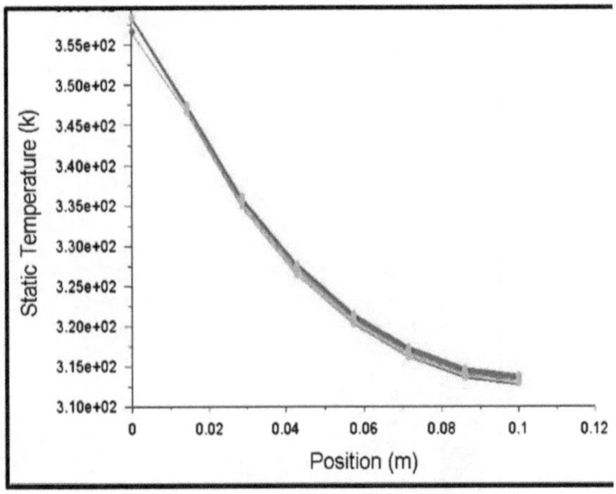

Fig. 8 Temperature distribution along the length of the triangular fin

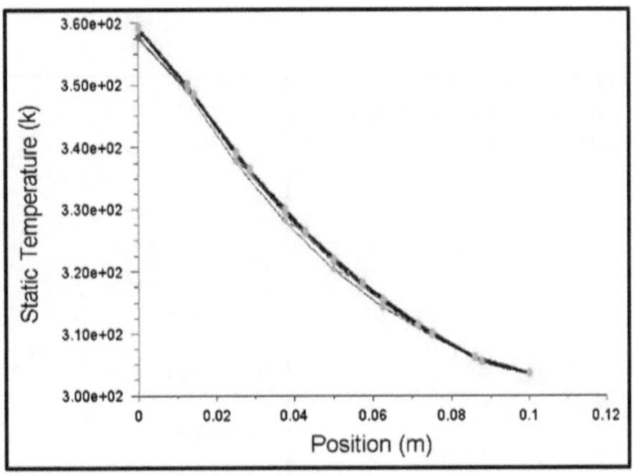

Various performance parameters obtained through theoretical calculations [8] are given in Table 2.

Various performance parameters obtained through computational analysis are given in Table 3.

Percentage error between theoretical and computational values are given in Table 4.

To consider the comparison of cost among different fin profiles their respective volume of materials are compared, which are given in Table 5.

By analyzing the results formulated in Tables 2 and 3, it can be seen that the theoretical and computational results are in close proximity to each other. The percentage error varies between 0.64 and 10 % which can be attributed to the

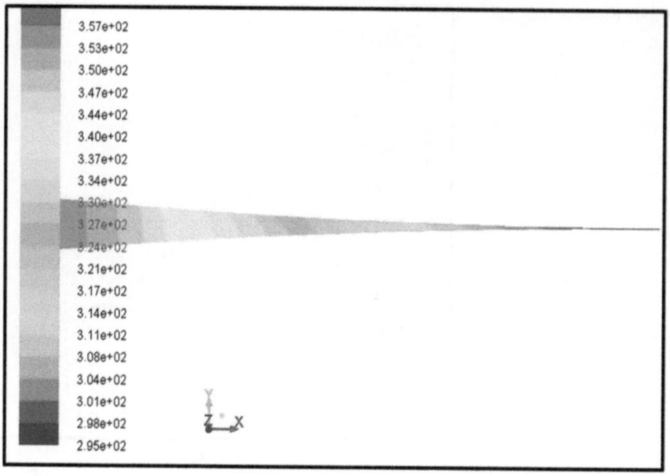

Fig. 9 Temperature contours for concave profile fin

Fig. 10 Temperature distribution along the length of the concave profile fin

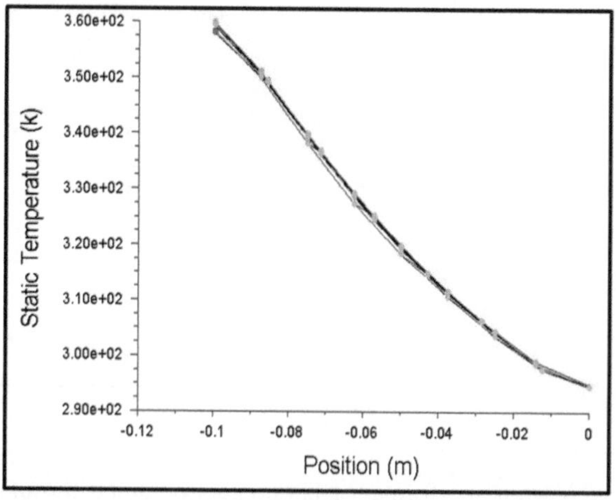

coarse mesh used for the analysis. The results can be further improved by refining the mesh depending on the computational and memory capacity of the system.

Also, both the theoretical and computational results depict that the rectangular profile fin has the highest efficiency whereas the concave parabolic profile fin has the least efficiency amongst the analyzed profiles.

Further, from Table 5, it is clear that the material requirement for the rectangular profile fin is maximum leading to additional cost as compared to the triangular and concave parabolic profiles which require least material.

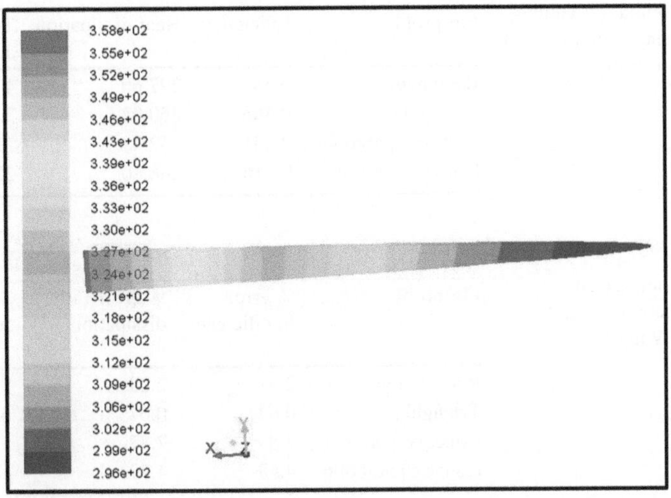

Fig. 11 Temperature contours for convex profile fin

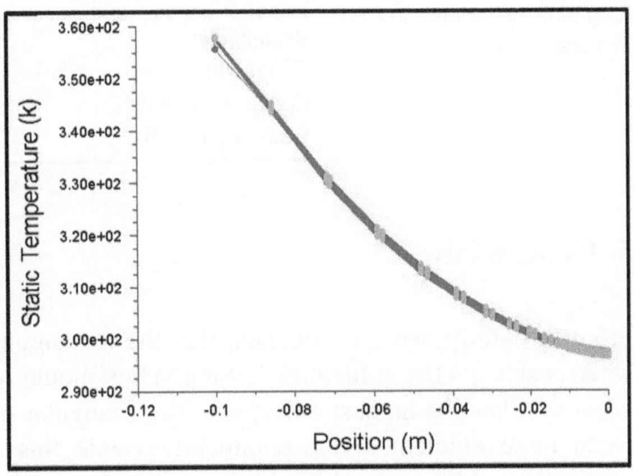

Fig. 12 Temperature distribution along the length of the convex profile fin

Table 2 Theoretical values of Performance Parameters of different Fin Profiles

Fin profile	Efficiency	Heat dissipation	Temperature at the tip
Rectangular	0.520	291.1	44.0
Triangular	0.468	261.8	28.4
Concave parabolic	0.418	234.0	20.0
Convex parabolic	0.492	275.6	20.0

Table 3 Simulation values of Performance Parameters of different Fin Profiles

Fin profile	Efficiency	Heat dissipation	Temperature at the tip
Rectangular	0.532	297.60	40.0
Triangular	0.465	260.02	30.0
Concave parabolic	0.451	252.4	22.0
Convex parabolic	0.516	288.80	22

Table 4 Comparison between Theoretical and Simulation Values

Fin profile	% error in efficiency	% error in heat dissipation	% error in temperature at the tip
Rectangular	2.31	2.23	9.09
Triangular	0.64	0.67	5.63
Concave parabolic	7.89	7.87	10.0
Convex parabolic	4.88	4.79	10.0

Table 5 Material Requirement for different Fin Profiles

Fin profile	Volume of material ($\times 10^{-4}$ m^3)
Rectangular	8.0
Triangular	2.691
Concave parabolic	2.691
Convex parabolic	5.328

5 Conclusion

From this study, we can conclude that the rectangular fin profile should be used when cost is not the influential factor and fins should be used for small applications because it has the highest efficiency. The triangular profile fin should be installed when more efficient and economically viable fins are required for large-scale applications. Moreover, the selection of fins is greatly determined by the cost and efficiency of the fins but the manufacturing of the fins is also an influential factor in case of complex profiles such as concave and convex parabolic profiles.

6 Future Work

This work can be further extended by incorporating the effect of heat transfer through radiation and forced convection. Also the effect of boundary layer interference between the adjacent fins on heat transfer can be considered.

Acknowledgments We gratefully acknowledge the computational facilities and resources provided by Department of Mechanical Engineering, NIT Hamirpur. We also express our sincere gratitude toward Dr. Somesh Sharma and Dr. Rakesh Sehgal for giving us valuable suggestions and guidelines during the course of work.

References

1. Fourier JB, Théorie analytique de la chaleur, Paris, 1822 (English translation by A. Freeman, The Analytical Theory of Heat, Dover Publications, Inc., New York, 1955)
2. Harper RR Brown WB (1923) NACA Technical Report 158
3. Franco A (2009) An analytical method for the optimum thermal design of convective longitudinal fin arrays. Heat Mass Transf 45(12):1503–1517
4. Armstrong J, Winstanley D (1988) A review of staggered array pin fin heat transfer for turbine cooling applications. ASME J Turbomach 110:94–103
5. Van Fossen GJ (1981) Heat transfer coefficients for staggered arrays of short pin fins. ASME paper no. 81-GT-75
6. Chyu MK, Hsing YC, Shih TI-P, Natarajan V (1998) Heat transfer contributions of pins and endwall in pin-fin arrays: effects of thermal boundary condition modeling. ASME paper no. 98-GT-175
7. Bejan A, Morega AM (1993) Optimal arrays of pin fins and plate fins in laminar forced convection. J Heat Transf 115:75–81
8. James W, James WR, Kraus Allan D (2001) Extended surface heat transfer, 10th edn. Wiley, USA, pp 1–59

Dynamic Modeling and Simulation of Flapping Wings UAV

Balasubramanian Esakki, Vasantharaj Rajagopal and Rusendar Babu Srihari

Abstract Flapping wing UAVs, or Ornithopters, are becoming popular compared to rotary wing because they are less noisy and durable while maintaining same maneuverability. Recent advances in small-scale flapping-wing microaerial vehicles have extended the capabilities of flight control for a number of applications, such as intelligence, surveillance, and reconnaissance activities. Ornithopters have not advanced to autonomous control because their lift and thrust are generated from the same flapping mechanism which couples the forces, unlike a fixed wing which has separate lifting and propulsions systems. In this paper, a state space model of an Ornithopter is studied and a pole placement control is implemented for its stability. In addition, the coupled motion dynamics of Ornithopter has been studied and a computed torque control algorithm is implemented. Furthermore, an Ornithopter is designed and the real-time test is also carried out in the outdoor. In future, an autopilot system with the developed control algorithm will be implemented for the autonomous flight.

Keywords Ornithopter · Computed torque control · State space model · Autonomous flight · Pole placement

B. Esakki (✉) · V. Rajagopal
Vel Tech Dr. RR and Dr. SR Technical University, Chennai, India
e-mail: esak.bala@gmail.com

V. Rajagopal
e-mail: mail2vasantharaj@gmail.com

R. B. Srihari
Vel Tech High Tech Dr. Rangarajan and Dr. Sakunthala Engineering College, Chennai, India
e-mail: naveenhari8@email.com

R. P. Bajpai et al. (eds.), *Innovative Design, Analysis and Development Practices in Aerospace and Automotive Engineering*, Lecture Notes in Mechanical Engineering, DOI: 10.1007/978-81-322-1871-5_15,

1 Introduction

Micro Air Vehicles (MAVs) are emerging category of flights that can be used in numerous applications, such as intelligence, surveillance, and reconnaissance activities. Especially flapping wing vehicles have attracted by the researchers compared with fixed-wing and rotary-wing vehicle. The fixed-wing vehicle is lacking hovering capability and poor maneuverability. The rotary wing vehicles are noisier and designs of such systems are complex. Flapping wing vehicles (Ornithopters) are the practical solution to achieve good maneuverability, reduced noise, improve stealth property and some of the biological-inspired vehicles have hovering capability. Compared with fixed and rotary wing vehicles, Ornithopters are still in the emerging trend in terms of analytical and experimental research. Ornithopter autonomy is intricate since the kinematic, aerodynamics, guidance, and navigations are much more complex in fixed and rotary wing vehicle. Unlike fixed and rotary wing vehicle, the Ornithopters have capability to produce lift and thrust simultaneously by twisting the wings throughout the flapping stroke. The design and development of flapping mechanism is being a challenging problem and there were more efforts made to mimic the birds and also the insect-like wing motion using rigid mechanisms [1–5]. However, stable and hovering model of Ornithopters are rare in the field of flapping wing vehicles. The control algorithm is complex and autopilot system is necessary. The hovering control of flapping wing was developed in [6] which is used to navigate the vehicle at certain altitude. The attitude control was improved with an aid of light sensors and rotational velocity detectors [7]. The detailed controllability analysis of flapping wing vehicle can be in seen in [8]. The sensory-based control [9, 10], velocity, and altitude control [11] of flapping wing vehicle and low-cost sensor vehicle control [12] have been developed. In this paper initially two-dimensional dynamic models of Ornithopters are considered and corresponding state space model will be derived. The pole placement control algorithm is implemented to achieve the control of position and orientation (pose) of an Ornithopter. Considering the simplified biological wing analogy, three-dimensional model of Ornithopters considering rolling and flapping motion is studied. The computed torque control algorithm is being implemented to control the coupled motion of the system. Simulation results are presented to validate the control algorithms. Furthermore, an Ornithopter is designed and the real-time test also carried out in the outdoor. In future, an autopilot system with the developed control algorithm will be implemented for the autonomous flight.

2 System Modeling

This paper describes the two-dimensional model and coupled oscillator model to study the dynamics of an Ornithopter.

Fig. 1 Two-dimensional model of ornithopter

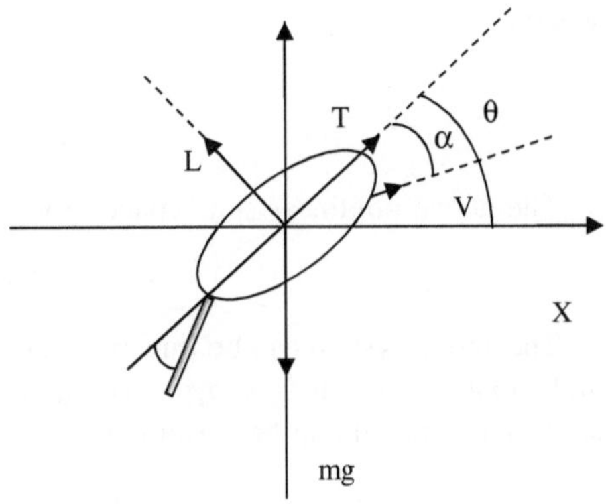

2.1 Two-Dimensional Model of an Ornithopter

The Fig. 1 depicts the two-dimensional representation of Ornithopter structure. The motion of the Ornithopter at a velocity V at an angle θ is considered to resolve the thrust force T and lift force L components in the two dimensions. The angle of attack α can be varied to achieve required thrust as well as lift forces. Also, the tail angle φ is used to change the orientation of Ornithopter.

The equations of motion can be written as [11],

$$\begin{aligned} m\ddot{x} + C_x\dot{x} &= T\cos\theta - L\sin\theta \\ m\ddot{y} + C_y\dot{y} &= T\sin\theta + L\cos\theta - \text{mg}. \\ \dot{\theta}I + C_\theta\ddot{\theta} &= -C_t\delta_E \end{aligned} \tag{1}$$

The parameters of an Ornithopter considered are $m = 230$ g and wingspan of 1.2 m. The damping values C_x and C_y are varied based upon the desired damping characteristics. The rotational damping C_θ and torque constant C_t also can be varied based upon the rotational inertia.

In order to analyze the system a state-space model of the system is necessary. A Nonlinear state-space representation of 2D dynamics is formulated as,

$$x = \begin{bmatrix} x_1 \\ x_2 \\ x_3 \\ x_4 \\ x_5 \\ x_6 \end{bmatrix} = \begin{bmatrix} x \\ \dot{x} \\ y \\ \dot{y} \\ \theta \\ \dot{\theta} \end{bmatrix} \quad \Longrightarrow \quad \dot{x} = \begin{bmatrix} \dot{x}_1 \\ \dot{x}_2 \\ \dot{x}_3 \\ \dot{x}_4 \\ \dot{x}_5 \\ \dot{x}_6 \end{bmatrix} = \begin{bmatrix} \dot{x} \\ \ddot{x} \\ \dot{y} \\ \ddot{y} \\ \dot{\theta} \\ \ddot{\theta} \end{bmatrix} = \begin{bmatrix} x_2 \\ \frac{u_1}{m}\cos x_5 - \frac{b_x}{m}x_2 \\ x_4 \\ \frac{u_1}{m}\sin x_5 - \frac{b_x}{m}x_4 - g \\ x_6 \\ -\frac{k_t}{I}u_2 - \frac{b_0}{I}x_6 \end{bmatrix} \tag{2}$$

where,

$$u = \begin{bmatrix} u_1 \\ u_2 \end{bmatrix} = \begin{bmatrix} T \\ \delta_E \end{bmatrix}.$$

The above nonlinear state-space representation is given by,

$$\overset{\bullet}{x}_t = S(x(t), u(t), t). \tag{3}$$

The above system can be linearized at an equilibrium point with initial state x_0, and input u_0, which is subject to small perturbations and corresponding linear state-space model can be formulated as,

$$\begin{aligned} \overset{\bullet}{x}_0(t) + \overline{x}(t) &= S(x_0(t),^\bullet u_0(t) + \overline{u}(t), t) \\ &= S(x_0(t),^\bullet u_0(t), t) + \frac{\partial S}{\partial x}\overline{x} + \frac{\partial S}{\partial u}\overline{u} + \ldots.., \end{aligned} \tag{4}$$

The above can be rewritten as:

$$\overset{\bullet}{\overline{x}}(t) = \frac{\partial S}{\partial x}\overline{x} + \frac{\partial S}{\partial u}\overline{u} = A(t)\overline{x} + B(t)\overline{u}. \tag{5}$$

In order to simulate the vertical hovering condition, θ is approximately considered to be equal to 90° and the angle of attack, α, is approximately equal to zero is substituted in (2).

The linear state-space model is formulated as

$$\overset{\bullet}{X} = AX + BU \tag{6}$$

where,

$$A = \begin{bmatrix} 0 & 1 & 0 & 0 & 0 & 0 \\ 0 & -\frac{b_x}{m} & 0 & 0 & -g & 0 \\ 0 & 0 & 0 & 1 & 0 & 0 \\ 0 & 0 & 0 & -\frac{b_y}{m} & 0 & 0 \\ 0 & 0 & 0 & 0 & 0 & 1 \\ 0 & 0 & 0 & 0 & 0 & -\frac{b_\theta}{I} \end{bmatrix} \quad \text{and} \quad B = \begin{bmatrix} 0 & 0 \\ 0 & 0 \\ 0 & 0 \\ \frac{1}{m} & 0 \\ 0 & 0 \\ 0 & -\frac{k_t}{I} \end{bmatrix}.$$

The Eq. (6) will be used for developing a controller to understand and describe the dynamics of a two-dimensional Ornithopter.

2.2 Coupled Motion Dynamics

In order to determine the dynamic equations of motion of an Ornithopter, the kinetic energy and potential energy of the system as shown in Fig. 2 have to be determined. The proposed 2-DOF system which has flapping motion ψ and rotational motion φ is to develop the dynamics [13].

Fig. 2 Coupled motion

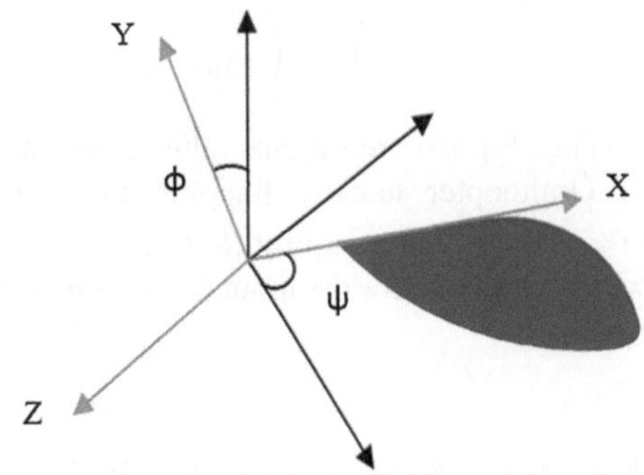

The kinetic energy given by,

$$T = \frac{1}{2} I \, \dot{\psi}^2 + \frac{1}{2} J \, \dot{\phi}^2 \tag{7}$$

where J is the inertia of the wing holder and I is the inertia of the wing in the wing frame.

The potential energy is given by,

$$V = \mathrm{Mgz}_1 \cos\psi + \frac{1}{2} K_1 \psi^2 + \frac{1}{2} K_2 \phi^2 + \mathrm{Mgz}_2 \cos\phi \tag{8}$$

where, M is the mass of the wing. z_1 represents the amplitude of flapping motion and z_2 represents the amplitude of rotational motion. K_1 represents the flapping motion spring constant and K_2 denotes the torsional spring constant.

Applying Lagrange's principle the equation of motion of coupled dynamics can be written as:

$$M(\Omega)\,\ddot{\Omega} + C(\Omega, \dot{\Omega}) + G(\Omega) = \tau \tag{9}$$

where, $M(\Omega)$ is the inertia matrix given by,

$$M(\Omega) = \begin{pmatrix} \sin^2\psi I_x + \cos^2\psi, I_z, +J & \cos\psi, I_{yz} \\ \cos_{yz} & I_y \end{pmatrix}$$

$C(\Omega, \dot{\Omega})$ is the centrifugal force which is given by

$$C(\Omega, \dot{\Omega}) = \begin{pmatrix} \sin 2\psi (I_x - I_z)\, \dot{\phi}\, \dot{\psi} - \sin_{yz} \dot{\psi}^2 \\ \frac{1}{2} \sin 2\psi (I_x - I_z)\, \dot{\phi}^2 \end{pmatrix}.$$

The gravitational force term $G(\Omega)$ is given by

$$G(\Omega, \dot{\Omega}) = \begin{pmatrix} K_1\phi + K_2\psi \\ \mathrm{Mgz}_1 \sin\psi + K_1\psi + \mathrm{Mgz}_2 \sin\phi + K_2\phi \end{pmatrix}.$$

The Eq. (9) represents the coupled motion three-dimensional dynamics of Ornithopter namely flapping as well as rotational motion of Ornithopter $\Omega = (\phi, \psi)^T$ will be useful to formulate a control algorithm. The control algorithm will be designed for the input torque τ to achieve the desired flapping and rotational motion.

3 Control System and Analysis

3.1 Control of Two-Dimensional Model

The position and orientation of two-dimensional Ornithopter is controlled by pole placement technique. The linear system is asymptotically stable when the real parts of system matrix are negative which are nothing but poles. Basically pole placement technique [14] designs feedback control wherein real part of the poles of placed in different locations according to the stability. This technique offer good stability without big overshoots and oscillations. The poles are chosen in such way that they should not be too close to each other otherwise it causes numerical stability. In addition, the poles cannot be too far from the origin which may cause the controller too hard in terms of necessary control inputs.

Poles are placed arbitrarily in the imaginary axis and corresponding control gains are given below.

$$\begin{aligned} K = [&8.5229 \quad 1.8788 \quad 0.3333 \quad 0.0501 \quad -1.5324 \quad -0.0407 \\ &0.0073 \quad 0.0020 \quad 0.0000 \quad 0.0000 \quad -0.0025 \quad -0.0001]. \end{aligned}$$

Simulation results are presented to describe the behavior of two-dimensional linear state-space model of an Ornithopter.

3.2 Control of Coupled Dynamics Model

The computed torque control law [15] can be written as:

$$\tau = MV + C(\Omega, \dot{\Omega}) + G(\Omega) \tag{10}$$

where,

$$V = \ddot{q}_d - 2\lambda\, \dot{\tilde{q}} - \lambda^2 \tilde{q}.$$

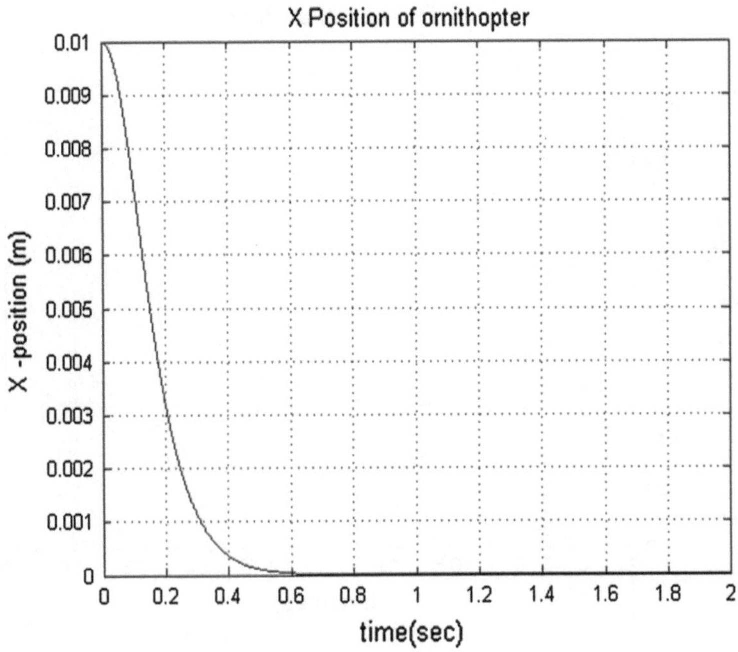

Fig. 3 X position of ornithopter

Substitution of control torque (10) into the coupled dynamics given in (9) yields

$$\dot{\tilde{q}} + 2\lambda\dot{\tilde{q}} + \lambda^2\tilde{q} = 0 \tag{11}$$

where, the error is given by, $\tilde{q} = q - q_d$.

The controller will compensate for the nonlinearity in Eq. (10) with the aid of quadratic equation and formulates second-order error differential equation given in Eq. (11). It shows that the system will be asymptotically stable.

4 Simulation Studies

Simulation has been carried out for the two-dimensional dynamics of an Ornithopter placing the poles at different locations using pole placement method. The initial conditions are considered to be (0.01, 0.02 m and 0.05 rad). The poles are placed at $V = [-60, -80, -12, -22, -25, -20]$. The simulation results show that (Figs. 3, 4 and 5), the position and orientation reaches the equilibrium in short span of 0.6 s. The coupled oscillator model dynamics was controlled with the aid of computed torque control method. The rotational motion and flapping motion of the wing was controlled within 5 s which is shown in Figs. 6 and 7.

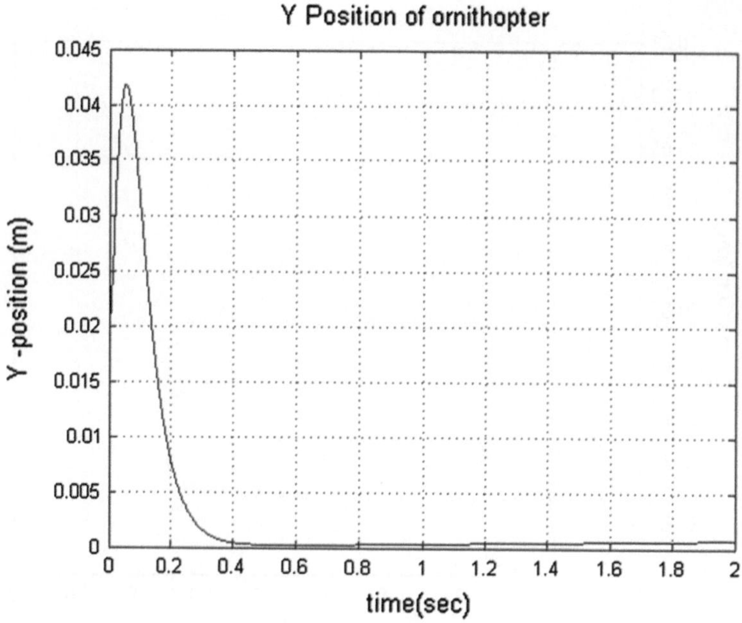

Fig. 4 Y position of ornithopter

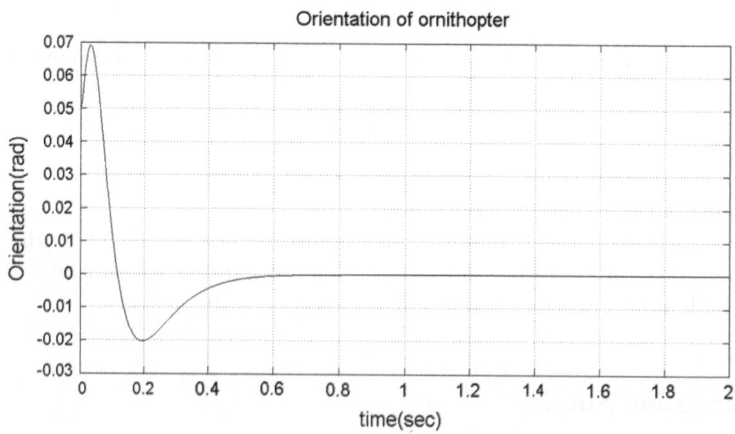

Fig. 5 Orientation of ornithopter

5 Design of an Ornithopter

The Construction of the wings and tail are made by assembling the carbon rods with the appropriate dimensions by placing the mylar sheets over them. The carbon rods and the Mylar Sheets exhibit excellent mechanical and thermal

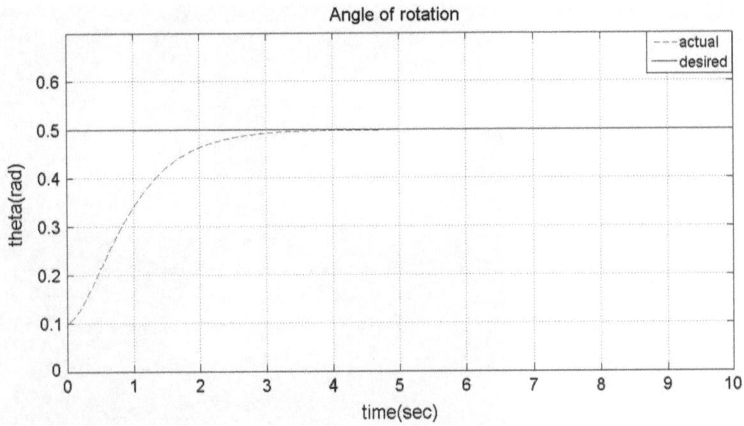

Fig. 6 Rotational motion of ornithopter

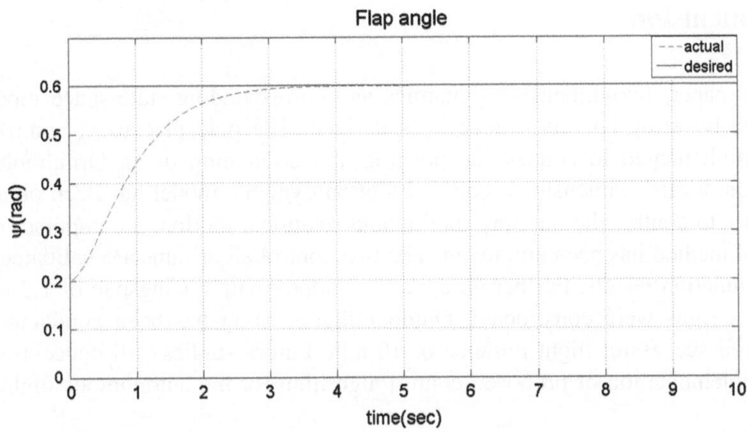

Fig. 7 Flapping motion of ornithopter

properties which are very useful for designing Ornithopters. The carbon rods were placed in the respective place and construction of Ornithopter structure has been made using mylar sheets in the wings as well as in the tail which is shown in Fig. 8. The wingspan of 120 cm and weight of 230 g Ornithopter have been designed. The flight tests were carried out in the outdoor environment and 15–20 min endurance was achieved.

Fig. 8 Full scale ornithopter

6 Conclusion

In this paper, two-dimension dynamics and corresponding state-space model at vertical hovering condition have been derived. The pole placement control law was implemented to control the position and orientation of an Ornithopter. In addition, a three-dimensional coupled motion dynamic model has been obtained. In order to control the flapping motion and rotational motion, a computed torque control method has been employed. The two-control algorithms are validated with the simulation results. Furthermore, an Ornithopter with a wingspan of 1.2 m and 230 g weighs were constructed. Outdoor flight testing has been conducted and achieved successful flight durance of 10 min. Future studies will concentrate on the implementation of proposed control algorithm for the autonomous flight.

References

1. Zbikowski R, Galinski C, Pedersen BC (2005) Four-bar linkage mechanism for insect like flapping wings in hover: concept and an outline of its realization. J Mech Des 127(4):817–824
2. Banala SK, Agrawal SK (2005) Design and optimization of a mechanism for out-of-plane insect winglike motion with twist. J Mech Des 127(4):841–844
3. McIntosh SH, Khan ZA, Agrawal SK (2006) Design of a mechanism for biaxial rotation of a wing for a hovering vehicle. IEEE Trans Mechatron 11(2):145–153
4. Raney LD, Slominski CE (2003) Mechanization and control concepts for biologically inspired micro aerial vehicles. In: AIAA guidance, navigation and control conference, 11–14 Aug 2003
5. Fearing R, Chiang K, Dickinson M, Pick D, Sitti M, Yan J (2000) Transmission mechanism for a micromechanical flying insect. Proceedings of the IEEE international conference on robotics and automation, pp 1509–1519

6. Deng X, Schenato L, Sastry S (2003) Model identification and attitude control for a micromechanical flying insect including thorax and sensor models. In: IEEE international conference on robotics and automation, pp 1152–1157
7. Schenato L, Wu WC, Sastry S (2004) Attitude control for a micromechanical flyinginsect via sensor output feedback. Rob Autom IEEE Trans 20(1):93–106
8. Schenato L, Campolo D, Sastry S (2003) Controllability issues in flapping flight for biomimetic micro aerial vehicles (MAVs). In: IEEE international conference on decision and control, pp 6441–6447
9. Kang CW, Park CG (2009) Attitude estimation with accelerometers and gyros using fuzzy tuned Kalman filter. In: The European control conference, Budapest, pp 3713–3718
10. Kingston B, Beard AW (2004) Real-time attitude and position estimation for small UAVs using low-cost sensors. In: AIAA 3rd unmanned unlimited technical conference, Workshop and exhibit, pp 6484–6488
11. Shigeoka KS (2007) Velocity and altitude control of an ornithopter micro aerial vehicle. Master's thesis, The University of Utah
12. Tayebi A, McGilvray S, Roberts A, Moallem M (2007) Attitude estimation and stabilization of a rigid body using low-cost sensors. In: IEEE conference on decision and control, New Orleans, pp 6424–6429
13. Khan ZA, Agrawal SK (2007) Design and optimization of a biologically inspired flapping mechanism for flapping wing micro air vehicles. In: IEEE international conference on robotics and automation, pp 373–378
14. Ogata K (2010) Modern control engineering. Prentice Hall, Englewood Cliffs
15. Sankaran J (1997) Real-time computed torque control of flexible-joint robots, MASc thesis, University of Toronto

Off Line Performance Measures of Two Wheeler Engine: In Automatic Objective Method

G. Mugunthan, E. Balasubramanian, R. Madhavan and Jebin Sing

Abstract This paper deals about the performance measures such as of kick force and gear shift force of two wheeler engine without human intervention. The automatic objective method has been designed in a solid modelling package which cater multi-variant of engine through adjustable exhaust system, modular clamping and load system, and also by servo-based force application and load measurement. Simulation studies were performed by controlled force application to kick lever of engine there by measuring the kick force required to start an engine and number of kick required. A geared stepper motor setup has been designed for gear shift with load cell for load measurement and also clutch actuation setup is incorporated. A servo setup to accelerate engine to specific speed and also fuel consumption verification is carried out by applying desired load through drive shaft engine. The proposed work replaces the human intervention effort in the case of manual test rig which increases the quality and productivity in the Automobile sectors. At the outset the simulation studies will eliminate the error on the various performance measures of the two wheeler engine which helps in increasing the product life-cycle management.

Keywords Kick starter · Engine acceleration · Engine gear shifting system

1 Introduction

During the past decades, tremendous efforts have been made to optimise the parameters of kick starter system, acceleration system, and gear shifting system to improve the riders comfort. To measure the various performance of engine like power,

G. Mugunthan (✉) · E. Balasubramanian · R. Madhavan · J. Sing
Department of Mechanical Engineering, Vel Tech Dr. RR & Dr. SR Technical University, #42 Avadi-Vel Tech Road, Avadi, Chennai 600062, Tamilnadu, India
e-mail: mugunthan1987@gmail.com

E. Balasubramanian
e-mail: esak.bala@gmail.com

R. P. Bajpai et al. (eds.), *Innovative Design, Analysis and Development Practices in Aerospace and Automotive Engineering*, Lecture Notes in Mechanical Engineering,
DOI: 10.1007/978-81-322-1871-5_16,

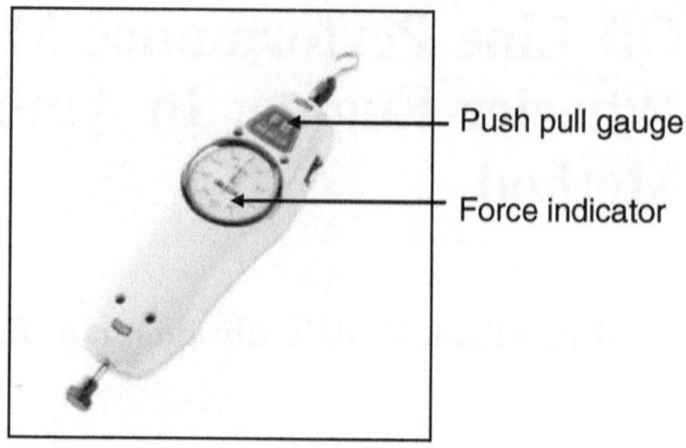

Fig. 1 Push pull gauge for force measurement

torque, fuel consumption dynamometers were used [1–3]. Various researches were carried out to improve all the above-mentioned fictional parameters [4–10]. Force required for actuation of kick lever for starting the engine and gear lever for gear shifting is measured subjectively using push pull gauge as shown in Fig. 1.

On the other hand, manual measurement of kicking force and gear shifting force leads to bottoming error, i.e. lever stopped by mechanical stopper [11–14]. It is difficult to stop manual operation to avoid bottoming error and another drawback is data logging which depends upon mechanical system which accumulated with measurement error as well. In addition, to perform the functions effectively without much error, a high-skilled man power was required to perform the operation.

It is evident from the above literature that, normally testing methods in kick starter system, acceleration system, and gear shifting system as an individual subsystem is successful in objective way. However, as the combination of three in an engine level stand still as a subjective manner. In this method, experienced testers used to check comfort both in research and development environment and also mass production environment. In order to measure the dynamic response of the two wheeler at engine level, an extensive micro controller-based data logging system is necessary. However, they are available with high cost and also many unknown factor will influence the riders comfort such as rider's action, environmental conditions, and also road parameter which are difficult to control. This paper overcomes the above-mentioned difficulties with an aid of engine performance simulator. Section 2 discuss about the conceptual modeling of engine performance simulator with four subsystems. Each subsystems tests in an objective manner has been extensively described in Sect. 3. Hardware in the loop simulation results are carried and examined in Sect. 4. The conclusion has been made at the end to show that riders comfort is improved with a help of engine performance simulator model.

2 Conceptual Model of Engine Performance Simulator

The conceptual model of engine performance simulator comprises of four subsystems as shown in Fig. 2.

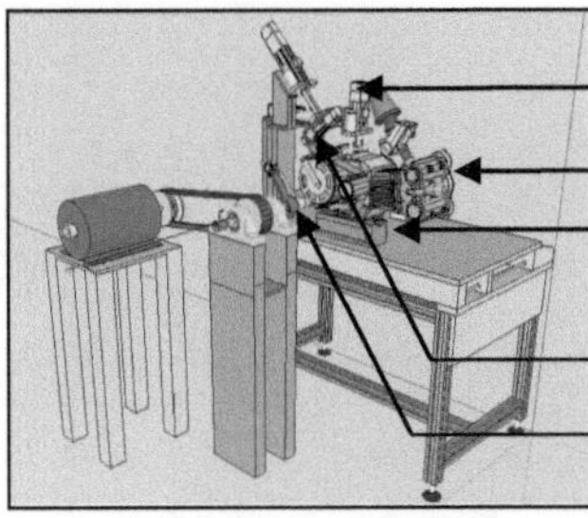

Fig. 2 Engine performance simulator

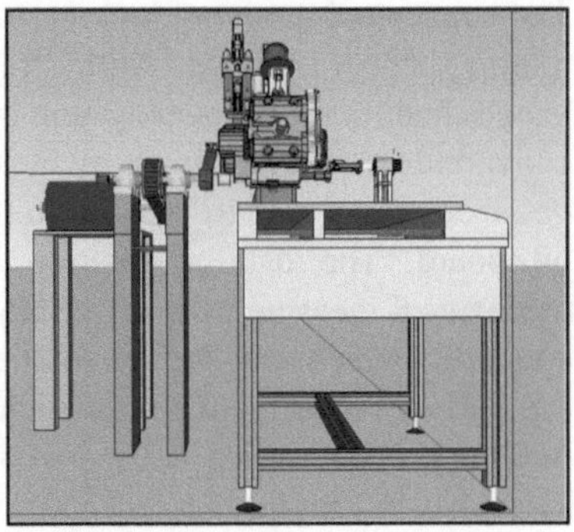

Fig. 3 Side view of engine performance

1. Engine clamping system
2. Kick force measuring subsystem
3. Acceleration subsystem
4. Gear shift force measuring subsystem.

In this engine performance simulator, power is given by the external actuator to transfer motion for crank the engine and gear shifting. Compared to the existing methods, the proposed method avoids the human intervention at the time of testing which reduces the error in the analysis. Figure 3 shows the Front view of engine performance simulator.

2.1 Engine Clamping System

For making stability during testing, special engine holder is provided at the top and fixture for orientation and vibration arresting at the bottom of the engine which is shown in Fig. 4. The proposed system should have the capability to withstand high inertia forces for the speed of 1,500 rpm or more and also to hold 250 kg of load or more.

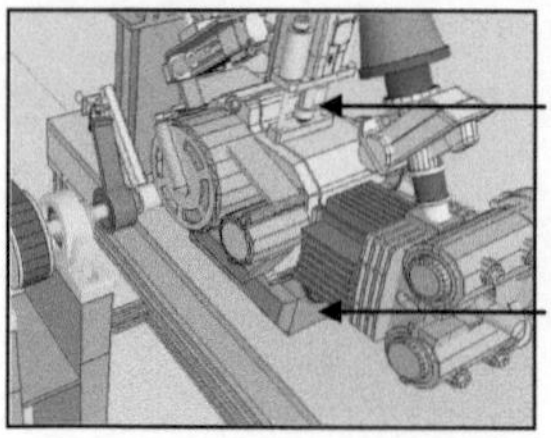

Fig. 4 Engine clamping system

2.2 *Kick Force Measuring Sub System*

The conceptual model of kick performance measuring subsystem consists of drive motor, damper equipped with strain gauges and rotary encoder [4]. The strain gauges and rotary encoder system is used to measure the analog data which will be converted into digital signal by means of Analog to Digital Converter (ADC). The obtained digital signals are sent to the computer via parallel port PCI 6221 or USB NI board. The data acquisition includes embedded specialized modules for mechanical measurement of certain sizes. Each channel has its own measurement converter which can be set to the sampling frequency in the range 0.1–9,600 samples/sps. The typical data such as speed, force, and rotary displacement are measured. This device is used to apply the load through a servo-based motor to kick lever of engine. In order to avoid parallax error the drive shaft of driver is placed in collinear with kick lever axis of engine [15–17].

The damper with strain gauge is mounted to measure the force experienced in the kick lever of engine. The complete calibration has been carried out to avoid source of variation from the sensor. In order to obtain the extracted or reversible load on the kick lever, a rotary encoder is mounted at end of the shaft. The accuracy of the encoder can be selected as minimum as one second. The above said measurements will be carried out using kick force measuring subsystem which is shown in Figs. 5 and 6. Number of actuation happened to start engine is another parameter need to be logged. Engine RPM is encoded to check the starting of engine and throttling has to be done to prevent engine off (detailed in Sect. 2.3).

2.3 *Acceleration Subsystem*

In order to start engine and for gearing action-acceleration is must. A linear controller encoded electrical drive is used to pull throttle cable as shown in Fig. 7. In parallel engine RPM is encoded to maintain engine at desired speed. Depending upon the type of engine, speed for ideal speed, and speed at each gear will be varied.

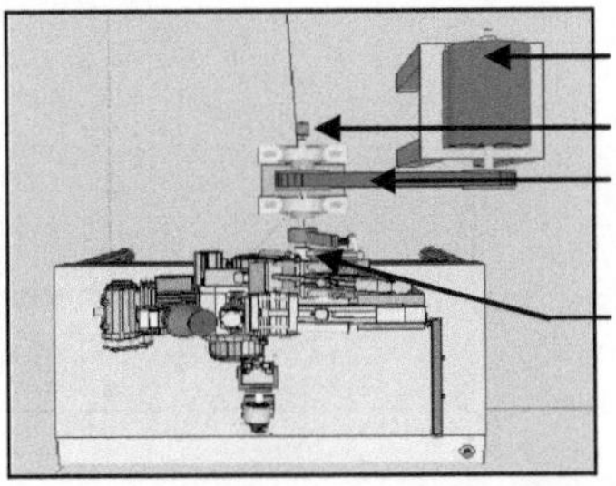

Fig. 5 Kick force measurement

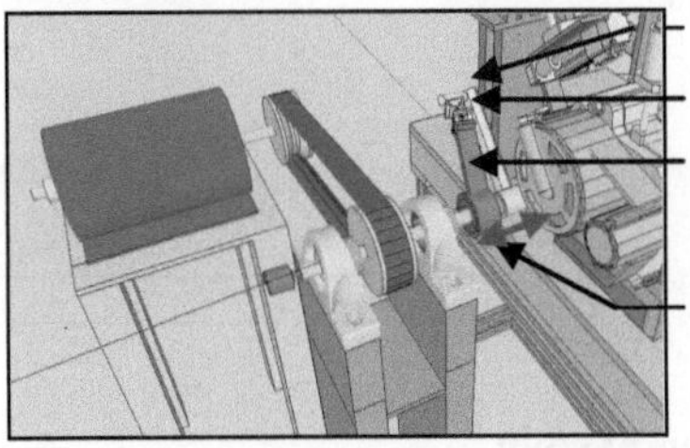

Fig. 6 Kick force measurement system

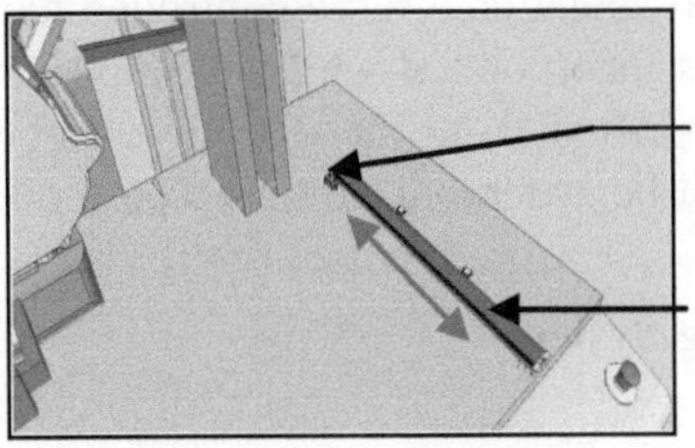

Fig. 7 Acceleration subsystem

2.4 Gear Shift Force Measuring Sub System

It consist of three major elements such as clutch actuation system, Gear change system, and data logger.

2.4.1 Clutch Actuation System

For smooth gear shifting operation, drive line connected to the driven line has to be disengaged in the engine. Automatic clutch disengagement happens by actuating pneumatic cylinder through solenoid as per the design requirement as shown in Fig. 8.

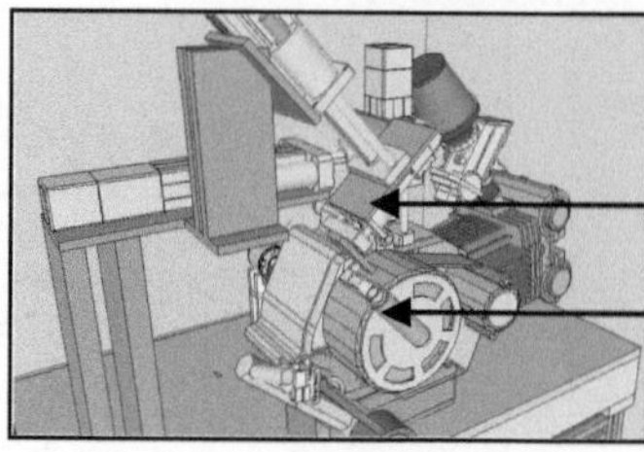

Fig. 8 Clutch actuation system

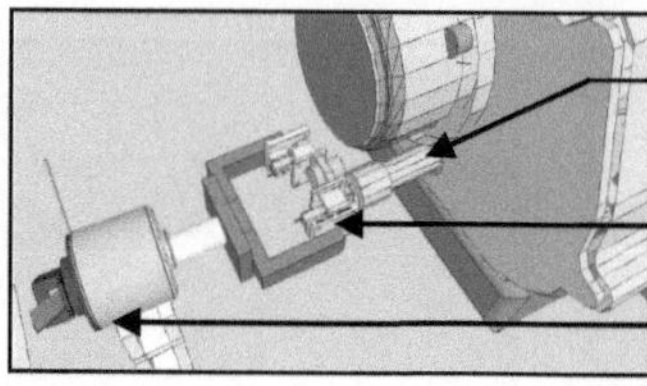

Fig. 9 Wheel wobbling measurement

2.4.2 Gear Change System

For engine gear shaft actuation a servo controlled motor coupled with load cell on both the direction is incorporated which is shown in Fig. 9. Bidirectional motor acts to move gear shaft in both the direction. Engine speed, gear shifting, and clutch actuation will be synchronized based on the test requirement.

2.4.3 Data Logging System

The strain gauges (both direction) and rotary encoder (servo motor) system is used to measure the analog data which will be converted into digital signal by means of ADC. The subsystems like acceleration system, gear shifting, and clutch actuation is synchronized to avoid errors. The obtained digital signals are sent to the computer via parallel port PCI 6221 or USB NI board. The data acquisition includes embedded specialized modules for mechanical measurement of certain sizes. Each channel has its own measurement converter which can be set to the sampling frequency in the range 0.1–9,600 samples/s. The typical data such as speed, force and rotary displacement are measured. This device is used to apply the load through a servo based motor to gear shaft of engine. In order to avoid parallax error the drive shaft of driver is placed in collinear with kick lever axis of engine.

3 Diverse Testing of Engine

Various engine testing namely kick force measuring, gear shift force measuring test has been carried out using the designed engine performance simulator. The engine was clamped to achieve the proposed testing as shown in Fig. 10.

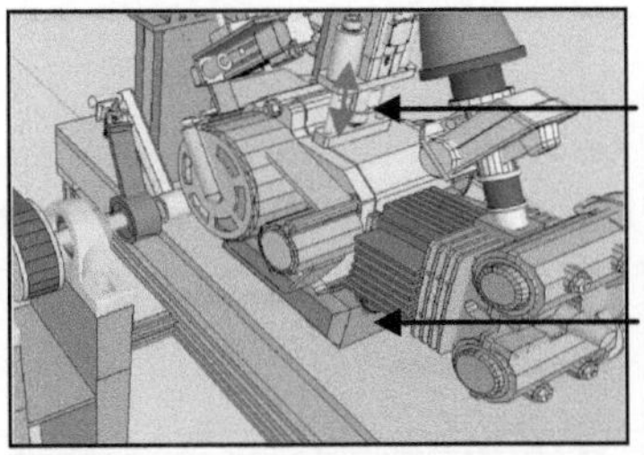

Fig. 10 Engine clamping operation

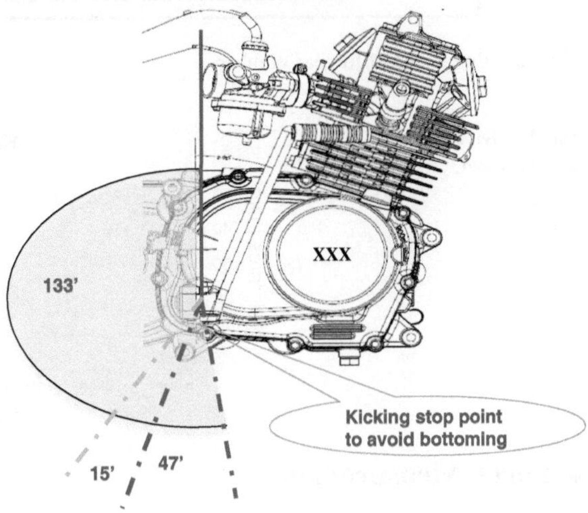

Fig. 11 Kick force measuring system

3.1 Kick Force Measurement Test

Based on clamping conformation signal, actuation for kick lever of engine will be started. Kick lever actuation of about 148′ travel of kick lever has to be monitored for load to angle, actuation should be stop before 47′ to avoid bottoming of lever which leads to measurement error, as shown in Fig. 11 (Total kick lever travel allowed—195′, angle may vary depending on engine design). In parallel, data for number of actuation to start the engine need to be logged. Throttle position should be maintained in order to keep engine alive.

3.2 Gear Shift Force Measuring Test

After getting conformation of engine start, engine rpm need to be maintained and monitored. Based on the desired engine rpm clutch actuation and gear shifting actuation need to be done as shown in Fig. 12 (engine rpm to gear shift value vary based on the engine design).

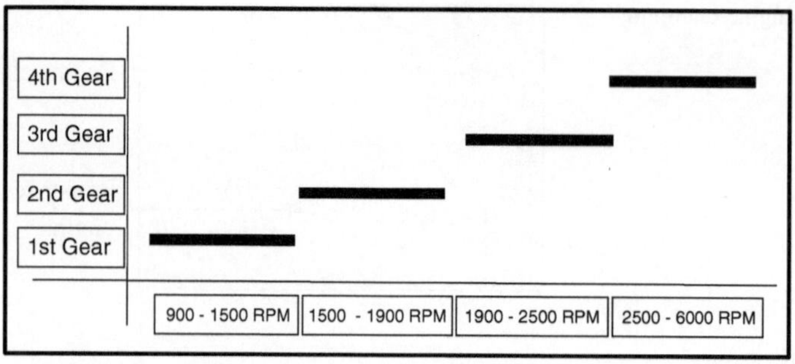

Fig. 12 Engine RPM to gear number

Fig. 13 Kick force measurement

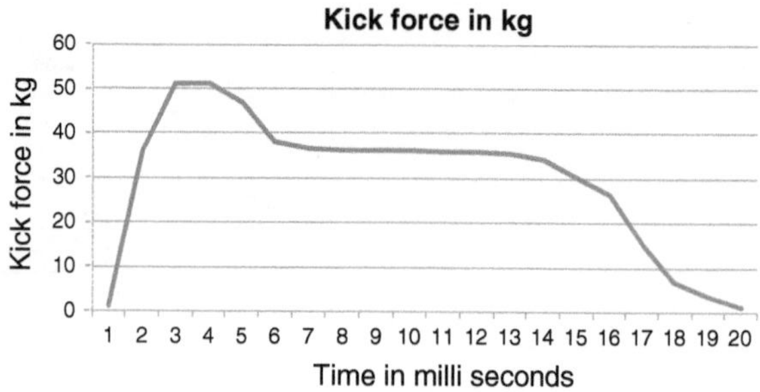

4 Force Measurement

4.1 Kick Force Measurement

In order to measure the kick force with respect to design specification at engine level, different speed of actuation can be simulated through the system. In view of this, a rapid load to kick lever has been applied and the pattern is shown in Fig. 13. Initially, force proposed to be high and tends to reduce as the kick lever starts rotating.

Similarly, maximum kick force for the different engine is proposed in trend chart as shown in Fig. 14 and number of kick required to start engine is shown in Fig. 15.

4.2 Gear Shift Force Measurement

In order to measure the gear shift force with respect to design specification at engine level, different speed, and degree of actuation can be simulated through the system. In view of this, a rapid load to gear shift lever has been applied and the pattern is shown in Fig. 16.

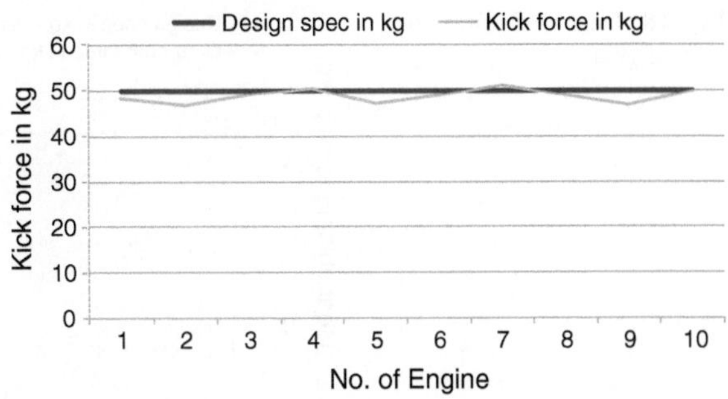

Fig. 14 Kick force at different engine

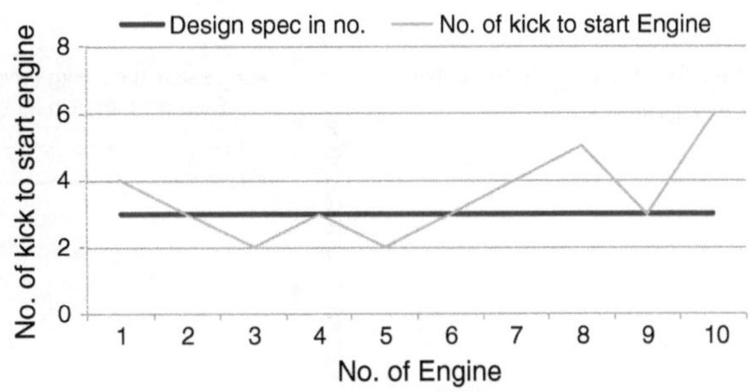

Fig. 15 No. of kick to start engine

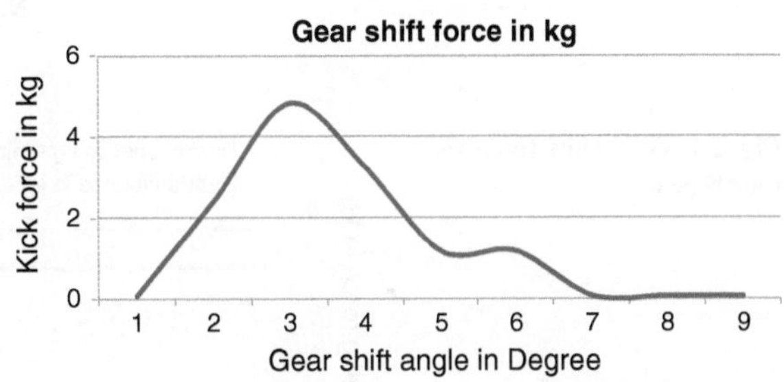

Fig. 16 Gear shift force measurement

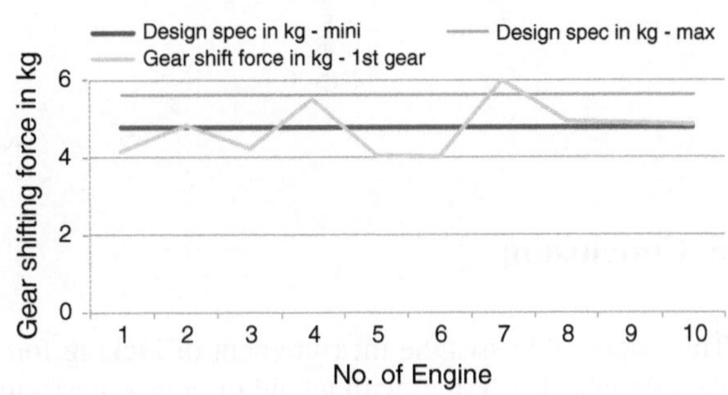

Fig. 17 Gear shift force for first gear

Similarly, maximum gear shift force for the different engine is proposed in trend chart as shown in Fig. 17 for first gear, Fig. 18 for second gear, Fig. 19 for third gear, and Fig. 20 for fourth gear.

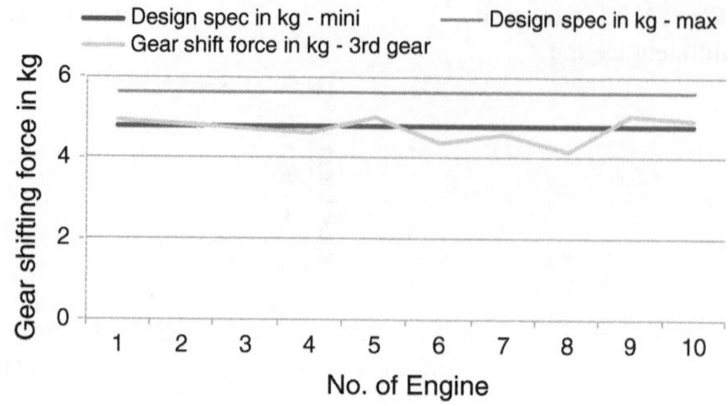

Fig. 18 Gear shift force for second gear

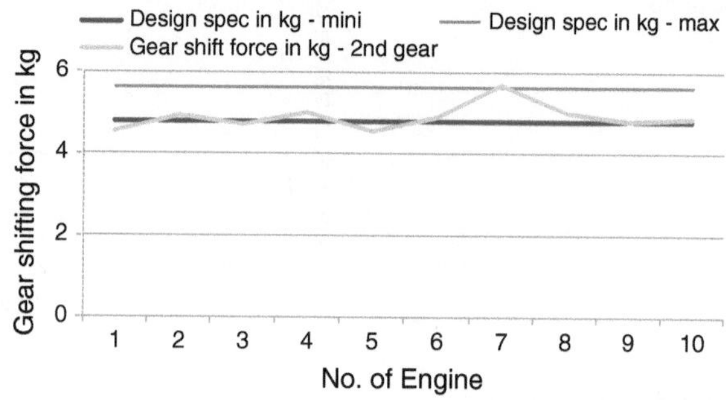

Fig. 19 Gear shift force for third gear

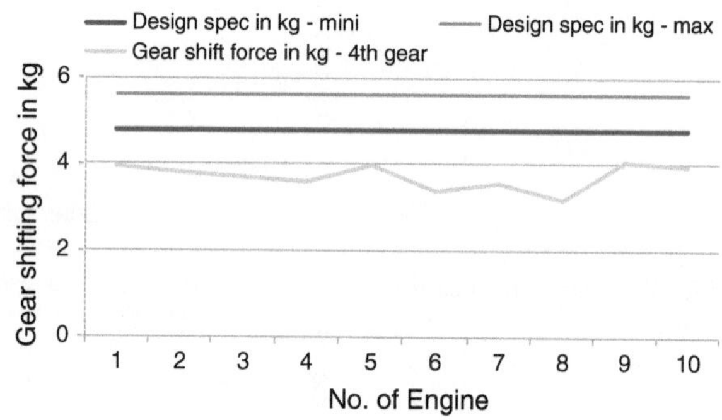

Fig. 20 Gear shift force for fourth gear

5 Conclusion

This paper addresses the measurement of kicking force and gear shifting force of the two wheeler system with an aid of engine performance simulator. In addition, number of kick required to start an engine is also carried out. The comprehensive hardware in loop simulation has been carried out to evaluate the performance of the engine without human intervention. Various tests in the normal speed conditions were carried out which provides great insight toward the characteristics of engine behavior in real-time scenario. It is concluded that simulation studies will eliminate the error on the various performance measures of the two wheeler engine which helps in increase in productivity and in turn improves product life cycle.

Acknowledgements Authors would like to express our sincere thanks to Mr. Krishnamachari, M/S City Electrical Pvt. Ltd, Bangalore for the support.

References

1. Super Flow 602 Owner's Manual (1998) Computerized Engine and Vehicle Test Systems, Version 2
2. Rayat HS (2000) The Automobile. S. Chand & Co. Ltd, New Delhi
3. Giri NK (2005) Automobile mechanics, 7th edn. Khanna Publishers, Delhi
4. Narang GBS (2005) Automobile engineering, 12th edn. Khanna Publishers, New Delhi
5. Carne JG, Simonis JC (1987) Modal testing and analysis. ASME 115(1):97–103
6. Ganesan V (1994) Internal combustion engines. Tata-McGraw Hill Publishing Co, New Delhi
7. Obert EF (1988) Internal combustion engine analysis and practice. International Text Book Co, Scranton
8. Mathur ML, Sharma RP (2008) A course in internal combustion engines. Dhanpat Rai & sons, India
9. Heywood J (1998) Internal combustion engine fundamentals. McGraw-Hill, USA
10. Taylor CF (1993) The internal combustion engines in theory and practice. Wiley, New York
11. Lechner G, Naunheimer H (1999) Automotive transmissions. Springer, Berlin
12. Achtenová G. (2004) Simulation model of automated gear-box. In: Proceedings of AED conference, Glasgow, pp 52–56
13. Achtenová G, Vávra J (2004) New control algorithm for automatic and automated gear-boxes. In: Proceedings of AVEC conference, Arnhem, The Netherlands, pp 52–56
14. Bracard S, Rougé M, Laguitton O (2006) Diesel engine and automated transmission: powertrain global optimisation under multiple constraints. (F2006P318T), Driver/vehicle interface, information and assistance—Lyon, SIA/Yokohama, Fisita
15. Bulucea CA, Nicola DA, Brandusa C, Cismaru DC, Manolea Gh, Popescu MC (2009) Embodied energy and environmental impact in electric transportation systems. In: Proceedings of the 4th IASME/WSEAS international conference on energy and environment, Cambridge, pp 232–239
16. Drighiciu M, Petrisor A, Popescu MC (2009) A Petri nets approach for hybrid systems modeling. Int J Circuits System Signal Process 3(2):55–64
17. Bettes H (2010) A quick view of chassis dynamometer testing. AERA Engine Builders Association—http://www.aera.org/ep/downloads/ep10/EP04-2010_34-39.pdf

Optimization of Variable Speed Turbo Coupling in Boiler Feed Pump

R. Pandithurai and I. Ambrose Edward

Abstract In Tuticorin Thermal Power Plant, the thermal energy is converted to mechanical energy and it is converted to electrical energy. The thermal power plant operates on modified rankine cycle with reheating and regeneration. This cycle is a closed cycle. In order to circulate the de-mineralized water in the rate of 700 tons per hour at 175 Kg/Cm^2. The de-mineralized water is circulated by boiler feed pump it is coupled with the constant speed motor. In this operation, 15 years back the throttle control was used to control the circulation of de-mineralized water to boiler drum. This throttle control method is inefficient method. To overcome this throttle control is replaced by speed control. The variable speed turbo coupling is used for speed control. The transmission efficiency is more at same percentage of speed. By analyzing the performance curve of the variable speed turbo coupling the most efficient power transmission scoop position is identified. Our project is based on; (1) Analysis of the scoop position at which the power transmission is high by using the performance curve. (2) Analysis of the scoop position at which the boiler feed pump is operating on maximum period in any one unit (TTPS). (3) Analysis to the maximum duration-operated scoop position by which avoiding energy losses due to operating at inefficient energy transmission efficiency position (scoop position) using optimization method genetic algorithm (GA).

Keywords Boiler feed pump · Gear ratio · Genetic algorithm · Optimization · Power loss

---

R. Pandithurai (✉) · I. A. Edward
Velammal College of Engineering and Technology, Madurai, India
e-mail: rspandithurai@gmail.com

R. P. Bajpai et al. (eds.), *Innovative Design, Analysis and Development Practices in Aerospace and Automotive Engineering*, Lecture Notes in Mechanical Engineering,
DOI: 10.1007/978-81-322-1871-5_17, © Springer India 2014

1 Introduction

Thermal power plant rankine cycle (Fig. 1).

1.1 Process of Rankine Cycle

1–2 Work input by CEP pump
2–3 Isentropic heat addition process LPH1, LH2, LPH3
3–4 Work input by boiler feed pump
4–5 Isentropic heat addition process HPH4, HPH5, Economizer
5–6 Latent heat addition process in furnaces
6–7 Super heating process
7–8 Isentropic heat expansion in HP TURBINE
8–9 Reheating process (heat addition process)
9–10 Isentropic Heat expansion in IP TURBINE
10–11 Isentropic Heat expansion in LP TURBINE.

2 Variable Speed Turbo Coupling

2.1 Main Parts

- Primary wheel and Secondary wheel (Fig. 2).
- Shell.
- Scoop tube position.
- Herring bone gear.
- Working Oil.

2.2 Operation of Variable Speed Turbo Coupling R16 K1

- Variable speed turbo coupling is used between the industrial Machine and constant speed motor for infinitely speed control.
- They are incorporated mutual housing.
- The lower section of the variable speed turbo coupling is the oil tank which contains working oil servo torque 10.

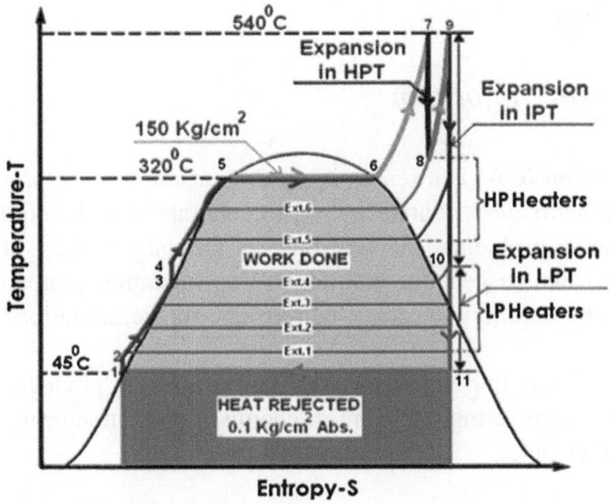

Fig. 1 Rankine cycle

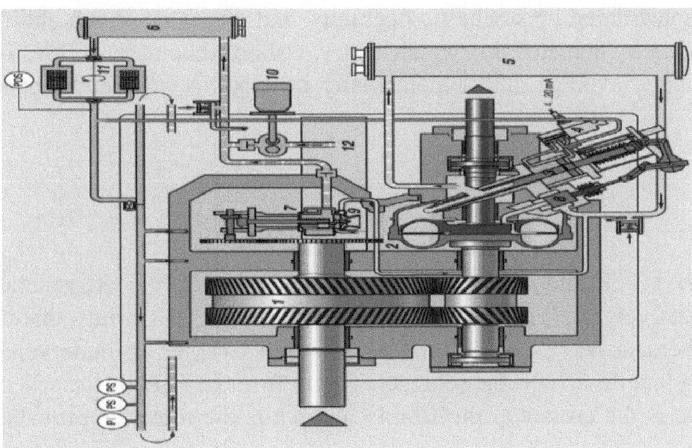

Fig. 2 Variable speed turbo coupling r16 k1

- The torques is transmitted hydrodynamically from impeller to turbine due to pressure drop and accelerates the fluid.
- The speed of the secondary shaft is less than the primary shaft.
- To transmit the power the slip is required.
- The speed of the secondary shaft is controlled by positioning the scoop tube which controls the working oil filling in the oil chamber.

3 Methodology

3.1 The Genetic Algorithm

A genetic algorithm (GA) is a search heuristic that mimics the process of natural evolution. This heuristic is routinely used to generate useful solutions to optimization and search problems. GAs belong to the larger class of evolutionary algorithms (EA), which generate solutions to optimization problems using techniques inspired by natural evolution, such as inheritance, mutation, selection, and crossover.

Genetic algorithms find application in bioinformatics, phylogenetics, computational science, engineering, economics, chemistry, manufacturing, mathematics, physics, and other fields.

3.2 The Fundamental Theorem of Genetic Algorithms

A GA is constructed by stochastic operators, and its robust search ability is based on the theorem depicted in, which states, "short schemata of low order with aptitude above average, and exponentially increase its number by generations," this is:

$$m(H,t+1) \geq m(H,t)\frac{f(H)}{f_{\text{avg}}}\left[1 - p_c\frac{\delta(H)}{l-1} - O(H)p_m\right] \tag{1}$$

where $m(H, t + 1)$ and $m(H, t)$ are the schemata number H in the generation $t + 1$ and t respectively, $f(H)$ is the average aptitude value of the strings that is included on the schemata H, f_{avg} is the total population's average aptitude value, l is the total string length, d(H) is the schemata length from H, $O(H)$ is the schemata order from H, pc is the crossover probability and pm is the mutation probability.

3.3 Genetic Algorithm Operators

A basic GA that can produce acceptable results in many practical problems is composed of three operators:

- Reproduction
- Crossover
- Mutation.

Table 1 Result of optimization

S. No	Before optimization	After optimization
Unit load (Mw)	210	210
Power output (Kw)	4000	3972.65
Power loss (Kw)	700	672.89
Output speed (rpm)	4498	4448

4 Result

4.1 Calculation Off Payback Period

$$\begin{aligned}\text{Cost of power} &= \text{Rs. } 2.5 \text{ per unit}\\ \text{Cost saving per day} &= \text{power saving per day} \times \text{cost of per unit}\\ &= 663.6 \times 2.5\\ &= \text{Rs. } 1{,}659/\text{day}\\ &= \text{Rs. } 6{,}55{,}535 \text{ lakhs/year}\end{aligned} \tag{2}$$

See Table 1.

5 Conclusion

The herring bone gear is the major power transmission part of the variable speed turbo coupling, which will increase the input speed of the coupling. In our project, By varying the gear ratio various output power and power loss corresponding to the particular various scoop position and from the output power and power loss graphs we have concluded 3.54 gear ratio is the most efficient among the range of 3.51–3.58. In this point, we can get the power saving of 27.65 Kw and cost about Rs. 6,55,535 lakhs/year.

It is not required to replace the gears immediately, as these spares have served more than 20 years, it can be scheduled in next renewal.

After replacing the gear ratio, the efficiency of coupling will be increased from 85.14 to 87.5 %, and output speed is reduced as 4448 rpm.

Cyclic Life Estimation of Turbine Rotor Blade Fitted in a Twin Spool Turbojet Engine

N. Sreenivasaraja

Abstract The main intent of the project is to establish cyclic life of high pressure turbine rotor blade fitted on a twin spool turbojet engine meant for fighter class application. An attempt is made to identify the gas bending loads, centrifugal loads, and thermal gas loads due to various rpm conditions of the engine. To establish cyclic life of turbine blade, stress approach is followed. The three-dimensional model is generated using CATIA V5 R20 and the analysis of the model is carried out in ANSYS 12.1 by simulating pressure and thermal loads so that mean and alternating stresses and vibrational modes could be extracted. Material properties for room and operating temperature are obtained from literature and these are used for generating Goodman diagram and S-N diagram. Campbell diagram is used to obtain engine cross over for the blades in bending and torsional frequencies. Simulation of defects is another area explored in respect of a blade wherein notch is created and the same analysis as above is carried out on this blade also. The purpose is to identify the reduction in the frequency and variations in engine cross over points in the Campbell diagram and to compare with the serviceable and unserviceable blades. The loss of frequency and change in engine cross over locations has given an idea about the effect of unserviceability on blade and its manifestation during engine operation.

Keywords Crack propagation · Thumb rule · Low cyclic fatigue · Bending stress · Campbell diagram

N. Sreenivasaraja (✉)
Aeronautical Engineering (Part time), Karpagam University, Coimbatore, India
e-mail: sreeraja2010@gmail.com

R. P. Bajpai et al. (eds.), *Innovative Design, Analysis and Development Practices in Aerospace and Automotive Engineering*, Lecture Notes in Mechanical Engineering,
DOI: 10.1007/978-81-322-1871-5_18,

1 Introduction

Gas turbine engine which is meant for fighter class aircraft is supposed to have very high reliability and no failure of components are permitted despite the fact that the engines undergo deep throttling very often. Therefore, the design aspects should be reasonably full-filled, so that no component failure occurs at any point of time during its operation so as to arrive at the ear-marked cyclic life of the components.

This has developed the necessity to understand the design efficiency of high pressure turbine rotor blade of a gas turbine engine. The project is on the cyclic life establishment of high pressure turbine rotor blade of a flying engine, which has almost done its life flawlessly [1]. Since the turbine rotor blades are the worst stressed components in an engine due to loadings from thermal conditions and mechanical conditions, the effect of temperature on the rotor blade is also considered. Analysis is carried out by applying pressure distribution and thermal distribution in order to obtain the stress and frequency values for various rpm conditions of the engine. The values obtained from the above analysis are taken as an input to draw the Campbell diagram.

1.1 Twin Spool Engine Arrangement

The high pressure turbine drives the high pressure compressor and low pressure turbine drives low pressure compressor. Thumb rule for twin spool engine is low spool pressure ratio should be less than that of high spool pressure ratio [2].

2 Turbine Material

Superalloys are a group of alloys that exhibit excellent mechanical properties at high temperatures. The nickel-based Superalloys were invented in the 1940s primarily for gas turbine applications because of their longtime strength and toughness at high temperatures. Axial flow turbines are meant for expanding high pressure and high temperature gas and in the process turbines are made to rotate which in turn transmits the required torque to both compressor and accessories mounted on the engine through the gearbox [3, 4].

3 Modeling and Analysis

By knowing the design details from the design calculations, modeling is done by using CATIA V5 software. All the important corner points of the turbine base drawing are identified and a sanity check is carried out for its dimensions (Table 1).

Table 1 Material properties of SU-718

Property	Value	Unit
Density	8.19	g cm^{-3}
Isotropic Elasticity		
Young's modulus	2.08e + 05	MPa
Poisson's ratio	0.3	–
Strength		
Tensile yield strength	800	MPa
Tensile ultimate strength	1,100	MPa

3.1 Meshing

In the FEM analysis of high pressure turbine rotor blade meshing is the initial step that is to be followed after the model is being imported for the purpose of analysis [5]. If there are substantial differences between the two, the analysis should continue with a more refined mesh 219,555 node and 128,022 elements and a subsequent comparison until convergence are established.

4 Low Cycle Fatigue

If the amplitude of the cyclic stresses applied to a component is very high (LCF), the accumulated strain energy per reversal will be significantly higher than for an HCF loading cycle [6, 7]. For this reason, under LCF, a component will spend a very small proportion of its life in the primary crack initiation stage and the majority of its life in the secondary crack propagation phase of fatigue failure [8].

5 Structural and Vibrational Analysis

5.1 Static Structural and Centrifugal Stress Analysis

In order to estimate the life of the high pressure turbine rotor blade a variety of analysis has been conducted on the blade. Centrifugal force on blade will be exceedingly high but it is rarely the primary cause of failure[9, 10]. By applying the boundary conditions for the turbine under bending mode and 1 N at leading and trailing edge for torsional mode, the results can be used to find out the life of the turbine.

As a result, it is identified that the Von misses stresses are found at root of the blade as shown in Fig 1 (Figs. 2, 3, 4, 5 and 6). The steady stress for 10,000 rpm (100 %) of the engine is found to be 34.935 N/mm^2. The obtained values are taken as an input for drawing the S–N and Goodman diagram by which the cyclic

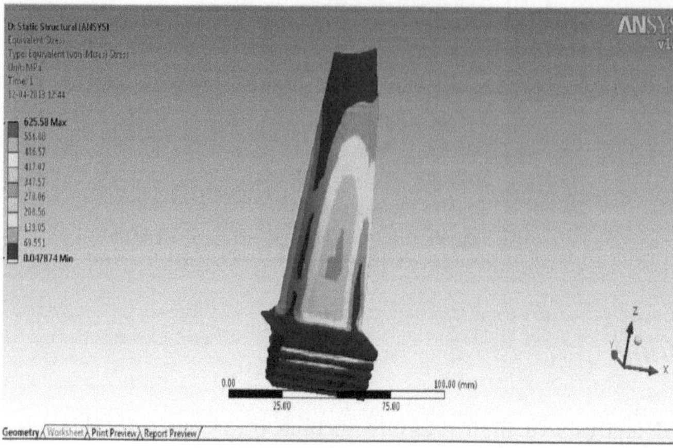

Fig. 1 Von misses stresses for bending loads

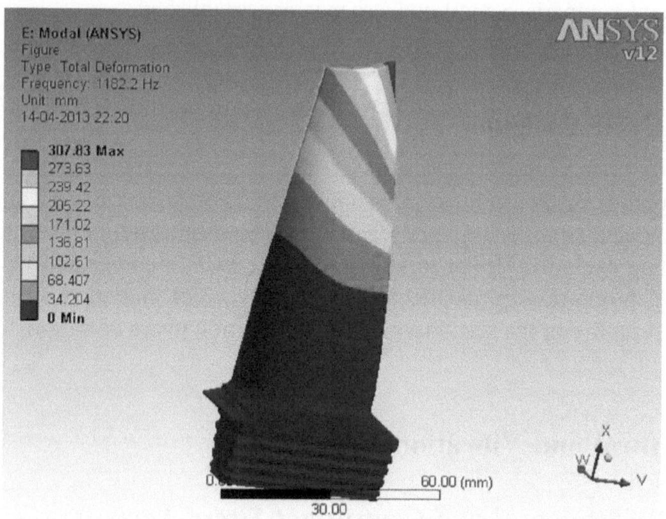

Fig. 2 Bending frequency at 10,000 RPM

life of the blade is predicted. The same exercise is repeated for all other rpm considerations and the gas bending loads values for various rpm condition are placed in the Table 2 (Tables 3, 4 and 5).

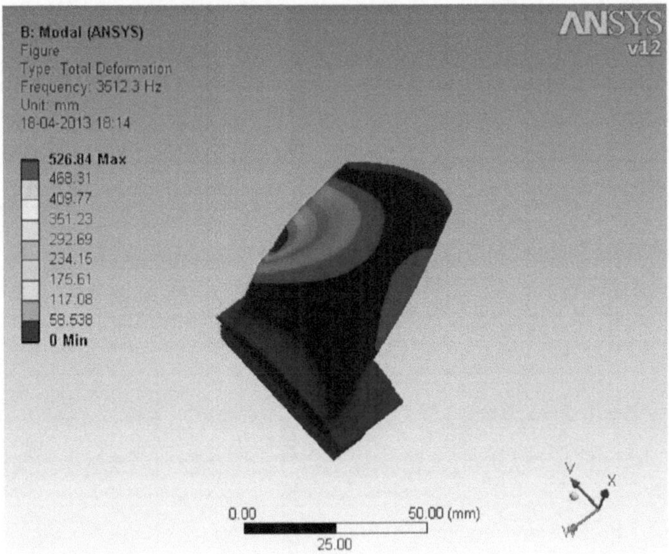

Fig. 3 Torsional mode at 10,000 RPM

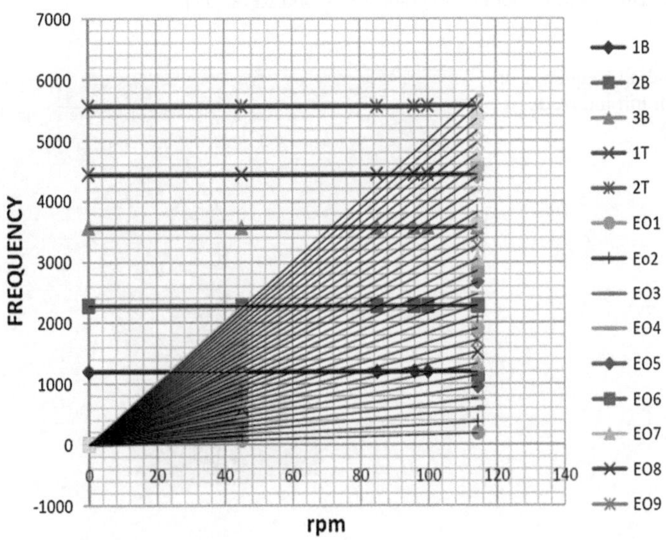

Fig. 4 Campbell diagram for turbine blade

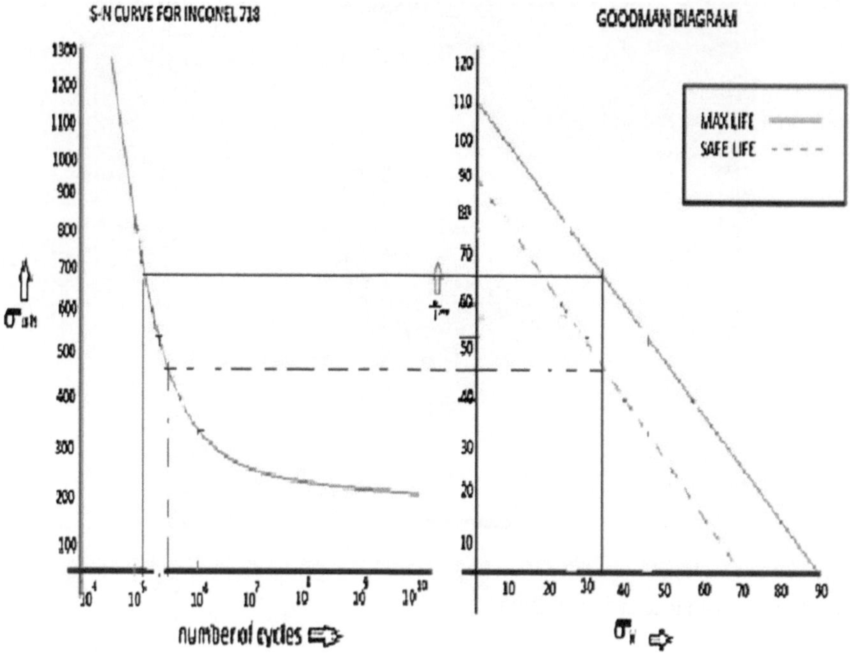

Fig. 5 S–N curve and Goodman diagram for 10,000 RPM [11]

Fig. 6 Bending frequency of single notch turbine blade at 10,000 RPM

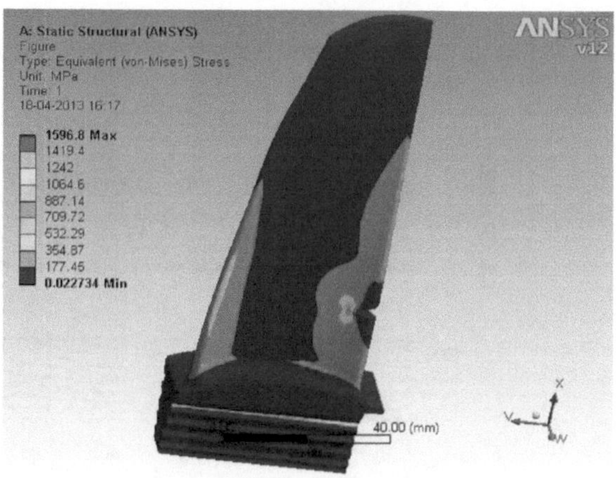

5.2 *Bending and Torsional Mode*

The Campbell Diagram shown in Fig. 7 is used to find the failure of the turbine at various RPM including bending and torsional mode.

Table 2 Bending and torsional frequencies

S. No.	RPM	Bending frequency (Hz)	Torsional frequency (Hz)
1	11,460	1184.1	3516.9
2	10,955	1183.5	3514.9
3	10,580	1183	3513.6
4	10,000	1182.8	3511.5
5	9,600	1181.8	3510.2
6	8,500	1180.5	3506.7

Table 3 Alternating stress and cyclic life

S. No.	RPM	Mean stress (N/mm^2)	Alternating stress (N/mm^2)	No. of cycles (N)
1	11,460	46.2	320	1×10^6
2	10,580	39.26	420	3.7×10^5
3	10,000	34.715	450	3.5×10^5
4	9,600	31.265	490	2.5×10^5
5	8,500	25.193	580	1.5×10^5
6	4,200	23.46	600	1×10^5

Table 4 Single notch turbine blade bending stress

S. No.	RPM	Stress (N/mm^2)
1	11,460	209.71
2	10,580	187.98
3	10,000	159.68
4	9,600	147.13
5	8,500	115.37
6	4,200	101.28

Table 5 Bending and torsional frequencies for single notch blade

S. No.	RPM	Bending frequency (Hz)	Torsional frequency (Hz)
1	11,460	1119.9	3507.9
2	10,580	1118.7	3504.5
3	10,000	1117.9	3502.4
4	9,600	1117.9	3502.4
5	8,500	1116.2	3497.5
6	4,200	1112.6	3487.6

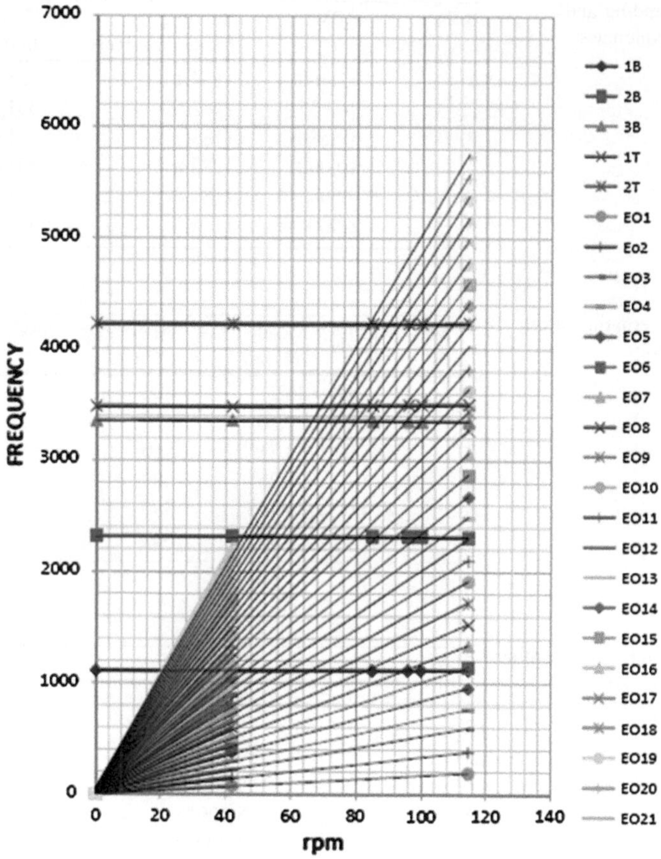

Fig. 7 Campbell diagram for single notch blade

6 Conclusion

This has given sufficient inputs regarding design aspects of the blade. It is although a rudimentary exercise as far as the design understanding is concerned; this has given an idea about the blade configuration, type of assembly on the disc, type of loads normally encountered on the blade, etc., during its operation. In future different materials are used to find the failure of the turbine at various RPM.

References

1. Mahobia GS, Sudhakar RG, Srini A (2013) Effect of salt coatings on low cycle fatigue behaviour of nickel—base super alloy GTM-SU-718. Elsevier 55:304–334
2. Philip B, Mahendra Babu NC (2010) Numerical estimation of fatigue life of aero engine fan blades. SasTech 9(2):47–54

3. Shepard DP (2004) Life extension methods in aero engines. Ommi 3(1):1–15
4. Ono Y (2004) High-cycle fatigue properties at cryogenic temperatures in INCONEL 718 nickel based superalloy. Mater Trans 45(2):342–345
5. Garcia de la Yedra A (2011) Thermo-mechanical fatigue behaviour and life prediction of C-1023 nickel based superalloy. Int J Eng Sci Technol 3(6):88–101
6. Zaretsky EV (2001) Determination of turbine blade life from engine field data. ASME 1–35
7. Peters JO (1999) Role of foreign object damage on thresholds for high- cycle fatigue in Ti-6Al-4 V. Metall Mater Trans 1–45
8. Pollock TM (2006) Nickel-based superalloys for advanced turbine engines: chemistry, microstructure, and properties. J Propul Power 22(2):361–374
9. Ahmad KAB (2009) An investigation of turbine blade failure in aircraft turbine engines. University of Malaysia, pp 1–11
10. Jelaska D (2001) Crack initiation life at combined HCF/LCF loading. University of Split, pp 1–8
11. Srinivasan N (1994) Micro structural control in hot working of Inconel-718 superalloy using processing map. Volume 25A, pp 1–10

CFD Analysis of Swirl Enhancement in a Direct Injection Diesel Engine with Vortex Generator in Inlet Manifold

G. Sivakumar and S. Senthil Kumar

Abstract Swirling enhancement is studied numerically for a direct injection diesel engine with vortex generators placed in the inlet manifold using commercial CFD code, ANSYS Fluent^{TM}. The three-dimensional geometry and mesh are created using preprocessor ANSYS ICEM. Simulations have been carried out to investigate the effect of vortex generator in inlet manifold for a single cylinder engine. Transient flow characteristics inside the engine cylinder for the intake stroke with RNG k-ε turbulent model are studied and the results are presented in terms of swirling velocity component, swirl number, and turbulent kinetic energy. It is observed that the modified inlet manifold creates uniform mixing inside the engine cylinder which is essential for effective combustion for diesel engine. Results show that it could be used as a simple and effective swirl creating mechanism in the in-flow of engine cylinder.

Keywords Swirling enhancement · Manifold · Diesel engine · Vortex generator · Mixing

1 Introduction

In the recent years, engine development researches are focusing mainly on reducing emission and effective utilization of energy from fuel. It is of great concern all over the world due to the pollution control which in turn affects the

G. Sivakumar (✉)
Department of Mechanical Engineering, Vel Tech Dr. RR and Dr. SR Technical University, Chennai, India
e-mail: sec2.veltech@gmail.com

S. Senthil Kumar
Department of Aeronautical Engineering, Vel Tech Dr. RR and Dr. SR Technical University, Chennai, India
e-mail: s.senthilms@gmail.com

R. P. Bajpai et al. (eds.), *Innovative Design, Analysis and Development Practices in Aerospace and Automotive Engineering*, Lecture Notes in Mechanical Engineering,
DOI: 10.1007/978-81-322-1871-5_19,

climate change. It is well known that combustion and emission characteristics are controlled by the flow created inside the cylinder during the intake and evolves in the compression stroke. Flow mixing enhancement inside the diesel engine has been studied both experimentally and numerically over several years. It is more important to create swirling inside the engine cylinder as it directly affects the combustion performance, output, and emission of the engine.

Swirling creation inside the engine was earlier studied by various researchers [1–3] by changing the parameters like manifold shape, combustion chamber configurations, and piston head shape. There are two ways of enhancing the combustion process thereby increasing the engine thermal efficiency. First one is by increasing the compression ratio as high as possible. But the problem is that high compression ratio may create knock, which should be avoided for good performance. Second one is by enhancing the turbulence in order to have a better mixing of fuel and air. However, two much turbulence leads to excessive heat transfer from the gases to cylinder walls [3]. Paul et al. [1] studied the swirl enhancement by using spiral-helical and helical-structured manifolds and presented swirling characteristics in terms of flow patterns, turbulent kinetic energy, and swirling number. They conclude that spiral-helical-shaped manifold creates higher swirl flow inside the cylinder than spiral manifold. Martins et al. [2] performed a CFD study with modified helical inlet port under different value lifts using commercial CFD code Fluent and found that the increase in valve lift decreases the swirl number. It indicates that the valve position also plays an important role in the in-cylinder flows.

Other approach was used to increase swirling inside the cylinder by changing the piston top surface with different bowl shapes and bowl positions by Murali et al. [4]. They experimentally studied, using particle image velocimentry (PIV), the effect of bowl shape on the top piston surface, and found that bowl shape on flat piston shows a good improvement. Jin et al. [5] carried out an experimental study to enhance swirling by using swirl control valve (SCV) with different SCV angles and presented flow patterns and swirl number for different SCV angles. It was that this modification enhances the swirling characteristics inside the cylinder. All the geometry shapes and mechanism tried by various researchers [1–6], have difficulties in manufacturing point of view and would increase the total vehicle cost. Hence, it is of necessity to have simpler mechanism which would give the same performance at the lower cost.

From the literature review, it is found that understanding the flow characteristics inside the cylinder, which are directly, linked with the engine performance and emission characteristics. And capturing the flow patterns inside the cylinder using experimental techniques is expensive and it is difficult to carry out parametric studies with different combinations of inlet manifolds, piston top surfaces. So the CFD is only effective analysis tool used by many authors [1, 2, 3, and 6] to study the flow behavior with various manifolds and piston top bowl shapes. It must be emphasized the 3D CFD simulation is necessary to bring more light to the understanding of flow dynamics in the cylinder.

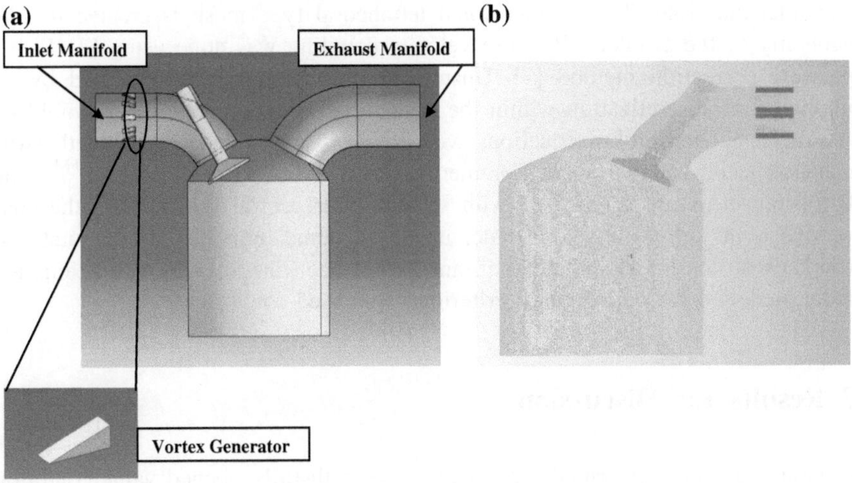

Fig. 1 **a** Geometry of the engine; **b** Mesh of the computational geometry

Table 1 Dimensions of the engine

Details	Value (mm)
Bore	104
Stoke	113
Inlet valve diameter	35
Exhaust valve diameter	45
Maximum inlet valve lift	13

Hence, the objective of this work is to study swirling enhancement by modifying the inlet manifold with a simple wedge-shaped vortex generators (VGs) attached circumferentially over the inner surface of the manifold at the entry position, as shown in Fig. 1a and compare with the base engine cylinder flow characteristics. The dimensions of the engine are given in Table 1 for a CI diesel engine.

2 Numerical Methodology

Numerical simulations under isothermal conditions have been carried out for a single cylinder CI engine with modified inlet manifold using vortex generators. Computational domain and mesh structure of the problem are shown in Fig. 1a and b, respectively. For simplicity, simulations are performed with fully opened valve condition at the inlet port and fully closed valve condition at the exhaust port. Pressure inlet boundary condition is specified for the starting face of the inlet manifold. No-slip boundary conditions are specified at walls of the engine cylinder and inlet manifold. Three-dimensional tetrahedral type mesh is created for the geometry of the problem. Pressure velocity coupling was done using SIMPLEC pressure correction method [7]. Unsteady Computations are performed by an implicit time discretization within the domain using incompressible Reynolds—Averaged Navier–Stokes equations with RNG k-ε turbulence model with swirl enhancement available with Commercial software ANSYS FLUENT™. The Diffusion terms are discretized with second-order central scheme and the convective terms are with second-order upwind scheme. For all the computations, residuals of continuity, momentum, and turbulence kinetic energy equations are monitored, and the convergence criterion value used equal to 10^{-4}.

3 Results and Discussion

Computations are performed for intake stroke with fully opened value condition for both the base engine and base engine with VGs attached at the inlet manifold, as shown in Fig. 1a. Flow characteristics in terms of velocity vectors and turbulent kinetic energy (TKE) are presented for a single cylinder engine at different cross-sections along the axis of the cylinder. Figure 2 shows the turbulent intensity (Left) and the velocity vector plots (Right) for a center plane (x = 72 mm, where x is the distance measured in the perpendicular to the axis of the cylinder)) cut parallel to the cylinder axis for the both cases, namely, inlet manifold with and without VGs. It is clearly seen from Fig. 2 (Top-Left), the turbulent intensity is not uniform and having many small vortices (see Fig. 2: Top-Right), which are not good for better combustion, whereas the inlet manifold with VGs having uniform turbulent intensity and not having many small vortices, which are good for better combustion, as observed in Fig. 2 (Bottom) and in Fig. 5.

Having better air-fuel mixture is depending on how big the size of vortex and that should too cover the whole size of the cylinder. In order to increase combustion speed and reduce ignition time, the large vortex inside the cylinder to be created with higher TKE.

To check the effect of VGs in turbulence enhancement, the variations of TKE at different planes perpendicular to the axis of the cylinder (say, z = 20, 40, 60, and 80 mm, where z is the distance along the axis of the cylinder measured from BDC) are shown in Figs. 3 and 4, for with VGs case and without VGs case, respectively. It can be seen from Fig. 3 that VGs create uniform turbulence throughout the cylinder, which are essential for good combustion and emission characteristics.

It indicates the importance of modification of the inlet manifold with VGs. TKE is not propagating uniformly through the cylinder without VGs case (base engine), as shown in Fig. 4. It may not be good for having better combustion process and more emission products may come out of the engine at the end of the combustion process. It should be mentioned that with VGs case, swirling vortices created in the

Fig. 2 Contours of turbulent intensity values at center plane axis (x = 72 mm) cut parallel to the cylinder for the inlet manifolds with and without VGs (*Top* without VGs and *Bottom* with VGs)

perpendicular and parallel planes, with respect to the cylinder axis, cover the whole space in that particular plane, which would make better air-fuel mixing so as to have good combustion efficiency and less emission outcomes, as observed in Fig. 3. This trend can also be noticed clearly in Fig. 5 with the path lines inside the cylinder for both the cases; with and without VGs.

In order to quantify the effect of turbulent creation inside the cylinder, it is necessary to calculate the swirl number. The swirl number is the ratio of the angular momentum to the axial momentum. Figure 6 shows the variation of swirl number along the stroke length for two inlet manifold configurations, with and without vortex generators. It is seen that the swirl number for with VGs is higher than that for without VGs, which shows the impact of VGs on swirling enhancement at all the positions along the stroke length. When the distance along the stroke length equals to 40 mm, the swirl number value reaches to 16, which shows the swirling enhancement by the VGs placed at the inlet manifold.

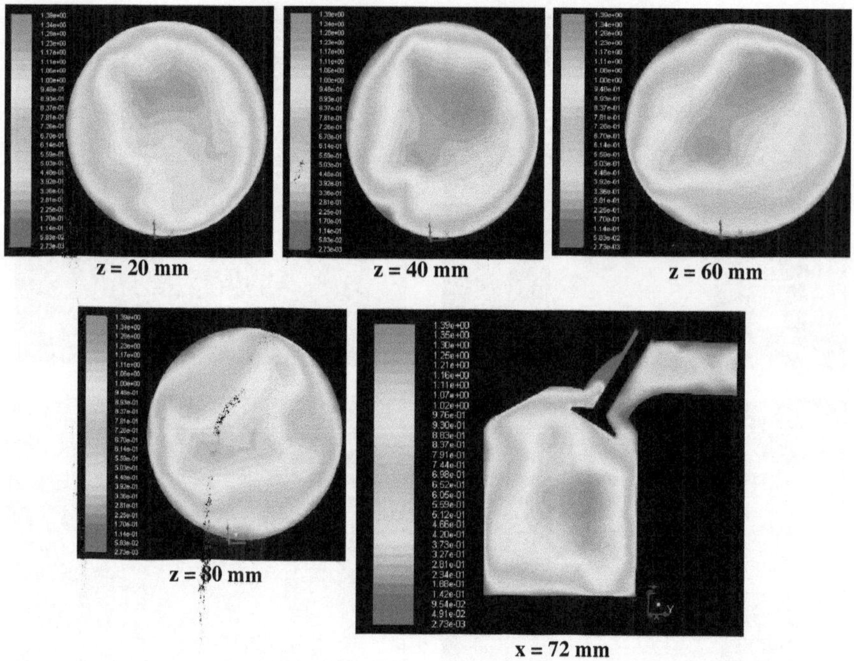

Fig. 3 Contours of turbulent kinetic energy at different planes in the cylinder, when the inlet manifold is modified with VGs

4 Conclusions

Three-dimensional numerical simulation of flow inside the engine cylinder with and without VGs at inlet manifold has been performed. It is found that the model predicts swirling very well which gives confident that swirling inside the cylinder could be enhanced using different manifold configurations. The flow patterns, TKE and swirl number show that VGs placed at inlet manifold could create good swirling characteristics for better combustion and emission characteristics. Hence, it is very essential to perform combustion and emission characteristics with the present VG configurations and it has to be quantified with experimental techniques. It would be the next stage of this project. Further study will be carried out for various manifolds at different locations along the inlet manifold and finally the manifold shape will be optimized for better mixing.

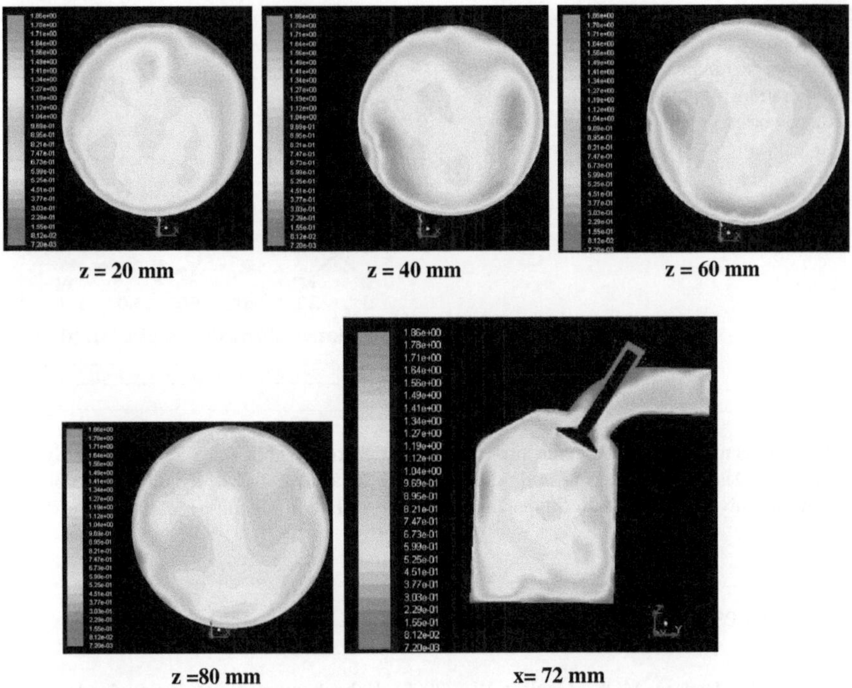

Fig. 4 Contours of turbulent kinetic energy at different planes in the cylinder for the inlet manifold without VGs

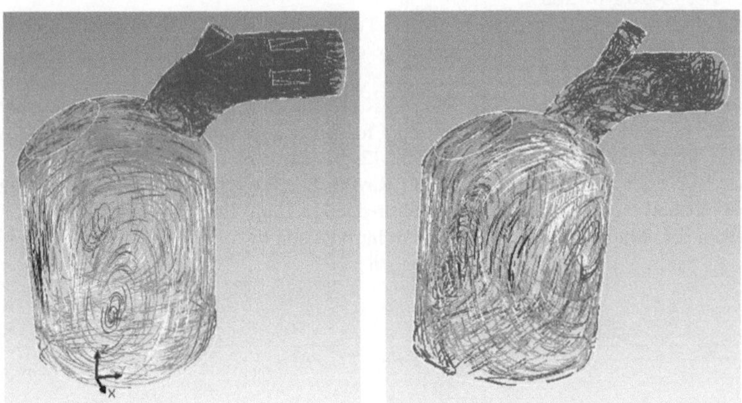

Fig. 5 Pathlines based on velocity magnitude; *Left* with VGs and *Right* without VGs

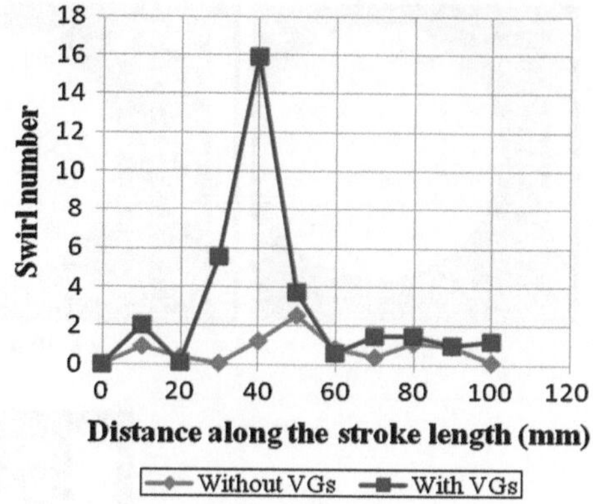

Fig. 6 Variations of the swirl number along the stroke length for two inlet manifold configurations; with and without vortex generators

Acknowledgements This research project was carried out in the Vel Tech-Dassault Systems Centre for Design, Engineering and Manufacturing located at the Research Park, Vel Tech Technical University, Chennai. Support for using the computing facility is acknowledged.

References

1. Velte CM, Hansen MOL, Okulov VL (2009) Helical structure of longitudinal vortices embedded in turbulent wall-bounded flow. J Fluid Mech 619:167–177
2. Paul B, Ganesan V (2010) Flow field development in a direct injection diesel engine with different manifolds. Int J Eng Sci Technol 2(1):80–91
3. Martins J, Teixeira S, Coene S (2009) Design of an inlet track of a small IC Engine for swirl enhancement, In: Proceedings of the 20th international congress of mechanical engineering, Gramado, 15–20 Nov 2009
4. Murali Krishna B, Mallikarjuna JM (2009) Tumble flow analysis in an unfired engine using particle image velocimetry, In: Proceedings of world academy of science, engineering and technology, vol 30
5. Lee J-W, Kang K-Y, Choi S-H, Jeon C-H, Chang Y-J (2000) Flow characteristics and influence of swirl flow interactions on spray for direct injection diesel engine, In: Seoul 2000 FISITA world automotive congress, Seoul, 12–15 June 2000
6. Prasad BVVSU, Sharma CS, Anand TNC, Ravikrishna RV (2011) High swirl-inducing piston bowls in small diesel engines for emission reduction, Appl Energy 88:2355–2367
7. Versteeg HK, Malalasekhara W (1995) An introduction to computational fluid dynamics: the finite methods. Longman Group Ltd, London

Numerical Study of Air-Intake Performance of a Scramjet with Strut Blockage at Various Angle of Attacks

S. Sivaraj and S. Senthil Kumar

Abstract Scramjet engine performance mainly depends on the establishment of supersonic flow in the air intake. The air intake geometry plays a dominant role in obtaining the required pressure recovery of the intake; thereby influencing the performance of the combustion chamber. This study is the numerical simulation of a scramjet inlet at Mach number of 6, in order to study the performance characteristics of the inlet at various angles of attack (AOA). The scramjet inlet used here is a mixed compression type with two external ramps. Inlet characteristics and the effect of fuel injection strut blockage on the performance of the inlet are studied. For the strut blockage of 20 % and at zero angle of attack, the pressure recovery of the inlet is found to be about 21 % and the inlet is in started condition. It is found that the pressure recovery decreases as the angle of attack increases for a fixed strut blockage which results in much spillage of air.

Keywords Scramjet · Hypersonic · Air-intake · Strut blockage · Kantrowitz limit

1 Introduction

Scramjet air intake employs external–internal compression intakes for obtaining required pressure recovery at the combustion chamber. In this generic configuration was studied that the external compression is provided by 2 ramps. The shock

S. Sivaraj (✉) · S. Senthil Kumar (✉)
Department of Aeronautical Engineering, Vel Tech Dr. RR and Dr. SR Technical University, Avadi, Chennai 600062, India
e-mail: siva.prop@gmail.com

S. Senthil Kumar
e-mail: s.senthilms@gmail.com

R. P. Bajpai et al. (eds.), *Innovative Design, Analysis and Development Practices in Aerospace and Automotive Engineering*, Lecture Notes in Mechanical Engineering,
DOI: 10.1007/978-81-322-1871-5_20,

emerging from the ramps converge at the cowl lip, which in turn turns the external flow through an oblique shock for internal compression inside the intake. The above system of oblique shocks provides less pressure loss and maximizes the mass flow in the combustion chamber for an efficient performance. But due to the viscous nature of flow oblique shocks emanating from the ramps do not meet exactly at the cowl lip but ahead of the lip which entails in different shock system called as coalesced shock. The coalesced shock hits the cowl lip and gets reflected, interaction of this reflected shock with the expansion waves from the expansion corners of the profile leads to the formation of shock train system and flow field around the cowl lip.

Van Wie et al. [1] conducted a small-scale supersonic inlet experiment to better understand the major factors that influence unstart and restart characteristics of simple inlets. The results of the experiment have shown that both hard and soft unstarts and restarts can occur, depending principally on the inlet geometry and the amount of inviscid flow entering the inlet. Unsteady effects observed within the test hardware were also shown to affect the unstart characteristics of the inlets. For the inlets investigated, the Kantrowitz limit was successful in predicting the restart contraction ratio (CR) with an accuracy of approximately 10 % if only the hard unstarts were considered and the Kantrowitz limit was based on the mass-averaged Mach number at the cowl lip. Depending on the test configuration, unstarts were classified into two broad categories as either “hard” or “soft.” The hard unstarts appear to occur when the flow at the inlet throat chokes. The soft unstarts occur as large-scale separation develops within the inlet. The ability of the classical Kantrowitz limit to predict the restart CR was assessed, and it was shown to be applicable for the hard unstart/restart configurations. The role of fluid injection upstream of the unstarted inlet was also assessed. The use of this injection may ultimately lead to improving the starting characteristics of inlets.

Schulte and Henckels [2] examined the application of bleed to improve the inlet efficiency using a generic flat plate model and a single shock/laminar boundary layer interaction. A bleed set-up has been presented yielding a maximum reduction of the boundary layer separation bubble and the thermal loads on the wall by requiring only a minimal bleed mass flow (about 60 % of the mass flow transported through the boundary layer). This bleed set-up, which is considered to be most suited regarding technical feasibility, can be used as the basis for an implementation of boundary layer bleed in hypersonic inlets.

Smart [3] conducted Wind-tunnel testing of a hypersonic inlet with rectangular-to-elliptical shape transition at Mach 6.2. These tests were performed to validate the use of a recently developed design methodology for fixed geometry hypersonic inlets suitable for airframe integrated scramjets. Results indicated that flow features within the inlet were similar to design and that the inlet typically captured 96 % of the available air flow. Typical mass-flow-weighted total pressure recoveries of 55 % were obtained for compression ratios of 14.8 throughout the test program. Assessment of the inlet starting characteristics indicated that the inlet self-started at Mach 6.2 despite the fact that it had an internal CR well above the Kantrowitz limit. These results demonstrate that high performance, fixed-geometry inlets can be designed to combine a nearly rectangular capture with a smooth transition to an elliptical throat.

Matthews and Jones [4] showed that re-expansion intakes have very low cowl drag at the expense of decreased efficiency and nonuniform exit flow. Scramjet engine performance increases monotonically with the amount of isentropic compression for intakes of this type. An optimum flow-turning angle is only found in the case when the external compression occurs exclusively through a single shock wave. Calculated stream-thrust averaged performance parameters for the REX1020 intake were considerably higher than measured when isolator skin friction was excluded from the analysis. This effect is sufficiently large that the value of the wind-tunnel data is greatly decreased if the intake model isolator is not a replica of the engine isolator. Viscosity also results in reduced CR (and lower performance) as the design must account for significant boundary-layer displacement at the intake throat.

Sun et al. [5] investigated numerically the flow field of the entire flow path and the aerodynamic characteristics of a hypersonic vehicle at a 7.0 free stream Mach number. The emphasis of the analysis is placed on the effects of the inlet operating conditions. Results indicate that the inlet started and unstarted operations have remarkable impacts on the flow pattern of the full flow path. When the inlet operates in a started state, the transverse pressure gradient generated by the forebody alters the air-capture characteristics and the entering flow quality of the inlet. The expansion of the nozzle jet flow is obviously affected by the external flow around the afterbody with a continuous cross-sectional shape changing along the flow direction. When the inlet operates in a unstarted state, substantial aerodynamic instabilities can be observed in the entire flow path of the vehicle. Due to the oscillations of the external compression shock system and the nozzle jet flow, the aerodynamic characteristics of the vehicle vary periodically with the lift to drag ratio changing from 0.25 to 2.09.

Mahapatra and Jagadeesh [6] presented an experimental study to show the effect of CR on shock interaction phenomena for a two-dimensional, planar scramjet inlet model. Experiments are conducted in a hypersonic shock tunnel, at Mach 8, at three CRs: 8.4, 5.0, and 4.3. CR here is defined as the ratio of the projected inlet area to the throat area. Investigations include Schlieren flow visualization around the cowl region and heat transfer rate measurement inside the inlet chamber. Various ramp/cowl shock interaction processes ahead of the inlet have been visualized using a high-speed camera. Edney type II interference pattern is observed for a CR of 4.3 with all its typical features resulting because of the fore body shock/cowl shock interaction. Peaks in the heat transfer rate measured inside the chamber show possible locations of shock impingement because of the shock interaction inside the inlet.

The objective is to study the performance characteristics of a scramjet inlet at Mach number 6 for various AOA. The obstruction to the flow due to the presence of the solid surfaces (fuel injection struts is known as "*blockage*" struts which have been used for fuel mixing), the fuel strut has blockage ratio of 20 % to the

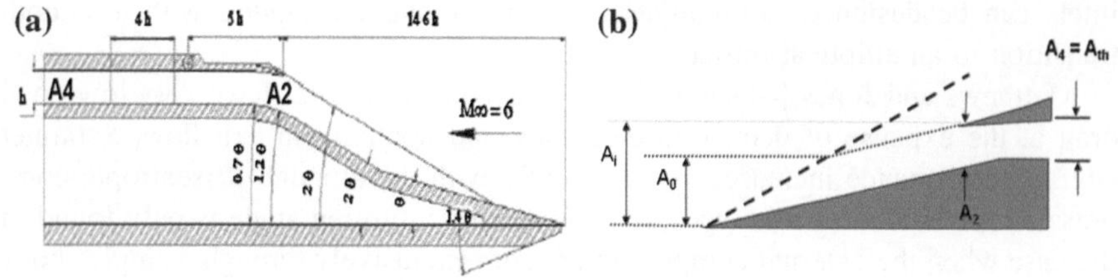

Fig. 1 **a** Geometry of the scramjet inlet [7]; **b** Inlet stations

flow at the strut locations. The effect of strut blockage on the performance of the inlet in terms of pressure recovery has to be studied.

The geometry of the problem is shown in Fig. 1. This is taken from the published work of Saravanan et al. [7] in order to validate the present computation with that of their experimental results. It has two ramp surfaces; the first ramp making an angle of 10.5° with the freestream and the second ramp making an angle of 21° with the freestream. There are two expansion corners in which one making an angle of 12° with freestream and the other making an angle of 7° with freestream. Cowl lip is located at the starting of the first expansion corner.

2 Kantrowitz Limit

Preliminary estimates of the internal CR that will self-start the inlet can be obtained from the Kantrowitz limit [8]. This limit is determined by assuming a normal shock wave at the beginning of the internal contraction and calculating the one-dimensional, isentropic internal area ratio that will produce sonic flow at the inlet throat. For a perfect gas the Kantrowitz limit can be calculated as follows:

$$\left(\frac{A_2}{A_4}\right)_{\text{Kantrowitz}} = \frac{1}{M_2}\left[\frac{(\gamma+1)M_2^2}{(\gamma-1)M_2^2+2}\right]^{\frac{\gamma}{\gamma-1}}\left[\frac{\gamma+1}{2\gamma M_2^2-(\gamma-1)}\right]^{\frac{1}{\gamma-1}}\left[\frac{1+\frac{\gamma-1}{2}M_2^2}{\frac{\gamma+1}{2}}\right]^{\frac{\gamma+1}{2(\gamma-1)}} \tag{1}$$

$$\left(\frac{A_0}{A_{\text{throat}}}\right)_{\text{Isentropic}} = \frac{1}{M_0}\left[\frac{2}{(\gamma+1)}\left(1+\frac{\gamma-1}{2}M_0^2\right)\right]^{\frac{\gamma+1}{2(\gamma-1)}} \tag{2}$$

where, A_2 is the cowl lip area, A_{th} or A_4, the throat area, A_0, the cross sectional area of the projected frontal area of the inlet and M, the Mach number, as sketched in Fig. 1b.

Based on the expression (1), the calculated CR of the inlet with no blockage (without struts) is 1.1667 and with 20 % blockage is 1.458 for the Cowl lip area of

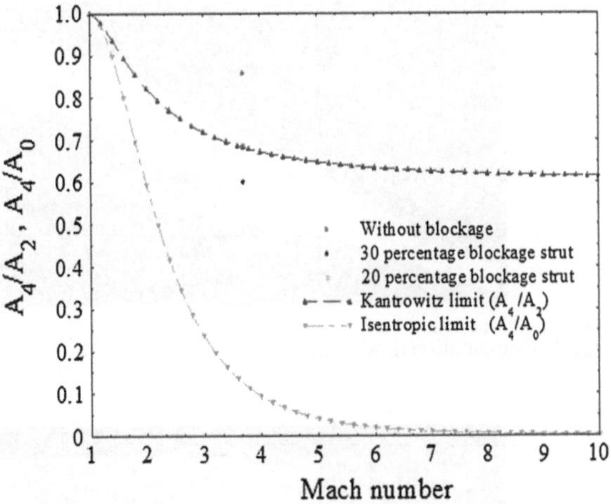

Fig. 2 Variations of contraction area ratio limits for different Mach number values

the inlet is 672.043 mm^2 and the throat area of the inlet is 576 mm^2. Mach number ahead of the cowl lip area is 3.7 for the given ramp angles and the CRs are calculated based on this Mach number.

Figure 2 shows the theoretical limit for the stating of the inlet, from the literature [8] it has been identified when the contraction area is above the limit (area enclosed by the Isentropic (2) and Kantrowitz (1) curves) the inlet is said to be self-started so the inlet without blockage and 20 % strut blockage are self-starting.

3 Numerical Methodology

Only the half symmetry model has been considered since the scramjet inlet is planar symmetric and a half symmetry model is sufficient to study the flow phenomenon. Computational domain of the problem is shown in Fig. 3. The boundary conditions used for this analysis are shown in Fig. 4. Computations are made within the domain using Compressible Reynolds—Averaged Navier—Stokes equations with Shear Stress Transport (SST) k-omega turbulence model available with Commercial software ANSYS FLUENT™. Implicit solver coupled with the Sutherland viscosity formulation is adopted. For all the computations, residuals of velocity and turbulence kinetic energy are monitored and suitable convergence criterion ($= 10^{-4}$) is set for the solutions to converge.

Pressure far field boundary condition is specified at the top and right side faces of the domain. No-slip boundary conditions are specified at walls of the ramps, struts, side plate, and cowl of the inlet. The tolerance level for the validation is 10^{-4}. 3D Hexahedron type structural mesh is preferred.

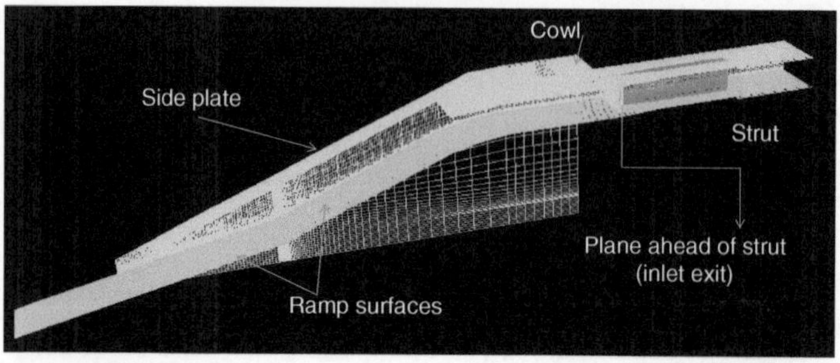

Fig. 3 Computational domain

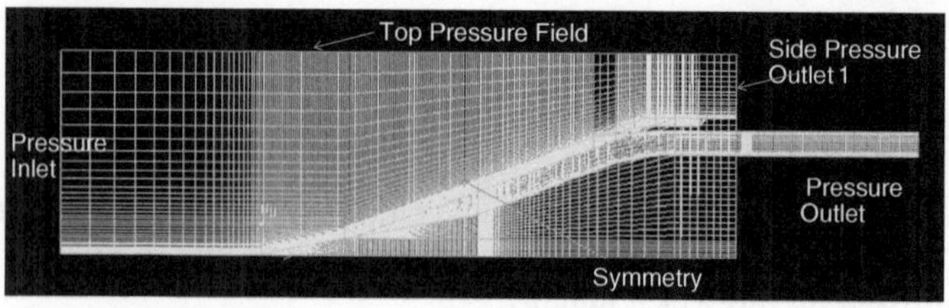

Fig. 4 Boundary conditions of the problem (2D cut view)

4 Results and Discussion

Computations have been carried at various AOA for 20 % strut blockage and Mach number 6. The validation has been carried out for 20 % strut blockage at zero angle of attack. Figure 5 shows the static pressure distribution at zero AOA on ramp-1 ($x/L < 0.3$) and ramp-2 ($0.3 < x/L < 0.5$) are similar with oblique shock values. But there is a pressure jump on the expansion corner (ramp 3 and 4, $0.55 < x/L < 0.61$) which is due to the formation of separation bubbles over this region. The cowl lip shock induces a stronger separation bubble due to boundary layer shock wave interaction on the ramp shoulder; the effects of flow separation are also stronger and carried forward to the expansion region. The pressure distribution beyond $x/L = 0.58$ is due to the highly viscous dominated flow on account of cowl lip shock impingement on the ramp, multiple shock reflections due to strut leading edge shock and the comparison of the pressure ratios of the ramp wall pressure to the free stream stagnation pressure ($P_{ramp}/P_{0\infty}$) shows reasonably good agreement between experimental results and it is shown in Fig. 5.

Figure 6 shows the Mach number contour of the inlet at zero AOA which clearly indicates the shock and its flow patterns. There is a drop in Mach number across the two oblique shocks emanating from ramps. When the cowl lip shock wave impinges on the boundary layer over the inlet, the large pressure rise across the shock wave acts as a severe adverse pressure gradient imposed on the boundary layer, thus causing the boundary layer to locally separate from the surface.

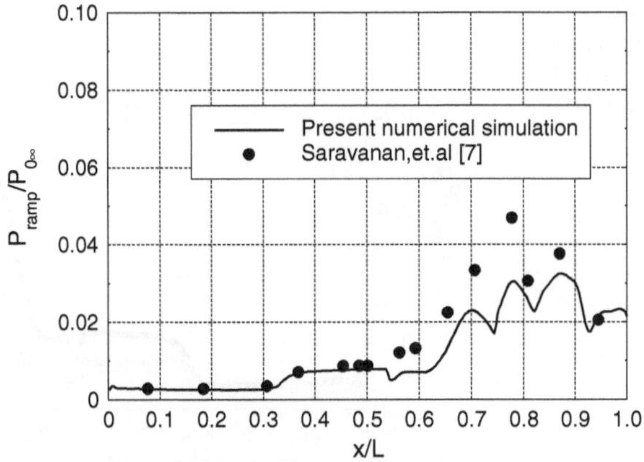

Fig. 5 Comparison of wall pressure distribution, $P_{ramp}/P_{0\infty}$ for 20 % strut blockage with the experimental results at zero AOA [7]

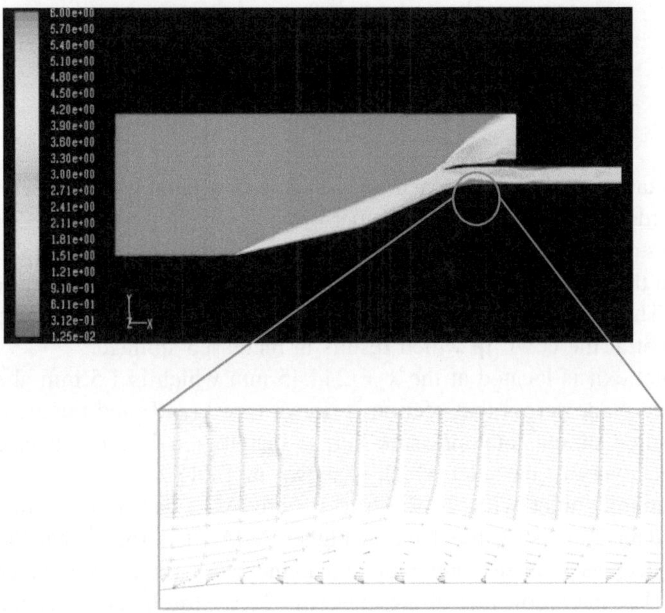

Fig. 6 Mach number contours for 20 % strut blockage at zero AOA

Because of the high pressure behind the shock, it feeds upstream through the subsonic portion of the boundary layer, the separation takes place ahead of the impingement point of the incident shock wave which leads to the formation of separation bubble and this is confirmed by the reverse flow in that region.

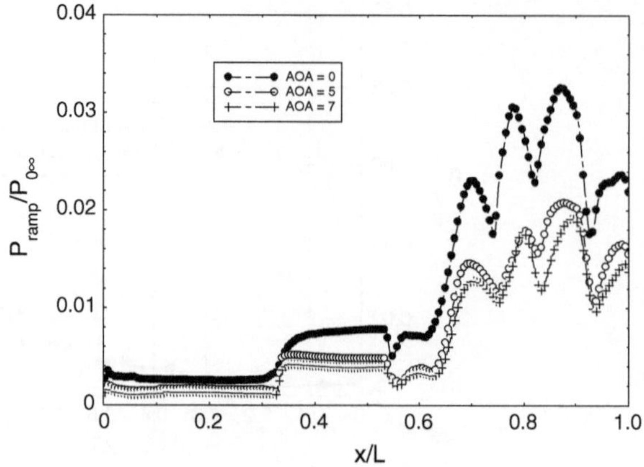

Fig. 7 Numerically simulated wall pressure ratio distributions, $p_{ramp}/p_{0\infty}$ for 20 % strut blockage at various AOA

Table 1 Computed pressure recovery for various AOA

S. No.	Angle of attack (°)	Pressure recovery (%)
1	0	20.7
2	5	17.13
3	7	15.13

Computations have been carried for the same flow parameters but with different AOA in order to see the effect on the performance of the inlet. Figure 7 shows the static pressure levels at various AOA for 20 % strut blockage. It is clearly seen in Fig. 7 that the pressure ratio ($p_{ramp}/p_{0\infty}$) at zero AOA is higher than those of 5 and 7 AOAs. This can be attributed that the shocks emanating from the ramp surfaces do not meet at the cowl lip which results in increased spillage.

The inlet exit is located at the x = 244.45 mm which is 1.5 mm ahead of the strut, hence plane across that section is drawn (see Fig. 3) and the area weighted average value of the total pressure across that region is used to calculate the pressure recovery of the inlet which is shown in Table 1.

The scramjet inlet with 20 % strut at zero AOA is having better pressure recovery than 20 % strut blockage at higher AOA. It is found that the pressure recovery decreases as the angle of attack increases for a fixed value of strut blockage. The mass flow rate of the inlet with 20 % strut blockage at zero AOA is higher which indicates no spillage when compared with those of 20 % strut blockage at higher positive angle of attack. The inlet with 20 % strut blockage at positive angle of attack has lesser mass flow rate and this is because the ramp shocks don't meet at the cowl lip which leads to the spillage of flow.

5 Conclusion

Numerical simulation of the scramjet inlet has been performed at Mach number 6 with 20 % strut blockage for various AOA to study the performance characteristics of the inlet. The effects of angle of attack with fuel injection strut blockage on the performance of the inlet have been studied. For the strut blockage of 20 %, the pressure recovery of the inlet is found to be about 21 %. It is found that the pressure recovery decreases as the angle of attack increases for a fixed strut blockage which results in much spillage of air.

References

1. Van Wie DM, Kwok FT, Walsh RF (1996) Starting characteristics of supersonic inlets. In: 32nd Joint propulsion conference and exhibit. doi: 10.2514/6.1996-2914
2. Schulte D, Henckels A (1998) Reduction of shock induced boundary layer separation in hypersonic inlets using bleed. Aerosp Sci Technol 4:231–239
3. Smart MK (1999) Design of three-dimensional hypersonic inlets with rectangular-to-elliptical shape transition. J Propul Power 15(3):408–416
4. Matthews AJ, Jones TV (2005) Design and test of a hypersonic isentropic-spike intake with aligned cowl. J Propul Power 21(5):838–843
5. Sun S, Zhang H, Cheng K, Wu Y (2007) The full flow path analysis of a hypersonic vehicle. Chin J Aeronaut 20(5):385–393
6. Mahapatra D, Jagadeesh G (2008) Shock tunnel studies on cowl/ramp shock interactions in a generic scramjet inlet. Proc IMechE Part G: J Aerosp Eng 222(G8):1183–1191
7. Saravanan R, Tennyson JA, Subramaniam S (2011). Scramjet air intake performance studies under two different start conditions at hypersonic Mach number, NCAE 2011, BIT Mesra, Ranchi, p 169
8. Kantrowitz A, Donaldson C (1945) Preliminary investigation of supersonic diffusers, NACA WR L-713

Numerical Study of Aerodynamic Characteristics of Triangular Flapping Wing for MAV

S. Senthil Kumar and Pranesh Rajan

Abstract Aerodynamic characteristics over a triangle-shaped flapping wing are analyzed numerically using commercial CFD code, Fluent^{TM}. The three-dimensional geometry and mesh are created using preprocessor GAMBIT. Transient laminar flow characteristics are studied and the results are presented in terms of pressure distribution, lift, and drag coefficients. It is observed that during the upstroke the lift coefficient reaches to the maximum positive value and its value falls down to the minimum negative value during the downstroke. However because of the symmetric flapping the average lift coefficient over a flap cycle is zero, which is consistent with the results obtained earlier.

Keywords Flapping wing · Dynamic mesh · MAV · Vortex shedding · Aerodynamics

1 Introduction

The potentials of Micro Aerial Vehicles (MAV's) have generated an enormous interest in this technology. There are significant technical challenges that are hindering the efficient utilization of MAV's technology. These include aerodynamics of low Reynolds Number flows, small-scale power generation and power storage, navigation and communication, propulsion and control as well as manufacturability. Generally, Micro Aerial Vehicles were classified as fixed wing and flapping

S. Senthil Kumar (✉)
Department of Aeronautical Engineering, Vel Tech Dr. RR and Dr. SR Technical University, Chennai, India
e-mail: s.senthilms@gmail.com

P. Rajan
Flight Mechanics and Control Division, NAL (CSIR), Bangalore, India

R. P. Bajpai et al. (eds.), *Innovative Design, Analysis and Development Practices in Aerospace and Automotive Engineering*, Lecture Notes in Mechanical Engineering, DOI: 10.1007/978-81-322-1871-5_21,

wing. Fixed wing aircraft get their primary lift generation as function of forward airspeed and amount of wing span and surface area. The angle of attack is a part of it but the amount of AOA needed is a function of the airspeed. For MAV flying in confined spaces very slow speeds are needed to avoid running into obstacles. The only way to do this is that increase the size of the wing for a fixed wing MAV. Due to this obstacle, it is necessary to make a larger MAV to fly in smaller spaces.

A Flapping MAV is one of the solutions for this problem. Flapping wings produce lift from flapping motion of the wing not from forward airspeed of the aircraft. Lift/thrust is controlled by the flapping frequencies and less dependent on air forward motion. This conceptually solves the issue for a fixed wing MAV trying to fly at very slow airspeeds or even hover. Therefore, the alternative is flight with flapping wings.

A large number of experimental and numerical works [1–11] have been carried out in recent years to investigate the basic physical phenomena including flow patterns and vortex structures in the wakes of flapping airfoil/wings. Most of the work been done to study the resultant aerodynamic forces acting on flapping airfoils/wings during the upstroke and downstroke of a flapping cycle. There are only a few literatures related to flapping wing to understand the overall aerodynamic performances of it.

Ashraf et al. [3] reviewed the recent progress of CFD analysis of flapping wing aerodynamics and discussed about the effects of flapping frequency, amplitude for thrust generation, and propulsive efficiency. Unsteady flow around an airfoil undergoing pure plunging motion was simulated using Fluent CFD code. Navier-strokes computations were performed for various parameters of frequency and amplitude.

Young et al. [4, 5] studied the flow over a 2D airfoil, oscillated sinusoidal in plunge, is simulated using a compressible Navier Stokes solver at Reynolds number 30,000. Close agreement is obtained between numerically simulated wake structures and experimental wake visualizations in the literature. Thrust of the airfoil and detailed wake structures are shown to be strongly dependent on both the Strouhal number and the reduced frequency of the plunge oscillation, at this Reynolds number. The findings are postulated to be the result of a vortex shedding frequency lock-in phenomenon, similar to that observed for transversely vibrating circular cylinders. Both experimental and numerical results indicate that the wake structure is strongly dependent on both k and h, for the Reynolds number range under study.

Steven Hoa et al. [6] reviewed the unsteady aerodynamics, aero elastic phenomena, and several flow control strategies for flapping wing flight, with an eye toward application of these theories to raising the performance of MAVs. Simple scaling arguments demonstrated that small and slow flying MAVs will operate in a fully unsteady flow regime and, as a result, quasi-steady aerodynamic analysis is not appropriate. Furthermore, the presence of highly flexible wings in an unsteady flow created a difficult aeroelastic analysis, but one that must be considered, as the flow-structure interaction proved important in determining overall lift and thrust.

Past researchers have uncovered a number of unsteady flow mechanisms that enhanced lift and thrust generation for flapping wings. These include clap and fling, rotational lift and wake capture, and, most prominently, the unsteady leading edge vortex. The unsteady leading edge vortex and the stiffness distribution of the wing were shown to dominate the lift and thrust production.

2 Numerical Methodology

The design process is carried out using 3D modeling and meshing software, GAMBIT 2.4.6. The coordinates of the design are taken on hypothesis basis. The size function is used to create fine mesh close to the wing surface and coarse mesh is used away from it to reduce the computational time. The planform of the flapping wing is assumed to be a triangle from our model. It is assumed and modeled to have thin wings as in the case of insect wings. However, the aero-elasticity that is observed in the case of insect wings is neglected for the ease of analysis.

Velocity inlet boundary condition is specified for the outer surfaces of the computational domain. No-slip boundary conditions are specified at walls. Three-dimensional tetrahedral mesh structures are used thought the computational domain. Dynamic mesh is used in order to take account of movement of wing so that a new grid structured has been generated at each time instants. Computations are carried out with unsteady laminar incompressible pressure based segregated solver with SIMPLEC pressure correction methods. The solver used is Commercial software FLUENT™. For all the computations, residuals chosen for continuity and momentum equations are 10^{-4} to check the solutions to converge and time step is 0.001. A sinusoidal flapping motion is assumed for the wing motion in terms of flapping frequencies. A user-defined function (UDF) is created in C language for incorporating flapping motion of the wing and it has been hooked in the solver.

3 Results and Discussion

3.1 Single Stroke Dynamics

The effect of flapping of the wing in unsteady regime is captured. The variations in the flow field for a given stroke amplitude and Reynolds number is captured in terms of static pressure and velocity magnitude. The variations are carefully studied to understand the underlying aerodynamics. To observe the flow field in a planar view a plane is cut behind the leading edge.

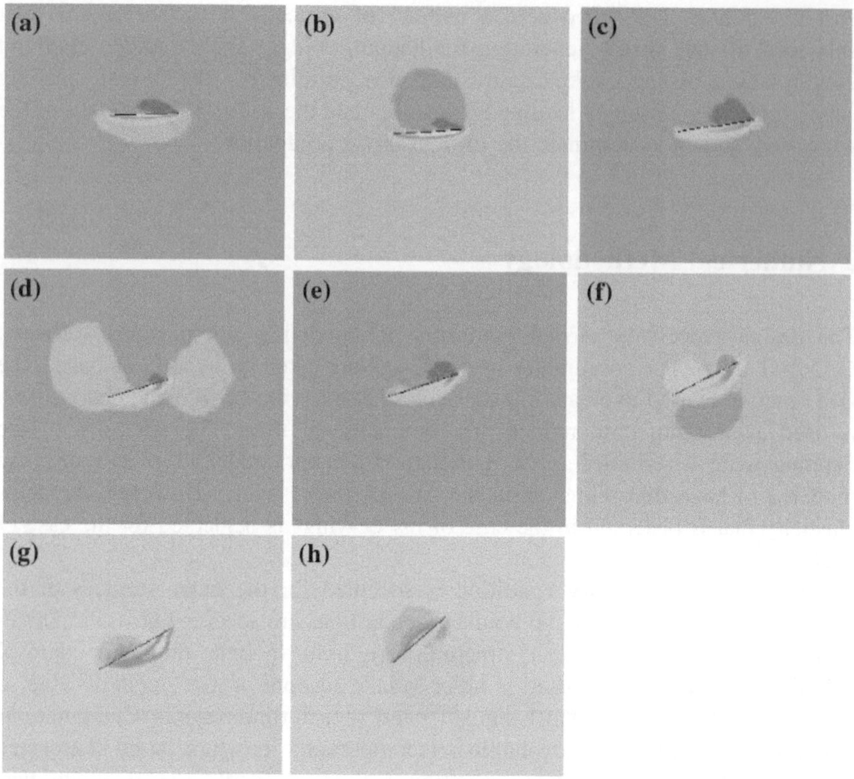

Fig. 1 Static pressure variation—upstroke. **a** t = 0.06 s, **b** t = 0.09 s; **c** t = 0.14 s, **d** t = 0.19 s, **e** t = 0.21 s, **f** t = 0.28 s, **g** t = 0.32 s, **h** t = 0.4 s

3.1.1 Static Pressure Variation During the Upstroke

Figure 1a shows that a disturbance is initialized just underneath the wing. This corresponds to a time instance of 0.06 s just the beginning of the upstroke. In Fig. 1b the disturbance propagates itself to the flow field nearby. Also the boundary layer is diffused throughout in the region. The vortex growth starts at this instant. There are two vortices one at the leading edge and other at the trailing edge. Both the vortices are of opposite rotation. The vortices continue to grow and one of the vortices tries to curl up the edge (see Fig. 1c). In the process a vortex is shed and a part of the vortex diffuses and combines with the shed vortex (Fig. 1d). The remaining vortex that is attached to the wing curls up (Fig. 1e, f) and gains strength (Fig. 1g). The vortex well remains attached till the beginning of the downstroke.

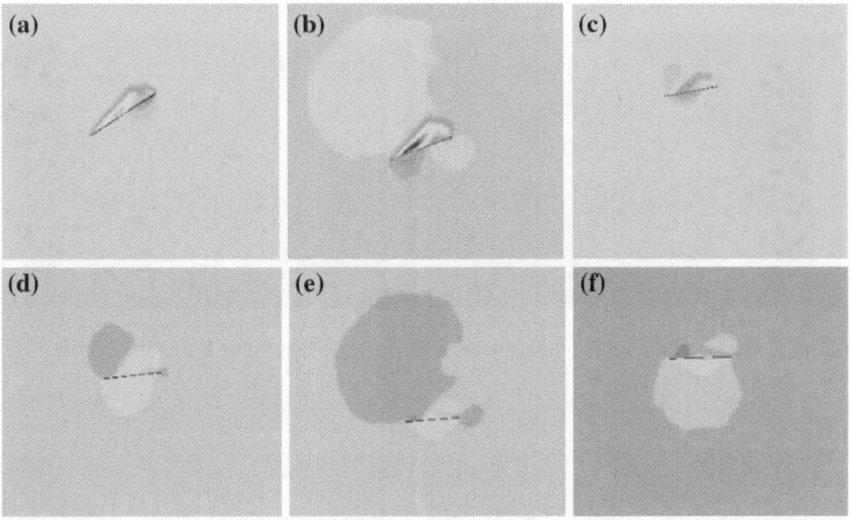

Fig. 2 Static pressure variation—downstroke. **a** t = 0.66 s, **b** t = 0.76 s, **c** t = 0.84 s, **d** t = 0.88 s, **e** t = 0.9 s, **f** t = 0.93 s

3.1.2 Static Pressure Variation During the Downstroke

With the beginning of the downstroke there is a clear visualization of the vortex shedding from the leading edge as shown in Fig. 2. It does not diffuse rather it is shed. In case of downstroke, there is vortex formation above the wing both at the leading and trailing edge. The same variations as observed in the upstroke are also observed here.

3.1.3 Variation of Velocity Magnitude

A zoomed in view of the variation in velocity magnitude is shown in Fig. 4a, b. This variation compliments that of the pressure variation and coincides perfectly with the results observed in that case of pressure variation. Some of the phenomena observed are (leading edge vortices alone are seen due to zooming), boundary layer diffusion, vortex of increasing strength, vortex curling up the tip, spanwise flow along the wing, vortex beneath the wing during upstroke, and vortex above the wing during down stroke.

Initially there is a diffusion of the boundary layer. Subsequently there is growth of vortex both in the leading edge and the trailing edge. The vortices are shed alternatively, i.e., when the trailing edge vortex is formed the leading edge are shed and vice versa. The attached leading edge vortex causes a negative pressure region and, hence, is responsible for enhancing lift production as shown in Figs. 3b and 4a.

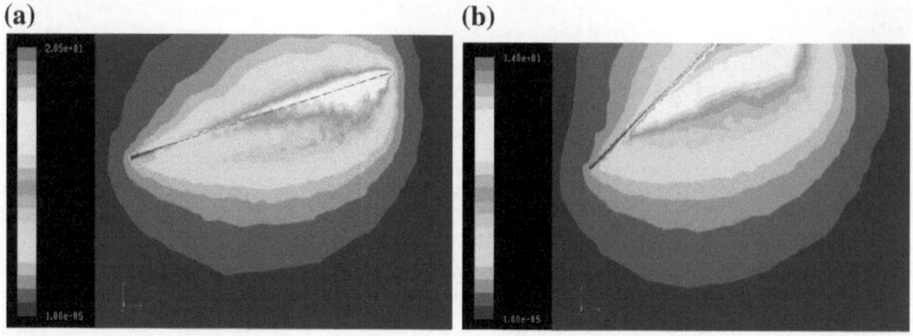

Fig. 3 Variation of the velocity magnitude. **a** Beginning of upstroke; **b** End of upstroke

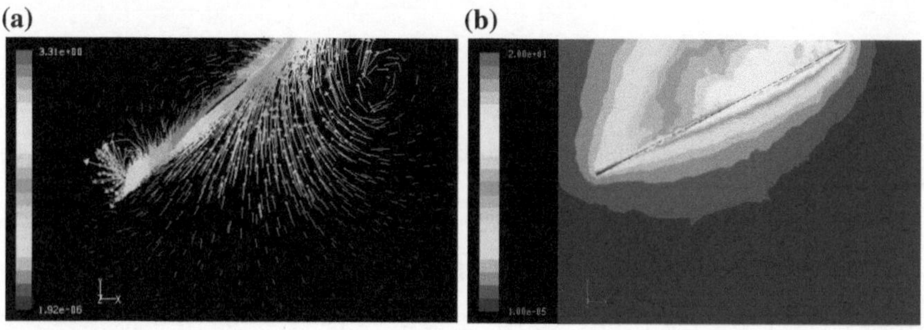

Fig. 4 **a** Velocity vectors at the end of upstroke colored by velocity magnitude; **b** Variation of the velocity magnitude at the beginning of downstroke

Also the vortices are of opposite direction resulting in von Karman Street when continued for a longtime interval. During upstroke the vortices are formed beneath the wing, as shown in Fig. 3b.

3.1.4 Unsteady Forces

To illustrate the unsteady behavior, we compute the time-dependent lift on the wing after an impulsive start. We examine the lift coefficient C_l, drag coefficient C_d as a function of time. The lift coefficient and drag coefficient are plotted for the stroke amplitude of 20° and the frequency of 30 Hz.

Figure 5a shows the lift coefficient, C_l at different time instants and the drag coefficient, C_d at different time instants are plotted in Fig. 5b. In the lift coefficient, the movement of flap wing starting the upstroke and arriving to the highest peak from the center position, C_l elevates to the maximum value and is positive. The movement of wing flapping again from the highest position to the center point, i.e., downstroke the lift coefficient, C_l falls to the starting value. Therefore, in downstroke the lifting force, L is negative. However, because of the symmetric flapping

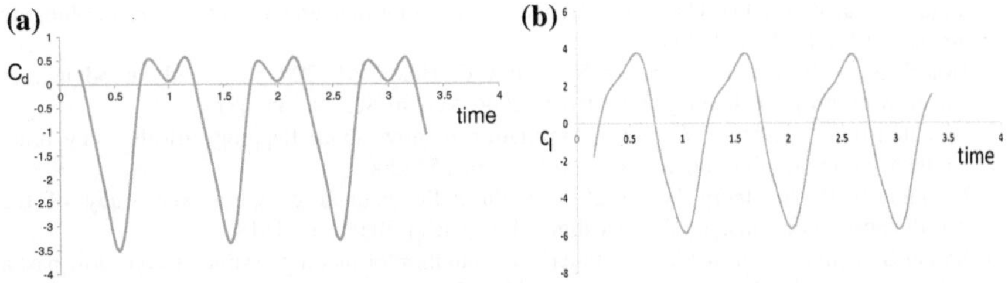

Fig. 5 **a** Drag coefficient plot; **b** Lift coefficient for stroke amplitude of 20° and frequency of 30 Hz

the average lift coefficient over a flap cycle is zero, which is consistent with the results, obtained earlier [3]. The resulting coefficient of lift and drag do not vary much with the varying stroke amplitude for small values of angle of attack.

4 Conclusions

Numerical simulation of flow over a flapping wing is carried out for a fixed flapping frequency. Transient laminar flow characteristics are studied and the results are presented in terms of pressure distribution, lift, and drag coefficients. Based upon the results, it appears as aerodynamic properties (i.e., pressure, velocity, C_l, and C_d) change according to the range of stroke amplitude. It is observed that during the upstroke the lift coefficient reaches to the maximum positive value and its value falls down to the minimum negative value during the downstroke. However, because of the symmetric flapping the average lift coefficient over a flap cycle is zero, which is consistent with the results obtained earlier. Further analyses will be performed for various flapping frequencies and wing shapes. Without understanding how flapping affects lift it would be impossible to design a flapping wing MAV. Findings show that a simple, sinusoidal flapping mechanism and triangular wing has the potential to generate a sufficient lift, but requires further investigation.

References

1. Shyy W, Aono H, Chimakurthi SK, Trizila P, Kang C-K, Cesnik CES, Liu H (2010) Recent progress in flapping wing aerodynamics and aeroelasticity. Prog Aerosp Sci 46(7):284–327
2. Dickinson MH, Gotz KG (1993) Unsteady aerodynamic performance of model wings at low Reynolds numbers. J Exp Biol 174:45–64
3. Ashraf MA, Lai JCS, Young J (2007) Numerical analysis of flapping wing aerodynamics. 16th Australasian fluid mechanics conference, Gold Coast, Australia, pp 1283–1290
4. Young J, Lai JCS (2007) Mechanisms influencing the efficiency of oscillating airfoil propulsion. AIAA J 45:1695–1702

5. Young J, Lai JCS (2004) Oscillation frequency and amplitude effects on the wake of plunging airfoil. AIAA J 42:2042–2052
6. Hoa S, Nassefa H, Pornsinsirirakb N, Taib Y-C, Hoa C-M (2003) Unsteady aerodynamics and flow control for flapping wing flyers. Prog Aerosp Sci 39:635–681
7. Ramamurti R, Sandberg W (2001) Simulation of flow about flapping airfoils using finite element incompressible flow solver. AIAA J 39:253–260
8. Ramamurti R, Sandberg WC (2002) A three-dimensional computational study of the aerodynamic mechanisms of insect flight. J Exp Biol 205:1507–1518
9. Sarkar S, Venkatraman K (2006) Numerical simulation of incompressible viscous flow past a heaving airfoil. Int J Numer Meth Fluids 51:1–29
10. Sarkar S, Venkatraman K (2006) Numerical simulation of thrust generating flow past a pitching airfoil. Comput Fluids 35:16–42
11. Tuncer IH, Platzer MF (2000) Computational study of flapping airfoil aerodynamics. AIAA J Aircr 37:514–520

Numerical Analysis of Suppression of Laminar Bubble at Low Reynolds Number Using Different Protrusions

S. Senthil Kumar, R. P. Vijay Ganesh, S. Karthikeyan and J. Karthik

Abstract Flow over airfoil at low Reynolds numbers encounters separation bubble which can be generally controlled by active or passive methods. In this project work passive method is used. The effectiveness of bubble burst control plate at different locations (5 and 7.5 % of the chord of airfoil) on an NACA 63_1-012 airfoil section for laminar separation bubble burst delay, stall separation, lift augmentation, and drag reduction will be analyzed computationally using commercial CFD code ANSYS FLUENT. The effectiveness of burst control plate is governed by its height, width, geometry, and distance between leading edge of the airfoil and trailing edge of plate. Flow patterns are analyzed and they are compared with experimental results. The results suggest that the stall angle of the original Airfoil (without the burst control plate) occurs at 10°, and it is successfully postponed to 13° (Airfoil with Triangular plate) and 14° (Airfoil with Rectangular plate) when the burst control plate is attached onto the Airfoil. The lift forces were calculated for a range of angle of attack from 0 to 15°. The overall analysis results demonstrated the application of the burst control plate on the NACA 63_1-012 Airfoil can be effective means of bubble burst control and Airfoil stall suppression in low speed flows.

Keywords Airfoil · Laminar bubble · Low Reynolds number · Burst control plate · Flow separation

S. Senthil Kumar (✉) · S. Karthikeyan · J. Karthik
Department of Aeronautical Engineering, Vel Tech Dr. RR and Dr. SR Technical University, Avadi, Chennai, India
e-mail: s.senthilms@gmail.com

R. P. V. Ganesh
Department of Aeronautical Engineering, Hindustan University, Padur, Chennai, India

R. P. Bajpai et al. (eds.), *Innovative Design, Analysis and Development Practices in Aerospace and Automotive Engineering*, Lecture Notes in Mechanical Engineering,
DOI: 10.1007/978-81-322-1871-5_22,

1 Introduction

Laminar bubble formation has significant role on the performance of air vehicles at low Reynolds numbers. Various applications operating in the low Reynolds number range are windmills, High Altitude Long Endurance Reconnaissance aircrafts, Micro Aerial Vehicles, and so on [6]. The bubble is formed due to the boundary layer separation and reattachment process in the adverse pressure gradient region, which is called as a Laminar Separation Bubble (LSB). It changes the airfoil shape as it forms on the upper surface of the airfoil and affects the aerodynamic characteristics of it. LSBs are found to increase drag which leads to reduce aerodynamic performance [7]. The laminar boundary layer is separated when the flow experiences an adverse pressure gradient. Due to this, the flow detaches from the wall surface and attaches at a point in the downstream, which creates a recirculatory region with the boundary layer. After the reattachment point, the flow gets more momentum and changes into turbulence in the downstream, which destroys the reverse flow near the wall.

Donald et al. [1] investigated the effectiveness of the nose slot at higher values of lift; a second suction slot was added near the mid-chord station of the model. The combination of a nose and a mid-chord boundary layer suction slot has been shown to be more effective in increasing the maximum lift of NACA 63_1-012 Airfoil section than a nose slot operating independently.

Sandham [2] carried out laminar separation bubble simulations numerically. The 3D simulation results show that breakdown of vortices are responsible for full transition to turbulence. Short laminar bubbles were not captured adequately in the two-dimensional simulations. He suggested that further simulations should be performed to understand the bursting mechanism which converts short separation bubbles to long bubbles. Chetan et al. [3] investigated experimentally a laminar separation bubble production on a flat plate. They carried out the flow visualization and pressure measurements in order to show the typical time-averaged characteristics of a bubble. From their results, they proposed the maximum height for a bursting criterion.

Ramesh et al. [4] investigated experimentally the dynamics and control of the laminar separation bubbles which are typically present on the suction surface of an aerofoil at a large angle of attack. A separation bubble is produced on the upper surface of a flat plate by appropriately contouring the top wall of the wind tunnel. The effect of reduction in reference speed on the bubble aspect ratio (ratio of height to length of the bubble) was studied. They found that both length and height of the bubble increase with decrease in speed. However, further analysis showed that the height increases at a greater rate than the length.

Mohseni et al. [5] performed the time-dependent unsteady calculations of low Reynolds number flows over an *EPPLER 387* Airfoil in both two and three dimensions. Various instantaneous and time-averaged aerodynamic parameters including pressure lift and drag coefficient are calculated in each case and compared with the available experimental data. In this simulation they demonstrate that there was a significant difference between two and three-dimensional pressure coefficient distributions over the Airfoil surface. The present three-dimensional simulations were shown relatively good agreement with experimental results.

Saxena [6] investigated laminar separation bubble at low Reynolds Number and its effect on airfoil characteristics. At low Reynolds number, the presence of laminar separation bubble is inevitable. Airfoil (SD7003) with varying Reynolds number at fixed angle of attack. The length of bubble is from 5 % to as much as 15 % of the airfoil chord ant their height is of order of 10 % of their length. The analysis shows that the major difficulty is in making correct and reproducible experimental measurements due to sensitivity of low Reynolds number flows to measurement techniques and methodology.

Aholt and Finaish [7] examined the controlling a Laminar Separation Bubble (LSB) over an airfoil at low Reynolds numbers using active external body force generated means, such as a plasma actuator. In this study, it was found that a localized body force was capable of significantly reducing the size of an LSB, or even eliminating it entirely. The actuator was found to marginally increase lift, while dramatically reducing the drag associated with the LSB. The resultant aerodynamic efficiency of the airfoil was comparable to that attained by the same airfoil at high Reynolds numbers, where the LSB was not observed.

Jahanmiri [8] showed the results of laminar separation bubble in different aspects, namely, basic concepts and characteristics, instability and control mechanism of LSB. However, a better understanding of the physical mechanisms and controlling the formation and structure of separation bubbles are still not complete. The laminar bubble separation control and dynamics are still far from complete despite numerous experimental and numerical studies have been carried out by various researchers.

Hence, the objective of this study is to investigate the stall characteristics of an NACA 63_1-012 with burst control plates, namely, triangular and rectangular plates attached at different locations on the top surface of the airfoil to suppress the laminar-separated bubble formed as the angle of attack increased.

2 Numerical Methodology

Commercially available CFD code, ANSYS Fluent™ is used for the flow field calculations and the governing equations are solved under incompressible pressure-based SIMPLEC method and standard k–ε two equations model is adopted as turbulence model. The k–ε model is one of the most common turbulence models used for this kind of problems. The convective flux are derived by a second order upwind scheme while the diffusive flux coefficients are obtained by central differencing. All the numerical tests are performed with convergence threshold residuals equal to 10^{-4}. Various shapes of domains are used to compute the results and eventually, the domain shown in Fig. 1, gave accurate results. The length of the domain is chosen as 32.5c and the width as 25c, where c is the cord length.

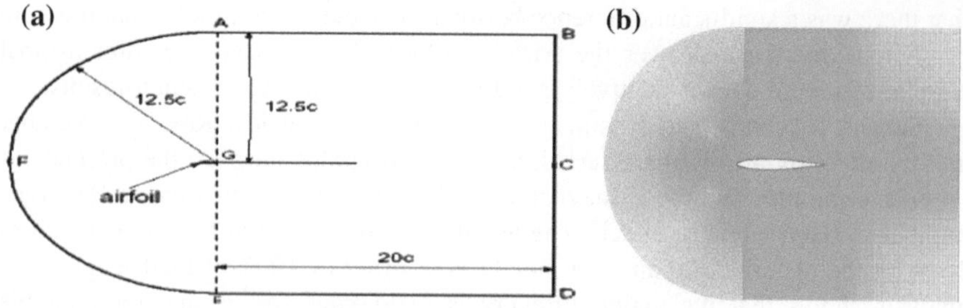

Fig. 1 **a** Computational domain; **b** Grid structure

Quad-type meshing elements are preferred. Figure 1b shows the grid structure of the computational domain. The following Boundary Conditions (B. C) are used for the simulations: Airfoil—No-slip B. C, Inlet—Velocity Inlet B. C (Edges BA, AFE, ED); Outlet—Pressure Outlet B. C (Edge BCD).

3 Results and Discussion

3.1 Validation: Coefficient of Lift Comparison for NACA 63_1-012

The analysis is carried out with air as working fluid and the Reynolds number of 1.3×10^5 for various angles of attack and it is compared with available experimental data [9]. The above process is called validation and this is done to ensure that the solution techniques and numerical method selected for the present problem predict the physics very well. The lift coefficients are calculated for various angles of attack and they are plotted in Fig. 2a. It shows that the results of numerical analysis have reasonably good agreement with the experimental data [9]. It is also noted that the maximum C_L got from numerical analysis is 0.849 (at 10°), which is approximately equal to the experimental one, i.e., C_L is 0.875 (at 11°).

Figure 2(b) shows that velocity vector plot for NACA- 63_1-012 at 10° angle of attack. Pressure at the leading edge is high, the leading edge is a stagnation region and the pressure is essentiality stagnation pressure. As the flow expands around the top surface of the aerofoil the surface pressure decreases dramatically, which is below the free stream static pressure. The pressure gradually increases, reaching a value slightly above free stream pressure at the trailing edge. This region of increasing Pressure is called adverse pressure gradient. It is seen from Fig. 2b that the flow is separated on upper surface of airfoil near trailing edge. When the flow is separated, the pressure at the back is lower than it would be if the flow were attached. Due to this reason a major increase in drag and a drastic decrease in lift.

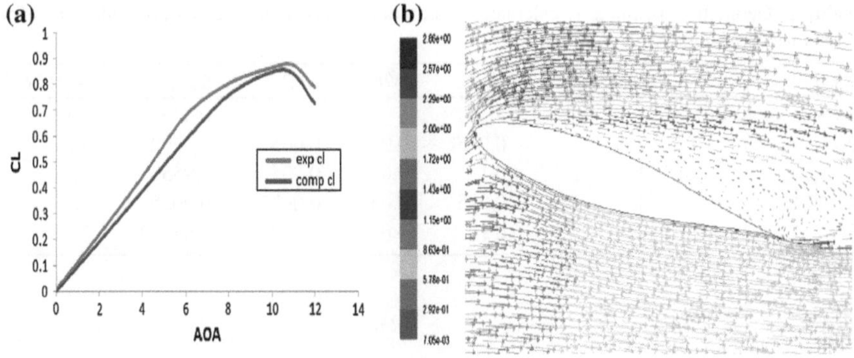

Fig. 2 **a** Comparison of the present numerical lift coefficient with experimental findings; **b** velocity vector plot for NACA- 63_1-012 at 10° angle of attack

Another word due to decrease in lift the stall will occur at higher angle of attack. Now the numerical analyses are performed with two burst control plates, namely triangular (Type 1) and rectangular (Type 2) plates.

3.2 Numerical Analysis of Airfoil with Triangular Plate

Computations are carried out to investigate the flow over airfoil with different triangular plate attached on the top surface of the airfoil for a Reynolds number of 1.3×10^5. The effectiveness of the plate is governed by its height, width, and the distance (x_p) between the leading-edge of the airfoil and the trailing edge of the plate. The various geometric parameters are shown in Table 1. The analysis of triangular plate is carried out for various angles of attack and it is compared with plain Airfoil. The lift coefficients are calculated for various angles of attack and plotted in Fig. 3. The lift coefficient of Airfoil with Triangular plate is increased as compared with that of plain Airfoil, and stall angle is also increased from 10 to 13°, as shown Table 1.

It is shown in Table 1 that the case 2 gives better performance as it delays stall by 3° and also gives better lift coefficient. Delay in stalling can be noticed in other cases (3, 4, and 5) also as they delay the stall by 2°. The effect of Triangular plate at the leading edge makes reasonable increment in lift and delay in stall, as shown in Fig. 3.

When the flow is separated, the pressure at the back is lower than it would be if the flow were attached. Due to this reason a major increase in drag and a drastic decrease in lift. Another word due to decrease in lift the stall will occur at higher angle of attack. The Triangular plate is placed over the upper surface of the Airfoil near the leading edge ($x_p/c = 0.075$, $w/c = 0.05$, $h/c = 0.005$). The effect of Triangular plate at the leading edge makes the reattachment of the separated flow to move upstream near the trailing edge, which makes reasonable increment in lift and delay in stall.

Table 1 Geometric parameters, calculated maximum lift coefficient, and stall angle for airfoil with various triangular plate cases

Case	Plate type	*Xp/C*	*W/C*	*H/C*	$C_{L\ Max}$	Stall angle
1 (original airfoil)	–	–	–	–	0.84917	10
2	1	0.075	0.05	0.005	0.9702	13
3	1	0.1	0.05	0.005	0.9536	12
4	1	0.075	0.05	0.0075	0.9025	12
5	1	0.1	0.05	0.0075	0.94902	12

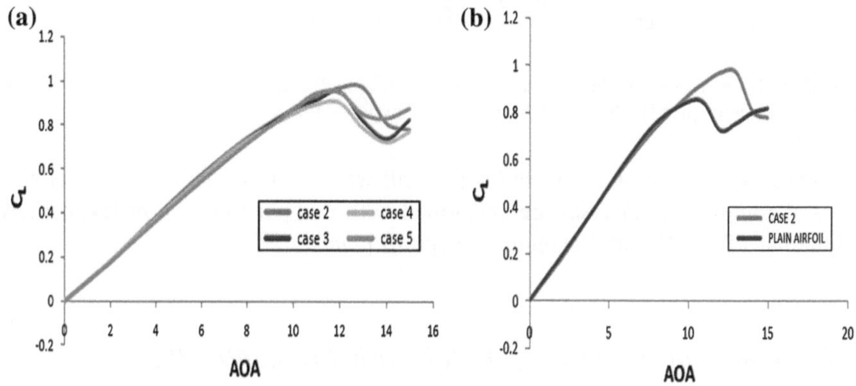

Fig. 3 Comparison of the lift coefficient of airfoil with triangular plate with that of plain airfoil

3.3 Numerical Analysis of Airfoil with Rectangular Plate

From the investigation on the burst control plate of type 1, the most effective location of the plate is found at upstream of $x_p/c = 0.075$ and it is believed that when x_p/c is higher than 0.075 the airfoil stall is triggered easily. Therefore the burst control plate of type 2 is set at $x_p/c = 0.05$ initially (as the reference location for the plate of type 2) and the effect of changing height, width and location of the plate will be studied in different cases accordingly. Note that the geometry of the plate of type 2 is a rectangular section with a constant chord wise height. Furthermore, the width of the plate of type 2 is much smaller than the plate of type 1.

In this particular case, the plate of type 2 is only 3.2 mm ($w/c = 0.016$) which is approximately 66 % smaller than the width of the plate of type 1 in the current investigation. With the plate configurations and dimensions as shown in case 9, the stall angle of the airfoil is successfully postponed to approximately 14° angle of attack compared to the original airfoil and the airfoil with the plate of type 1 (cases 2, 3, 4, and 5). The various geometric parameters are shown in Table 2. The analysis of Rectangular plate is carried out for various angles of attack and it is compared with plain Airfoil. The lift coefficients are calculated for various angles of attack and are plotted in Fig. 4. The lift coefficient of Airfoil with Rectangular

Table 2 Geometric parameters, calculated maximum lift coefficient, and stall angle for airfoil with various rectangular plate cases

Case	Plate type	*Xp/C*	*W/C*	*H/C*	$C_{L\ max}$	Stall angle
1 (original airfoil)	–	–	–	–	0.84917	10
6	2	0.05	0.016	0.005	0.9822	12
7	2	0.0751	0.016	0.005	1.03404	14
8	2	0.05	0.016	0.0075	1.11158	14
9	2	0.075	0.016	0.0075	1.16948	14

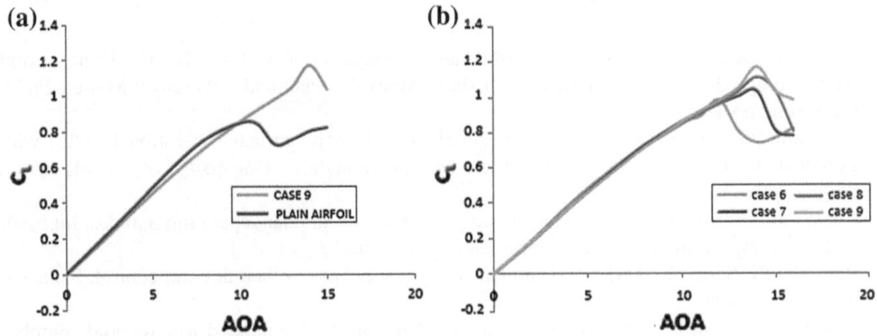

Fig. 4 Comparison of the lift coefficient of airfoil with rectangular plate with that of plain airfoil

plate is increased when compare to plain Airfoil lift coefficient, and stall angle is also increased from $\alpha = 10$ to 14°, as shown in Table 2.

Figure 4 shows the lift coefficient at various angles of attack for different cases (6, 7, 8, and 9). It can be seen that all the stalling curves are similar and but stall angle differs for each case. The case 9 gives better performance since it delays stall by 4° and also gives better lift coefficient. The other cases also give good performance over a plain Airfoil as it delays the stall by 3° but less maximum lift coefficient when compared with the case 9.

For the case 9, the Rectangular plate is placed over the upper surface of the Airfoil near the leading edge ($x_p/c = 0.075$, $w/c = 0.016$, $h/c = 0.0075$). The effect of rectangular plate at the leading edge makes the reattachment of the separated flow to move upstream near the trailing edge, which makes reasonable increment in lift and delay in stall.

4 Conclusions

The numerical analysis of laminar suppression bubble with different burst control plates has been performed on a NACA 63_1-012 airfoil at a chord Reynolds number of 1.3×10^5. The effectiveness of burst control plate is demonstrated computationally by changing the configurations of the plate such as height and width of the plate and distance between the leading edge of airfoil and trailing edge of plate. When the plate height (h/c) is increased from 0.005 to 0.0075 the stall angle of the airfoil is successfully delayed. Both types of plate (triangular and rectangular plates) are beneficial as shown in case 2 of type 1 and case 9 of Type 2 particularly. It is found that the small displacement of burst control plate (i.e., increase in x_p/c) does not delay the stall angle of the airfoil.

References

1. Gault DE, George B (1950) An experimental investigation of NACA 63_1-012 airfoil section with leading edge and midchord suction slots, Ames Aeronautical Laboratory Moffett Field, California, pp 1–23
2. Sandham ND (1999) Direct numerical simulation of short laminar separation bubbles with turbulent reattachment. In: Aeronautics and Astronautics, University of Southampton, Southampton, pp 1–28
3. Chetan SJ, Diwan SS (2006) On the bursting criterion for laminar separation bubbles. In: Sixth IUTAM symposium on laminar—turbulent, pp 401–407
4. Ramesh ON, Sourabh SD (2007) Laminar separation bubbles: dynamics and control. Sadhana, 32(1–2):103–109
5. Mohseni K, Sahin M (2008) Direct numerical simulation of separated low reynolds number flows around an eppler 387 airfoil. Cenareo, Rue des Freres Wright 29, B-6041 Gosselies, Belgium, pp 1–21
6. Saxena A (2009) The laminar separation bubble. Aerospace Engineering at the University of Maryland, personal report, pp 1–6
7. Aholt J, Finaish F (2011) Active flow control strategy of laminar separation bubbles developed over subsonic airfoils at low Reynolds numbers AIAA 2011-733. In: 49th AIAA Aerospace Sciences Meeting including the New Horizons Forum and Aerospace Exposition, January, Orlando, Florida
8. Jahanmiri M (2011). Laminar separation bubble: its structure, dynamics and control, research report 2011:06, Chalmers University of Technology, Sweden, pp 1–21
9. Abbott H, Albert ED, Stivers LS (1945) Summary of airfoil data. NACA Report No: 824

Numerical Simulation of Rayleigh-Bernard Convection in Enclosures Filled with Nanofluid

S. Senthil Kumar and S. Karthikeyan

Abstract Numerical investigations of Rayleigh-Bernard convection in enclosures of different modified bottom and top surfaces filled with Au-Water Nanofluid are presented. This paper describes a numerical predication of heat transfer and fluid flow characteristics inside enclosures bounded by modified bottom and top surfaces and two periodic straight vertical walls. Simulations are carried out for a Rayleigh number of 6×10^4 and two aspect ratios (0.25 and 0.5) with working fluid as water (base fluid). The same analyses are performed with the Nanofluid having Au nano-particles of same size in order to see the effect of Nanofluid on heat transfer. The Boussinesq approximation is used in order to take density change effect in the governing equations. The study investigates the effect of the nanoparticle volume fraction, and the aspect ratio on the heat transfer. The results are presented in terms of isotherms, streamlines local, and average surface Nusselt numbers. Results show that the flow and isotherms are affected by the geometry shape and by the presence of nanoparticles. It is also shown that for a fixed value of aspect ratio, the convective heat transfer is decreased for the Nanofluid when compared with that of base fluid due to an increase in thermal conductivity of the Nanofluid.

Keywords Nanofluid · Natural convection · Rayleigh-bernard convection · Wavy enclosure · Boussinesq approximation

---

S. Senthil Kumar (✉) · S. Karthikeyan
Department of Aeronautical Engineering, Vel Tech Dr. RR and Dr. SR Technical University, Avadi, Chennai, India
e-mail: s.senthilms@gmail.com

S. Karthikeyan
e-mail: karthi.innovators@gmail.com

R. P. Bajpai et al. (eds.), *Innovative Design, Analysis and Development Practices in Aerospace and Automotive Engineering*, Lecture Notes in Mechanical Engineering,
DOI: 10.1007/978-81-322-1871-5_23,

1 Introduction

Buoyancy-driven cellular flow structure in a bottom-heated horizontal wavy enclosures, noted as Rayleigh-Bernard convection, has been extensively investigated over the past four decades due to the important role it plays in fundamental fluid mechanics and heat transfer study in technological applications such as solar collectors, cooling of microelectronics equipments, heat transfer in nuclear reactors, electric machinery, double wall thermal insulation, natural circulation in atmosphere. Various researchers [1–5] have performed both experimental and numerical investigations on this area in order to understand the heat transfer characteristics. They analyzed with single phase flow heat transfer for the modified geometrical surfaces like sinusoidal/wavy cavity. Different flow patterns have been observed when changing the bottom and/or top surface of the enclosure. Geometry shape was governed by variables such as wavelength and amplitude, phase angle, interwall spacing. It was found that all these parameters significantly changes the flow and heat transfer characteristics inside it. Results showed that average heat transfer increases with the surface waviness when compared with flat wall cavity. However, the decrease of average heat transfer with the surface waviness was found by Adjlout et al. [6] in their natural convection study of an inclined cavity. Mahmud and Islam [7] performed laminar free convection simulation with entropy generation method inside an inclined wavy cavity. They found that the entropy generation was affected by the inclination angle of the cavity. Das and Mahmud [8] analyzed the free convection inside both the bottom and top wavy enclosure. They found the heat transfer rate increased when amplitude wave length ratio increased. Similarly, Abdelkader et al. [9] performed numerically an analysis of natural convection in horizontal wavy enclosures for different aspect ratios and Rayleigh numbers. They showed that the flow and heat transfer strongly affected by the wave amplitude.

Nanofluids have been found to enhance thermal properties like thermal conductivity of the fluid. Applications of Nanofluids, or suspensions of nanoparticles in liquids with heat transfer, have been studied by various researchers [10–14]. Stability and thermal conductivity characteristics of nanofluids were performed by Hwang et al. [10] and they concluded that the thermal conductivity of ethylene glycol was increased by 30 %.

Hence, the objective of this study is to investigate the natural convective heat transfer characteristics of the Nanofluid inside different wavy enclosures and to analyze the effects of the geometry parameters and nanoparticles. Heat transfer characteristics though the local and average Nusselt numbers are presented for different aspect ratios (amplitude-wavelength ratios) and for different top and bottom surface geometries. Flow and thermal fields are presented by streamlines and isothermal lines.

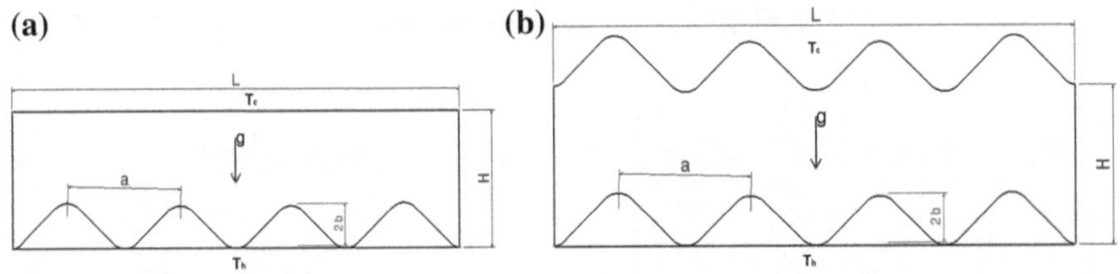

Fig. 1 Schematic diagram of **a** *bottom* wavy wall enclosure; **b** double side wavy wall enclosure

2 Mathematical Formulation

Two-dimensional horizontal wavy enclosures of height H and length L are considered for this study. Two configurations are chosen. In the first, the hot bottom wavy wall (BWW) and cold straight top wall are considered, as shown in Fig. 1a and in the second one, both the hot bottom and cold top walls are considered as wavy, which is called as double side wavy walled (DSWW) enclosure, as shown in Fig. 1b. Enclosures are filled with a water-based Nanofluid containing Au nanoparticle. The top wall is at a constant temperature T_c, while the bottom wall is at temperature T_h, where $T_h > T_c$. The two vertical walls are considered adiabatic. The flow is assumed to be steady, incompressible, and laminar. The nanoparticles are assumed to have same shape and size. The Boussinesq approximation is used in order to take into account density changes while other thermophysical properties are kept constant. The viscous dissipation effect is ignored. The flow behavior and heat transfer characteristics in the enclosures can be governed by the continuity, Navier-Stokes, and energy equations. These equations are not presented in this paper due to the constraint in number of pages.

The thermo-physical properties of the Nanofluid can be defined as follows [14]

$$\mu_{nf} = \frac{\mu_{bf}}{(1-\varphi)^{2.5}},\ \rho_{nf} = (1-\varphi)\rho_{bf} + \varphi\rho_s,\ (\rho\beta)_{nf} = (1-\varphi)(\rho\beta)_{bf} + \varphi(\rho\beta)_s,$$

$$(\rho c_p)_{nf} = (1-\varphi)(\rho c_p)_{bf} + \varphi(\rho\beta)_s,\ k_{nf} = k_{bf}\frac{k_s - 2\varphi(k_{bf} - k_s) + 2k_{bf}}{k_s + \varphi(k_{bf} - k_s) + 2k_{bf}}$$

where μ is the dynamic viscosity, k, the thermal conductivity, β, the thermal expansion coefficient, ρ, the density, C_p, the specific heat capacity, φ, the volume fraction of nanoparticles in the base fluid and the subscript *nf* denotes nanofluid, *bf* denotes base fluid and *s* denotes nanoparticles. The aspect ratio of the enclosure, as is equal to $2b/a$ where b in the half wavy height and a is the wavelength. The Rayleigh number based on enclosure height H, Ra $= g\beta H^3(T_h - T_c)/\upsilon\alpha$ and the Prandtl number, Pr $= \upsilon/\alpha$ where υ is the kinematic viscosity and α is the thermal diffusivity.

Fig. 2 Comparison of present simulation (*right*) results of isotherms (*top*) and streamlines (*bottom*) for Ra = 20,000, and Pr = 0.71 with those of published work (*left*) [9]

3 Numerical Methodology

The computational domains are discretized by uniform mesh spacing. Mesh size is determined based on conducting a series of computations to achieve independence of the solutions to the mesh. The governing equations with appropriate boundary conditions are solved under steady, incompressible conditions with the Boussinesq approximation, using the commercial CFD software, ANSYS Fluent^{TM}. The convective flux coefficients are derived by a second order upwind scheme while the diffusive flux coefficients are obtained by central differencing. The velocity and pressure fields are coupled using the SIMPLEC algorithm [15] and the discretized algebraic equations are solved iteratively using the Multigrid technique. All the numerical tests are performed with convergence threshold residuals for the momentum, continuity, and energy equations equal to 10^{-9}.

4 Results and Discussion

4.1 Grid Independence Study

The computations are performed for Rayleigh-Bernard convection in a straight-walled enclosure for Ra = 20,000 and Pr = 0.71 to validate the solution technique and to compare with the results of Abdelkader et al. [9]. The thermophysical properties of base fluid water and Au nanoparticles are taken from the literature [14]. The properties of the nanofluid are calculated using the formula given in the Mathematical formulation Section. In order to ensure that the solution is grid independent, the results are obtained on different uniform grid sizes (200 × 20, 400 × 40 and 800 × 80).

Figure 2 shows the streamlines and temperature fields of both the present results and those of Abdelkader et al. [9]. It can be seen that the hot fluid rises and the cold fluid falls, producing a circulatory motion within the enclosure that

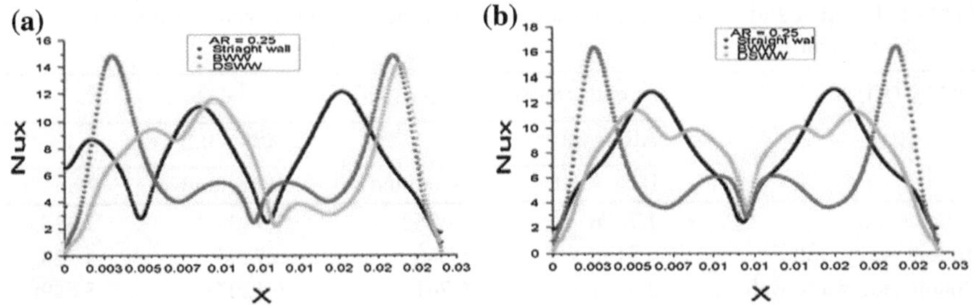

Fig. 3 Variation of local Nu number of the hot *bottom* wall of the different enclosures along the x-distance when AR = 0.25 **a** for base fluid and **b** for nanofluid

enhances the heat transfer through the enclosure. It is seen that the computed flow pattern and isotherms are similar to that found by Abdelkader et al. [9]. The average Nusselt number for the hot wall is also compared with that of the published [9] result. It may be noted that the result approaches the experimental profile as the number of grids increases. With 400 × 40, the solution agrees well with the results of Abdelkader et al. [9]. The grid spacing based on this grid size is fixed for all the computations in this study.

4.2 Effect of Wavy Surface

In order to see the effect of geometry shapes (i.e., wavy walled enclosure) on natural convective heat transfer, the simulations are carried out for both the bottom wavy walled and double side wavy walled enclosures with a Rayleigh number of 6×10^4 and the volume fraction of nanoparticles of 5 %. Figure 3a and b show the variation of the local Nusselt number values along the length of the enclosure in different enclosures of AR = 0.25 for the base fluid and Nanofluid, respectively. The average Nusselt number values of the hot bottom wall are given in Table 1 for different geometry enclosures with two aspect ratios (0.25 and 0.5) for both the base and Nanofluids. It is clearly shown in Fig. 3a and b that the wavy surfaces alter the convective heat transfer characteristics and that it is more effective when both the bottom and top walls are modified with wavy surfaces for both the fluids. It is found that for the BWW enclosure, the effect is reversed whereas for the DSWW enclosure, the average Nusselt number is increased when AR = 0.25. When AR = 0.5, the average Nusselt number value for the straight walled one is higher than those of both the BWW and DSWW enclosures. The results are consistent with those of Adjlout et al. [6].

Table 1 Computed average nusselt number values at the hot bottom wall for different geometries and different fluids

Geometry type	Average nusselt number ($\overline{Nu}$)			
	AR = 0.25		AR = 0.5	
	Base fluid	Nanofluid	Base fluid	Nanofluid
Straight wall	7.7606	7.3358	7.7606	7.3358
Bottom wavy wall	6.7198	6.566	5.7296	5.328
Double side wavy wall	8.0321	6.761	6.1207	5.6498

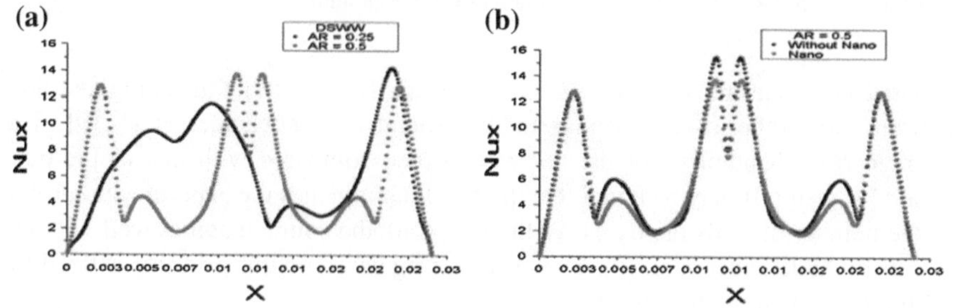

Fig. 4 Variation of local Nu number of the hot *bottom* wall of the DSWW along the *x*-distance **a** for the nanofluid with different aspect ratios **b** with different fluids when AR = 0.5

4.3 Effect of Aspect Ratio

The effect of the aspect ratio can be noticed from Fig. 4a and in Table 1 that for the same geometry shape, the Nusselt number decreases as the aspect ratio increases for both the base fluid and Nanofluid. For example with DSWW, as shown in Fig. 4a, it is evident that the local Nusselt number changes significantly. The average value of Nu for the base fluid is 8.0321 when AR = 0.25 and 6.1207 when AR = 0.5. The same trend is seen for the Nanofluid as well but with lesser value of Nu when compared with that of the base fluid.

4.4 Effect of Nanoparticles

The computations are performed with straight walled, BWW, DSWW enclosures for different aspect ratios and for a fixed volume fraction of Au nano particles of 5 %, to see the effect of nanoparticles when added to the base fluid. Figure 4b shows the comparison of the local Nu values of the base fluid with those of the Nanofluid for the double side wavy walled enclosure with the aspect ratio of 0.5. It is shown that the local Nusselt number of the Nanofluid is lesser than that of the base fluid. It is obvious that its average Nu value is also decreased from 6.1207 to

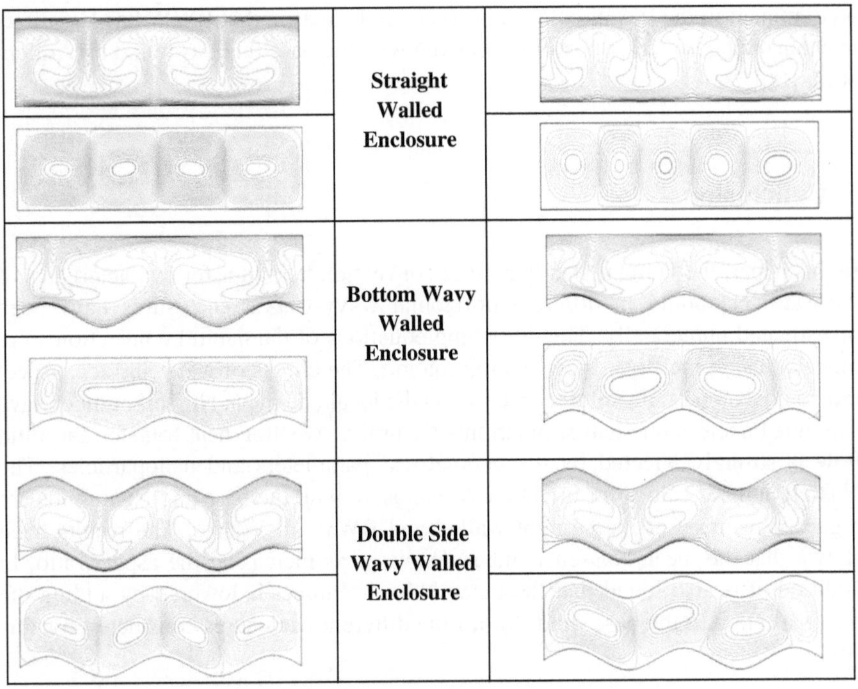

Fig. 5 Isotherms (*top*) and streamlines (*bottom*) of the base fluid (*left*) and the nanofluid (*right*) for different enclosures when AR = 0.25

5.6498 when AR = 0.5 (see Table 1), as expected. The same effect is seen with all the geometry shapes, aspect ratios. Overall, addition of nanoparticles in the base fluid results in a significant reduction in the convective heat transfer performance, which is consistent with the findings of various researchers [13, 14]. This can be attributed by the fact that adding nanoparticles to the base fluid increases its thermal conductivity thereby reducing the convective heat transfer.

Figure 5 shows the streamlines and isotherms of the base fluid and of Nanofluid for different enclosures, namely straight walled, BWW, and DSWW enclosures when AR = 0.25. It is shown that the same heat transfer enhancement mechanism has been found in the validation part in this paper that fluid adjacent to the hotter surface rises and the fluid adjacent to the cooler one falls, forming counter-rotating votices within the enclosures. The votices' shapes depend upon the geometry of the enclosure and working fluid, which is evident from Fig 5. The temperature and flow fields are symmetric with the base fluid for all the cases considered whereas they are asymmetric with the Nanofluid. It can be clearly seen that in the straight walled enclosure, four counter rotating vortices formed inside in it for the base fluid whereas five vortices formed for the Nanofluid. The same effect is also observed in the case of DSWW enclosure. As can be seen from Fig. 5, the effect of nanoparticles is predominant in the straight and DSWW enclosures, but less predominant in the BWW case. This could be the reason that BWW has less effect in enhancing convective heat transfer even with the base fluid, as found by Adjlout et al. [6].

5 Conclusions

In this paper, the laminar steady natural convection heat transfer characteristics of Au-water Nanofluid in different horizontal wavy-walled enclosures have been investigated numerically. The governing equations of the natural convection were modeled using the Boussinesq approximation. The effects of wavy surface, aspect ratio, and nanoparticle volume fraction on the local and mean Nusselt number have examined using numerical simulations. Natural convection heat transfer and fluid flow is strongly affected by the geometrical parameters and nanoparticles. The Nusselt number values for the DSWW enclosure with the base and Nanofluids are higher than those of the straight walled and BWW enclosures. The results have shown that the mean Nusselt number decreases as increasing the aspect ratio. In addition, it is also found that the mean Nusselt number is lowered by adding the nanoparticle into the base fluid, for all the different enclosures considered for this study.

References

1. Saidi C, Legay F, Pruent B (1987) Laminar flow past a sinusoidal cavity. Int J Heat Mass Transf 30:649–660
2. Wang G, Vanka SP (1995) Convective heat transfer in periodic wavy passages. Int J Heat Mass Transf 38:3219–3230
3. Nishimura T, Ohori Y, Kawamura Y (1984) Flow characteristics in a channel with symmetric wavy wall for steady flow. J Chem Eng Jap 17:466–471
4. Kumar BVR (2000) A study of free convection induced by a vertical wavy surface with heat flux in a porous enclosure. Numer Heat Transf A 37:493–510
5. Asako Y, Faghri M (1987) Finite volume solution for laminar flow and heat transfer in a corrugated duct. J Heat Transf 109:627–634
6. Adjlout L, Imine O, Azzi A, Belkadi M (2002) Laminar natural convection in an inclined cavity with a wavy wall. Int J Heat Mass Transf 45:2141–2152
7. Mahmud S, Islam AKMS (2003) Laminar free convection and entropy generation inside an inclined wavy enclosure. Int J Therm Sci 42:1003–1012
8. Das PK, Mahmud S (2003) Numerical investigation of natural convection inside a wavy enclosure. Int J Therm Sci 42:397–406
9. Abdelkader S, Mebrouk R, Abdellah B, Khadidja B (2007) Natural convection in a horizontal wavy enclosure. J Appl Sci 7 (3):334–341
10. Hwang Y, Lee JK, Lee CH, Jung YM, Cheong SI, Lee CG, Ku BC, Jang SP (2007) Stability and thermal conductivity characteristics of nanofluids. Thermochim Acta 455:70–74

11. Soleimani S, Sheikholeslami M, Ganji DD, Gorji-Bandpay M (2012) Natural convection heat transfer in a nanofluid filled semi-annulus enclosure. Int Commun Heat Mass Transf 39:565–574
12. Mohammed HA, Bhaskaran G, Shuaib NH, Abu-Mulaweh HI (2011) Influence of nanofluids on parallel flow square microchannel heat exchanger performance. Int Commun Heat Mass Transf 38:1–9
13. Stoian FD, Holotescu S, Stoica V, Bica D, Vekas L (2008) Comparative study of convective heat transfer in water and water based magnetizable nanofluid for thermal applications. J Optoelectron Adv Mater 10:773–776
14. Ternik P, Rudolf R (2012) Heat transfer enhancement for natural convection flow of water-based nanofluid in a square enclosure. Int J Simul Model 11:29–39
15. Patankar SV (1980) Numerical heat and mass transfer and fluid flow. McGraw-Hill, New York

A Conceptual Design of Versatile Furtive Craft

P. Gowtham, C. Gnanavel, V. Manoj Kumar and R. Naveen

Abstract The prudential work arose from the empirical study of persisting fighters that are prone to be habituated perspicaciously on land and sky. This agenda provoked us to an articulation of redundant conceptual design of a camouflage craft which will be versatile to air and water. This aberrant propaganda has been evolved to occlude the problems faced by the fighter crafts in naval dominance operations and by the submarines which are demolished due to limited recant performance. An effective evaluation in design is made with CATIA having solicitous consideration of buoyancy principle, aerodynamic effects, structural fatigue, design criteria, and propulsion. This articulate approach evolved with the meticulous development of submersible fighter craft having feasibility of operation in both air and water standards. This concept of submersible craft bolsters the defense sector by means of hybrid performance of fighter crafts and submarines. Dual propulsion mode involving jet propulsion to match fighter performance in air standard and adaptable electric propulsion was considered for submersible operation. Feasible handlings of the craft enriched with versatile performance of control surface maneuvering in dual modes were modeled.

Keywords Submersible craft · Buoyancy · Reynolds number · Hydrodynamics · Azimuthing thrusters · Hybrid vehicle · Drop down pods

---

P. Gowtham (✉) · C. Gnanavel · V. Manoj Kumar · R. Naveen
Department of Aeronautical Engineering, V. S. B. Engineering College, Anna University, Karur, India
e-mail: gowthamthinkpositive@gmail.com

C. Gnanavel
e-mail: gnanavelchandramohan@yahoo.com

V. Manoj Kumar
e-mail: icandoalways@gmail.com

R. Naveen
e-mail: navinorb@gmail.com

R. P. Bajpai et al. (eds.), *Innovative Design, Analysis and Development Practices in Aerospace and Automotive Engineering*, Lecture Notes in Mechanical Engineering,
DOI: 10.1007/978-81-322-1871-5_24,

1 Introduction

In the eighteenth century Lord Kelvin said that "Heavier-than-air flying machines are impossible". In 1903 Orville wright with Wilbur wright made the first controlled, powered, and sustained heavier-than-air human flight. It was a milestone in human heritage which introduced a new subject for study on Aeronautics, which comprises of many extents that should be evaluated with great care to make it most efficient and faster mode for transportation. Later in 1947 World War II aircrafts were used for combat missions. Living on a planet that has a surface largely covered with water which should be exploited well for the fortification of the defense. A new vehicle to achieve a stealthy underwater travel also came into existence called as submarines. More and more advancements started emerging in individual sectors of fighter crafts and submarine. Multiple application of aircrafts and submarines grew vastly. It not even took a century to fulfill the advanced requirement. The International defense sectors are in essential of technology development in its security enhancement with a comprehensive surveillance and the augmented rescue mission on emergency states. Our sagacious work arose from this speculative study of the aircraft and the submarine with consideration of several technologies that can be adoptable for accomplishing the said stuffs. In an approach of the design of a vehicle which is versatile for all the circumstances made us to the development of the concept of hybrid vehicle. In the study of fluid dynamics a prioritized concentration is made on its branches which are Aerodynamics and Hydrodynamics. This study made us to collate the aircraft and the submarine where similar properties are existing for both the standards that have upraised the study and made us to get involved in the meticulous study of aircraft and submarine to provoke a craft that is flexible for both the naval and the air standards. The design of the craft is made with a preliminary consideration of the airplane structure, crew withholding cabins, payload, and propulsion system that are versatile for air and naval standards. To make the design in a feasible form an initial analysis and computation was used. This design will be an amalgamation of the key technologies as the speed and range of an aircraft, the loitering capabilities of a boat, and the stealth of a submarine that is with the foreseen target mission being covert coastal insertion of the defense sector forces.

2 Mission Profile

For Aircraft

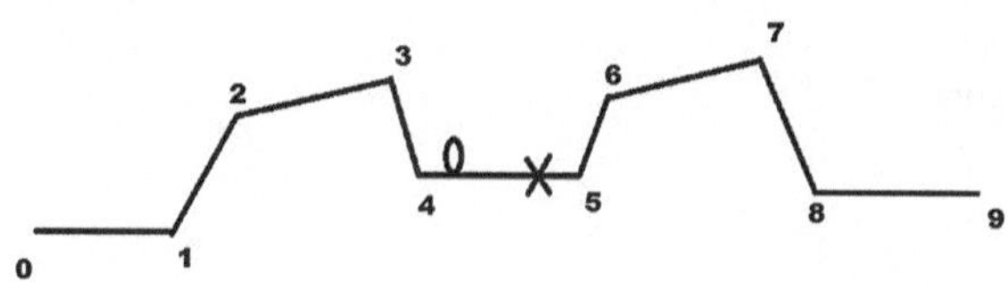

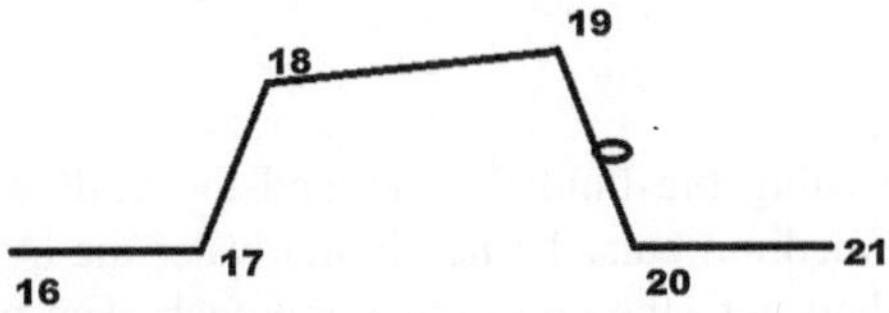

For Submarine

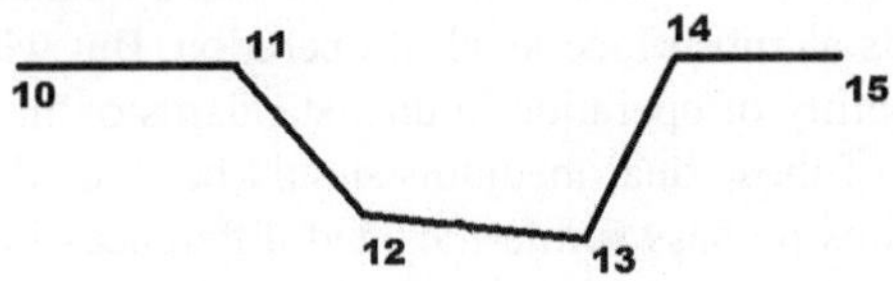

Combined over all mission profile

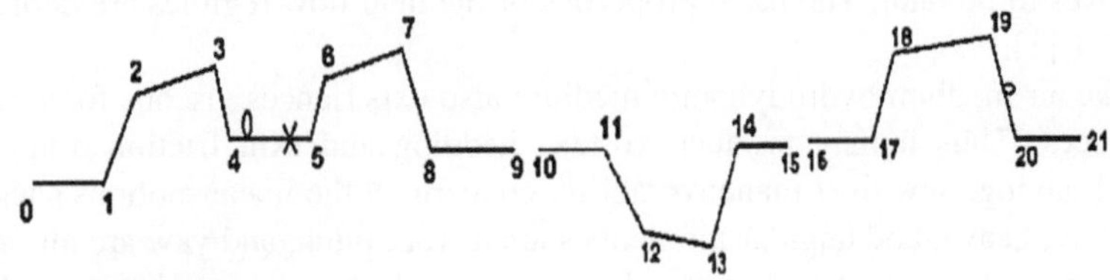

0–1	Take Off from land
1–2	Climbing
2–3	Cruising
3–4	Descent
4–5	Combating
5–6	Climbing
6–7	Cruising
7–8	Decent
8–9	Landing on water surface
10–11	Submerging
11–12	Deploy
12–13	Loitering time/transit
13–14	Retrieve
14–15	Return to water surface
16–17	Take off from water
17–18	Climbing
18–19	Cruising
19–20	Loitering
20–21	Landing on surface

3 Fluid Flow Regimes

In order to have a clear vision regarding the fluid flow around the craft while encountering dual standards of fluid medium considering air and water the hybrid fluid flow regime anticipated is studied with the conceptual approach. The basic standards and parameters of both the medium are hereby differentiated to our experimental platform as aerodynamics and hydrodynamics. And also by means of combinations of those dual standards at subsurface level of operation. But with a view to develop a craft having feasibility of operation in dual standards of air and water medium, the basic properties of those dual mediums should be studied and evaluated carefully. Both the mediums possess similarities and differences in the operational standards. A very basic evaluation denotes that water medium is nearly 1,000 times denser than that of the air medium. Thus without the comparisons of those hybrid fluid flow regimes of air and water designing and integration of this craft proves to be fatal. The basic properties of the dual flow regimes are depicted in Table 1 [1].

As like air medium hydrodynamic medium also experiences viscous forces and wave forces. Thus it may produce vortex shedding and skin friction drag. By means of having view over maneuverability concern all the linear motions such as surge, sway, heave, and angular moments such as roll, pitch, and yaw are all same for these hybrid standards. Thus the dual mode of the regime can be studied by means of analysis which is meant to be performed. It is carried out by means of various angles of attacks for both standards. Different Reynolds number variation is to be considered for the air and water medium. An extreme care should be taken to integrate hybrid variant performance in the dual standards of both structure and propulsive system of unit.

4 Propulsion Systems

The propulsion system considered for the design of our craft is studied with a meticulous approach so as to sustain both the naval and the air dominance with different fluid regimes. For the design of propulsion system several propulsion conceptual systems were studied for both the airborne and submerged propulsion where the airborne propulsion systems generally utilizes air to oxidize fuel and produce the thrust for the system and the submerged propulsion consists of Air Independent Propulsion (AIP) system generate electricity which in turn drives an electric motor for propulsion or recharges the boat's batteries for the propulsion of submarines [2]. Thus it made us to the development of the propulsion system that will be resourceful for the airborne and submersed missions where it should constitute the properties for its accomplishment of optimal performance. The basic characteristics of the anticipated propulsion system are those such as the thruster in the propulsion system should be 360° rotatable with optimum bollard pull in

Table 1 Properties of fluid in dual standards

Properties	Air medium	Water medium
Density	1.21 kg/m^3	1025 kg/m^3
Dynamic viscosity	1.82 × 10^{-5} Ns/m^2	1 × 10–3 Ns/m^2
Kinematic viscosity	1.51 × 10^{-5} m^2/s	1 × 10–6 m^2/s

maneuvering condition which enables feasibility of operation, the prompt conversion from the basic operation to the stored thruster position if possible that can be made to swing up into a housing in the hull if not required for the operation this will make the propulsion system more safer mode of operation and steering gear with a top bevel gear which familiarizes the 360° rotation which when installed makes the transition of craft between different fluid regimes more quicker [3]. When with these above considerations of the propulsion system quest made us to find the conceptual idea of our propulsion system is in concordance with the Rolls Royce modern developed concept of Azimuth Thrusters. They almost meet our necessities for the dual propulsion on airborne and the submerged states which when altered slighter will be worth efficient for our propulsion system. In the underwater propulsion all the electronic solution by incorporating a drop down azimuthing pods.

The batteries to be used for the underwater propulsion system significantly consume large weight which can be easily astounded by means of selecting high range submersible speed propulsion. The below figure portrayed will make us to understand this technique of high range submerged speed will effectively drive the overall weight [4] (Fig. 1).

Thus a single engine system with a propulsion unit optimized for air transit and acceptable performance in a submerged environment, which would provide a seamless integration between the two contending concepts, finally became the choice of the propulsion system for our craft design [5].

5 Landing and Takeoff

When we took the problem of landing and takeoff a configuration which could withstand both land and water touchdown was required for terrestrial operation. It was decided that a standard tripod design landing will be sufficient. Due to its simple design and proven operating principle it can withstand the loads implied over it during landing. For the water touch down the surface is designed so that the surface itself will be sufficient for withstanding the impact [6]. A small morphing technique will help in maintaining its stability. Standard cascade method of thrust reversal is employed safer landing to reduce the landing distance. The swing up azimuthing pods which can have its 360° can give a pull up force in opposite direction of the craft to reduce the landing distance on the water. When landing on

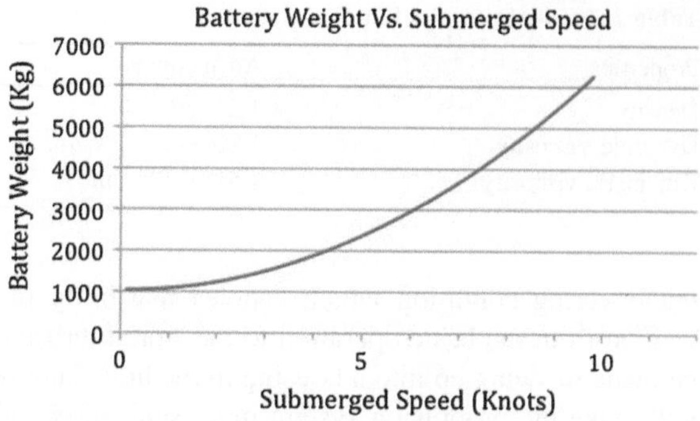

Fig. 1 Effect of submerged speed on battery weight

water due to the lesser density and surface tension force proffered on the fuselage it may try to sink but as soon as the broader surface contact of the wing happens, the surface tension offered by the seawater will help in retaining the craft on the sea surface which is the idea acquired by the study of surface tension. To detach from the surface of water the pontoons are evacuated and the flaperons are deflected downward.

6 Submersible Operation

The jet engine after travelling a long distance, the temperature of jet engine especially at the exhaust will be more. To avoid the quenching of the engine the engine will be fully sealed. The hot gases still present inside the engine can be expelled by an optional snorkel arrangement exhaust. From the study made on the submarine pontoons are required to made the craft to rule the under water. The pontoons are the empty tanks which will be filled by seawater by using a centrifugal pump [7]. To enhance the submarine operation, the azimuthing pods which have no deflection will be powered. The unused areas in the wing and fuselage, i.e., the empty spaces between the ribs and spars can also be wetted to make more effective submarine operation. Almost all empty spaces which can be effectively utilized to make the aircraft sink. Similar operation of the control surfaces can be achieved in under water also. The performance of the control surface and wing in the submerged condition will be nearly five times better than in the air.

$$\mathrm{L} = \frac{1}{2}\rho S C_{\mathrm{L}} \mathbf{V}^2$$

$$\mathrm{D} = \frac{1}{2}\rho S C_{\mathrm{D}} \mathbf{V}^2$$

Since the density of water is more there will be an increase in lift also. Considering the reduced velocity under water the above statement was made (Fig. 2).

Fig. 2 Rendered conceptual design

7 Structure

For the concept of operation of the versatile craft extreme care should be taken in the structural design. It is because the craft needs to operate in hydride medium of air and water. Hence structural design should be carried out considering the principle aerodynamics and hydrodynamic effects. The structure should be designed in such a way that it may be as light as possible for the combating mission and performance in the air standards. In the same hand, the craft should be heavy as possible for the submarine operational standard. For the design of the foil studying the hydrofoils and airfoils it is obvious that the symmetric airfoil will have a good performance. To make the craft to reach the depth the drag force created on the wing can be used to reach the surface of water from the subsurface the lift created can be used. To land on water and to loiter on the water surface a shape similar to a boat is required.

For the direction control of the furtive craft, a V tail configuration consisting of rudervator can be used which will also show the good directional control under water also.

8 Conclusion

An effective evaluation has been made on the submarine and aircraft provoked us to a unique design which will overcome the limitations of the recant performance of submarine and vulnerability of aircraft in the coastal areas. This sophisticated design of our proposed concept will pave the way to the ultimate enhancement of the Naval Air Forces in combating the existing fighters.

References

1. Raymer DP (2006) Aircraft design: a conceptual approach, Chap. 11, 4th edn. American Institute of Aeronautics and Astronautics, Inc., Reston, Virginia
2. Submarine Power and Propulsion—Application of Technology to Deliver Customer Benefit. In: Paper on submarine power and propulsion technology developments presented at UDT Europe 2008 in Glasgow, John Buckingham C Eng MIMechE Christopher Hodge CEng FIMarEST Timothy Hardy CEng MIMarEST, BMT Defense Services Ltd, Bath, United Kingdom
3. Azimuth Thrusters, Rolls-Royce, Rolls-Royce Marine AS, Propulsion—Ulsteinvik Sjøgata 98, N-6065 Ulsteinvik, Norway. http://www.rolls-royce.com/images/09.Azimuth_2p_06.05.11_tcm92-38824
4. East gate J, Goddard R (2011) Defence Engineering and Science Group. Submersible Aircraft Concept Design Study. In: 11th international conference on fast sea transportation (FAST 2011), Honolulu, Hawaii, USA
5. Roskam J (1999) Airplane design (Chaps. 4–7), Part V: Component Weight Estimation, Design, Analysis and Research Corporation, Lawrence, Kansas
6. Coltey DB, Meyer JA, Roseberry TC, Hicks J, Ahlgren EA Conceptual Design of a Submersible Tactical Insertion Aircraft, Auburn University, Auburn, Alabama
7. Pouvert PN Some Aspects of Submarine Design Part 1: Hydrodynamics under contract to Maritime Platforms Division Platforms Sciences Laboratory, Defence Science and Technology Organization, Department of Defense, Australian Government
8. Bureau Of Naval Weapons, Contract No. Noa(s) 59-6055c. Thrust augmented intermittent jet lift-propulsion system: final report, February 1960. Defense Advanced Research Projects Agency. Broad agency announcement: submersible aircraft, business opportunities database [online database]

Reduction of the Passage Between the Flame Tubes of the Combustor for a Millimeter Size Gas Turbine Engine in the Art of Micromachine Technology

Syed Alay Hashim, Nancy Manish, Deepam Mishra and Krishna Ahuja

Abstract The confluence of the market demands for the production of compact power source is not only in the electronics but also in the aeromechanical field. The micromachining technology has made feasible design and developments for the gas turbine engines in centimeter or millimeter size. The current design deals with microcombustor which keeps the performance on par with the modern combustor. In this research work, we tried to reduce the size of the combustor by reducing the passage between inner and outer flame tubes. Inside the combustor, the thermodynamics is almost same as large system but due to the millimeter size design, some changes are done in the mechanics. The design confirms almost the same gas velocity and the residence time (1 m/s) which is similar to a large scale combustor. At primary zone for keeping the flow close to the wall and for producing helical fluid path, the swirlers are designed in such a way that it can introduce the air at an angle of 30° from the axis of the combustor. This process enhances the turbulence intensity of the mixing process. A cylindrical bluish flame is established as we have chosen Liquefied Petroleum Gas (LPG) for this Micro combustor. This engine may be used with a lifting vehicle so that our interest is to minimize the combustor size as well as the combustion space with maximum heat release rate. Beside the mechanism of heat release rate, we are also interested in the rate at which chemical reaction takes place. Due to these reasons the process of vaporization in case of liquid fuel is eliminated, hereby introducing LPG as a working fluid. The turbine material and the pollutant emission point of view, equivalence ratio, and stage wise air induction is properly decided. Here, it is made

S. A. Hashim (✉) · N. Manish · D. Mishra · K. Ahuja
Department of Aeronautical engineering, Vel Tech Dr. RR & Dr. SR Technical University, Chennai, India
e-mail: aviator123hash@gmail.com

K. Ahuja
e-mail: kahuja67@gmail.com

R. P. Bajpai et al. (eds.), *Innovative Design, Analysis and Development Practices in Aerospace and Automotive Engineering*, Lecture Notes in Mechanical Engineering,
DOI: 10.1007/978-81-322-1871-5_25,

an annular combustor to keep lowest total pressure loss by decelerating the flow at the inlet dome of the combustion chamber. Cold and hot losses are also reduced in this design. One row of air holes on both the inner and outer domes are made in such a manner that it is blanketing the flame tubes inner surface with a relatively cool layers of air. In a nut shell this paper briefs the art of millimeter size combustor, including system design, integration, component design, fabrication, applications, and economics etc.

Keywords Millimeter size · 35 mm combustion passage · 30° swirlers and LPG engine

1 Combustor Functional Behavior

1.1 Physical Behavior

The combustion product can transfer efficiently into thermal energy and produces low total pressure loss with good recovery of kinetic energy. The combustion should be producing essentially flame stability, less NO_x and other emission. However, the scale reduction poses many new obstacles to be overcome. As such, the design space of a microcombustion chamber is highly restrained by the requirement for a sufficient residence time to allow complete combustion. Microcombustor development also faces unique challenges due to material and thermodynamic cycle constraints. These different challenges are shortly reviewed in the next paragraphs.

2 Other Restrictions

2.1 Time Accessibility Limitations

The flow rate decides the residence time of the combustion chamber.

$$\tau_{\mathrm{res}} = \frac{VP}{m^{\bullet} RT}$$

From the above equation it is clear that a combustor design which keeps longer flame tubes or a combustor has low volumetric flow rate can only increase the residence time. Enhancing static pressure or reduction in mass flow rate can only reduce the above flow rate. For producing high power density, mass flow rate per volume can be increased but operational static pressure is set based on the design parameters of the turbo machinery. The residence time cannot increase without compromising this power density. Indirectly, however, the residence time can also be increased by incorporating flow recirculation in the design. As this artificially lengthens the path of the gasses in the combustion chamber, it will result in a longer residence time without moderating the power density. Recirculation zones will be needed anyway for ignition of the incoming mixture and for high flame stability.

2.2 Annular Combustor Concepts

Based on the observed physical constraints and scaling effects, the strategy for a microcombustion chamber should be based upon the following three general concepts:

1. Slightly increasing the size of the combustor relative to the engine size to increase residence time.
2. Reaction time of the combustion can be reduced by premixing of the air and fuel.
3. Turbine inlet temperature limitations depend on mixture ratio, a lean mixture.

Air fuel mixing process takes countable time so the residence time of the matter inside the combustor is under the influence of mixing process. These things show that a combustor which keeps good mixing at the primary zone meets the residence time requirements. The turbine inlet temperature is under the influence of stable burning at low equivalence ratio. This can be achieved in two ways: the use of LPG or catalytic combustion of hydrocarbons. The latter solution will not be covered here as it seems to entail significant problems for micro scale applications. If we compare LPG to other gas turbine fuels, it has high-vaporization rate, good calorific value, higher diffusion velocity, low-reaction time, and appreciable flame speed. The hydrocarbon fuels need rich mixture in primary zone but LPG removes this due to its flammability limits. So that at upstream low-mixing ratio can be easily matched with the turbine inlet temperature. At this time LPG is investigated for this engine, successful results can be introduced for other gaseous fuel (propane) in future for the gas turbine engines.

3 Design Concepts

3.1 Design Process

The design of combustor for the gas turbine engines is a complex and difficult problem that is usually solved by reaching a reasonable compromise between the conflicting requirements. Design involves a broad range of technical discipline including combustion chemistry, fluid dynamics, heat transfer, stress analysis, and

Table 1 Combustor design parameter

Inlet temperature (T_{03})	300 K
Mass flow rate of fuel ($\dot{m}_f$)	0.0374 Kg/s
Inlet pressure (P_{03})	3 bar
Mass flow rate of air ($\dot{m}_a$)	0.5798 Kg/s
Fuel to air ratio	6.46×10^{-2}

metallurgy. Although there are many design parameters for a combustor. These are the most critical design parameters for the combustor design: combustor inlet conditions, air flow distribution and cooling air, combustor exit pressure, pattern factor parameter, flame tube length, and combustor exit area. The design of the combustion chamber was undertaken using the initial conditions as obtained from the test conducted on compressor. These parameters are listed in Table 1. The fuel chosen for the engine was LPG.

3.2 Evaluation of Reference Area

Many constraints are considered while designing the combustor of a gas turbine engine. Both the compressor and the turbine have major influence for the finalizing the combustor size. The combustor performance depends fully on the compressor exit conditions and also the combustor exit decides the performance of the turbine for maximizing its ability.

$$A_{\text{ref}} = 19,154.29\,\text{mm}^2.$$

3.3 Divisions of Air Flow

This air distributions to primary, secondary, and dilution zones have vital role to produce a compile and a stable combustion. Burner temperature profile and maximum heat release rate of the combustion should be acceptable for the flame tube material for the overall engine life. The primary zone accept 25 % of reference air for initializing the combustion, 30 % of air is introduced in secondary-zone for complete the combustion and the remaining 45 % is introduce to exit acceptable temperature for the turbine guide vanes. The variations in the annulus velocity are due to the flow movement through liner holes which is parallel to the axis of the combustor for cooling purpose. The percentage of airflow through inner casing and outer casing is calculated using the reference area, the airflow distributions are such as,

- In Inner Casing $\cong$ 15 % of airflow
- In Outer Casing $\cong$ 15 % of airflow.

3.4 Exit Temperature Calculations

Overall performance depends on the turbine inlet temperature so it is necessary to produce uniform temperature profile at the inlet of the turbine. To certify this appropriate temperature profile this should be established at the main burner exit. From these exit temperature profiles, the pattern factor of this combustor is calculated as:

$$PF \equiv \frac{T_{t\max} - T_{t\mathrm{av}}}{T_{t\mathrm{av}} - T_{t\mathrm{in}}} = 0.15.$$

3.5 Liner Length Concept

Law of Gas Dynamics is also used to determine the size of the burner with some ideas of scaling. Length of the burner is designed based on the length required for the complete combustion. The liner length is directly proportional to the pattern factor and also the liner diameter. The liner length is calculated by the following expression (Fig. 1).

$$PF = 1 - e^{\left(-0.05\,\mathrm{x}\,\frac{L_{\mathrm{linear}}}{D_{\mathrm{linear}}}\mathrm{x}\frac{\Delta P_{3-4}}{q_{\mathrm{ref}}}\right)^{-1}}$$

Liner length = 277 mm.

3.6 Combustion Vertex Design Concepts

The swirlers design should be in such a manner that it provides required turbulence only otherwise high-turbulence creates high pressure loss. In this design, swirlers turn the flow to 30 from the previous path. Total numbers of vanes are 88 with 11 mm in length and 4 mm in width.

3.7 Combustor Exit Flow Concept

Combustor exit area is determined based on the compressor pressure ratio and the mixture mass flow rate. Exit area is determined by the following equation,

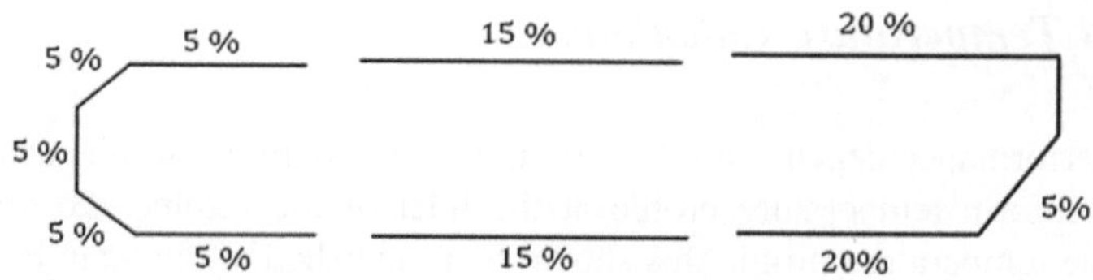

Fig. 1 Primary, secondary, and dilution zones air percentage

$$A^* = \frac{\dot{m}_3 \sqrt{T_3}}{P_3} \cdot \left(\frac{\gamma 3+1}{2}\right)^{\left(\frac{\gamma_3+1}{2(\gamma_3-1)}\right)} \sqrt{\frac{R_3}{\gamma_3}}$$

Using the above equation the combustor exit is calculated as,

$$A^* = 1.615193131 \times 10^{-3}\,\mathrm{m}^2.$$

3.8 *Fuel and Combustion*

In the early development of gas turbine engine, it was common belief that engine could use any fuel that would burn. This is true in theory, but not in practice. The turbojet engine is quit particular about the fuel used due to the high rate of the fuel flow, wide temperature, and pressure variations. Here LPG is stored as a liquid in the fuel tank but it becomes gas when it is released at ambient pressure, LPG is the mixture of propane (30 %), Butane (55 %), and Iso-Butane (15 %).

Mass flow rate of air (m_a) = 0.46 kg/s

By mass fraction, Total butane = 70 % and Propane = 30 %

$$\text{Moles of Butane} = \frac{70}{C_4H_{10}} = 1.21 \qquad \text{Moles of Propane} = \frac{30}{C_3H_8} = 0.682$$

Therefore total Moles = 1.21 + 0.682 = 1.8920

Butane by volume in LPG is 63.95 % and Propane by volume in LPG is 36.05 %.

3.9 *Air Fuel Ratio*

For a given fuel the amount of air which is required for the complete combustion is generally termed as Stoichiometric combustion. In combustion to this condition is not applicable always; mixture ratio depends on the loads on the engine. Near

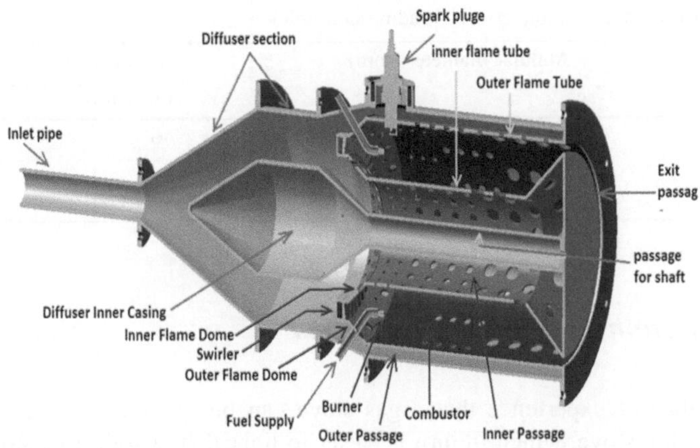

Fig. 2 Combustor cut section view

Stoichiometric the mixture releases all the latent heat of combustion. A lean mixture has low percentage of oxygen and the result of it is an incomplete combustion which forms CO and other unburned hydrocarbons in the exhaust. It is sometimes convenient to express the mixture strength in terms of an equivalence ratio. The equivalence ratio is the actual fuel/air ratio divided by the Stoichiometric fuel/air ratio. Thus for all the fuels, equivalence ratio is one denotes a Stoichiometric mixtures. For hydrocarbon fuel it is represented by C_XH_Y, the Stoichiometric relation is given in (Fig. 2).

$$C_xH_y + a(O_2 + 3.76N_2) \rightarrow xCO_2 + (y/2)H_2O + 3.76aN_2$$

Then the Stoichiometric air/fuel ratio would be

$$(A/F)_{\text{stioc}} = (m_{\text{ax}}/m_{\text{fuel}}) = 4.76a(MW_{\text{air}}/MW_{\text{fuel}})$$

For higher hydrocarbons such as propane and butane air/fuel ratio is around 15:1. It is also noticed that air/fuel ratio for other higher hydrocarbons such as kerosene, diesel, gasoline, ATF, and up to hexadecane are also around 15:1. Considering 1 mole of fuel (Table 2),

$$0.361C_3H_8 + 0.639C_4H_{10} + aO_2 + 3.76aN_2 \rightarrow bCO_2 + dH_2O + eN_2$$

$$\left(\frac{a}{f}\right)_{\text{stio}} = 15.5 \quad \text{and} \quad \left(\dot{m}_f\right)_{\text{stio}} = \frac{0.4606}{15.5} = 0.0297\,\text{kg/s}.$$

Table 2 Geometrical parameters of air admission holes

Combustor zone	Annular diameter (mm)	%	No. of holes	
			Inner liner	Outer liner
Primary	3	5	46	46
Secondary	6	30	72	72
Dilution	10	50	44	44

3.10 Injection and Ignition Concepts

It is a matter of experience that a gas stove can be ignited more easily than a kerosene wick stove, which in turn is easier to light than a solid fuel stove. Since ignition implies a flame in gaseous phase, one has to generate fuel vapour from liquid/solid before ignition take place. That is why it is easier to ignite gaseous fuel. Here spark plug of 150 cc bike is used for the ignition purpose. The spark plug is located in the primary zone to initiate the combustion and also protects itself by the combustion gases. The velocity of injection of LPG gas is calculated from this expression,

$$V_{\text{inj}} = \sqrt{\frac{2 \times (\Delta P)}{\rho_{\text{fluid}}}} \quad \text{and} \quad \dot{m}_f = \rho_f A_f v_f . C_D.$$

From the above equations the velocity of injection and the diameter of the fuel injector are calculated as 24.3 m/s and 0.6 mm.

4 Results and Discussion

All the components of the combustion chamber are assembled at their respective positions. Before testing rig cold air is passed to check the leakage in the system and at the same time the performance of the diffuser is also checked. Data for the experiment

I. Fuel injection pressure = 4 bar
II. Combustion pressure = 3 bar
III. Exit temperature = 500 K (result).

In the next experiments both the fuel and the air pressure are increased to 6 bar and the corresponding temperature is noted a 695 K. These results show that our estimated calculations are matched with the experimental values (Fig. 3).

Fig. 3 Components, assemble and testing views

5 Conclusion

This report gives an annular combustion chamber for the Microgas turbine engine applications. Here the passage between the flame tubes is reduced as possible as to produce a compact engine. A full scale combustor is the role model for this micro level design. This micro level design entails several new restrictions to the design space of such a device.

To reduce losses and to maintain the exit temperature the swirl vanes are designed in such a manner that airflow inside CC deviates at an angle 30° from their previous path, this helps in producing required turbulence for a complete mixing of air and fuel for complete combustion. Thus the designs of annular combustor for Microgas turbines are made simple and less complicated for fabrication. The material selection for this design for testing purpose only is SS304. Swirlers provide sufficient condition of flame stabilization. The model of this design is economically done and the assembly is also made easier.

Acknowledgment The authors are grateful to Mr. Muralidhara, Scientist Propulsion division, National Aerospace Laboratories Bangalore and Mr. Sunil Kumar Vishwakarma a passed out B.Tech (Aero) student for his unforgettable contributions in this project as well as Vel Tech University Chennai, India for providing facilities toward research at Vel Tech Research Park.

Influence of Temperature on Mechanical Characterization of Basalt/Epoxy Owen Fabric Composites

J. Alexander and B. S. M. Augustine

Abstract Glass-fiber-reinforced plastic (GFRP) composites are widely used in the manufacture of various components of aircrafts. This material can be replaced by using Basalt-fiber-reinforced plastic (BFRP) composites which are having superior material properties than GFRP. Aircrafts generally fly through different altitudes, through different environments, different temperature regions, etc. This effect will have the influence on the mechanical behavior of the materials. Therefore there is an imminent need to investigate the mechanical behavior of BFRP subjected to various temperature conditions. The objective of this chapter is to characterize the mechanical behavior of the BFRP when exposed to various temperatures like room temperatures like cryogenic to high temperature. The experimental results show that the bending strength,flexure modulus are improved at 50 °C post curing condition, and also the properties are much improved at aging condition. Impact energy has decreased at aging condition.

Keywords Basalt epoxy composites · Mechanical characteristics · Tensile strength · Bending strength · Impact strength · Compressive strength · Composite material · Thermal properties

1 Introduction

The continuous basalt fibers are manufactured from naturally occurring inert volcanic rock which are available throughout the world. These fibers are having excellent mechanical, thermal, and chemical properties which are suitable for various structural applications. The properties of BFRP are comparable with the

J. Alexander (✉) · B. S. M. Augustine
Department Aeronautical Engineering, Sathyabama University, OMR,
Chennai 600119, India
e-mail: vsjalexander@gmail.com

R. P. Bajpai et al. (eds.), *Innovative Design, Analysis and Development Practices in Aerospace and Automotive Engineering*, Lecture Notes in Mechanical Engineering,
DOI: 10.1007/978-81-322-1871-5_26,

properties of GFRP. GFRP are widely used for various structural applications including aero industry due to their higher specific strength, thermal properties, stiffness, better chemical properties. These performance also can be seen in BFRP, even better than GFRP [1–7]. Many researchers have investigated the properties of BFRP and their applications. Militky et al. [8] has tempered the basalt fibers at various temperature conditions and found the tensile strength and compared the same with glass fiber and proved basalt fibers are better. Lopresto et al. [9] has investigated the mechanical properties of BFRP at room temperature. It was demonstrated that the interfacial region of basalt-fiber-reinforced polymer was more susceptible than that of glass-fiber-reinforced composites after saltwater immersion and moisture absorption and an excellent interfacial shear strength was found [10, 11]. From the literature review, only limited investigations have been carried out on the basis of BFRP yet This chapter focuses on the variations of the strength of BFRP due to temperature variations and aging. The mechanical behavior of BFRP at various temperature is found out and this material is suggested to use various applications like Automobile, mechanical industries, and where ever various temperature variations are involved especially in the manufacture of aircrafts.

2 Specimen Preparation and Experimental Procedures

2.1 Materials

Basalt Owen fabric (plain weave with surface density 220 kg/cm^2) was imported from Incotelogy, Germany, and the epoxy LY556 and the hardener Aradur Hy 951 were purchased from Javanthy Enterprises, Chennai.

2.2 Fabrication of BFRP Laminates

BFRP laminates were fabricated by using compression moulding machine available at Indian Institute of Technology (IIT), Chennai. Fiber mats are cut into 300 mm × 300 mm size. 16 layers of fabrics were laid. Resin was taken 2 times that of fiber by weight. Hardener 1/10th weight of resin was added with the resin in order to activate the curing. This mixture was homogenized by using a stirrer and applied over the fabric. This is kept in the bed of the compression moulding machine and compressed hydraulically at a pressure of 30 bar and room temperature for about 3 h duration. Then the laminate is taken out of the machine and cured at room temperature for about 24 h. Volume fractions of the fiber and matrix were determined by using burnt test as per ASTM D-2584 standard. Tensile test, bending test, shear test, and impact test specimen were cut from this laminate and a set of specimen were kept inside a hot Owen at 50 °C at around 2 h, another set of specimen at 50 °C 3 h duration, another set at 80 °C for 2 h, another set at 80 °C for 3 h, and another sets at −22 °C in a cold freezer which are available at the chemical laboratory, Sathyabama University, Chennai. For each test three specimens were taken. One set (3 specimen) of all types of test specimen were undergone through aging. The aging process is given below 30 min at 50 °C, 10 min at room temperature for cooling, 30 min at 60 °C, 10 min at room temperature for cooling, 30 min at 70 °C, 10 min at room temperature for cooling, 30 min at 80 °C, and 10 min at room temperature for cooling.

2.3 Mechanical Characterization

Tensile tests were conducted to evaluate the strength (maximum stress) of the BFRP using a UTM (TUE-CN-1000) with a cross head speed of 2 mm/min according to ASTM D 3039/D 3039 M-00. Three samples were taken from each combination and the results were read through Data Acquisition System supplied by Fine Spavy Associates and engineers pvt. Ltd, Miraj from the computer attached to the UTM. Then the results were averaged. This showed the tensile strength of the polymer composites. In the same way flexural test was conducted as per ASTM D 790-03 for all temperature conditions by using UTM. The characteristics of specimen are tabulated in Table 1.

3 Results and Discussions

All the tests data were recorded through bar chats. The results are discussed here below. The recorded data comprises of average of three or more tests. The mechanical properties of BFRP at various temperature conditions were compared with the mechanical properties at room temperature.

It was found that at 50 °C treatment led to statistically significant increase of tensile strength (Fig. 1) but at 80 °C the tensile strength decreased, even at low temperature strength decreased. Probably, the changes of these properties are based on the changes in the crystalline structure of fibers and the molecular changes of resins. At low temperature the BFRP has become brittle and has lost its strength. But it is observed that the tensile behavior is best at the aging condition. The Young's modulus is very high at 50 °C curing condition (Figs. 2 and 3).

The failure modes of BFRP at various loading conditions are, fiber breakage at tension, fiber buckling, and delamination at compression. The failure modes are very complicated at bending. In flexural test, Interlaminar shear strength (ILSS) predicts the bond strength between the fiber and matrix. If the transverse loading exceeds ILSS, failure will occur in between the fibers and matrix. The behavior of interface is based on characters of fiber and the matrix. Temperature is also one of

Table 1 Standards and dimensions of specimen for all kinds of tests

Test	Standard of test	Thickness (mm)	Width (mm)	Gauge length (mm)
Tensile	ASTM D3039/D3039 M-00	2.6	25	100 (280 length)
Impact	ASTM D256	3	10	
Flexural	ASTM D 790-03	2.6	13	40
Shear	ASTM D2344/D 2344 M–00	10	20	60

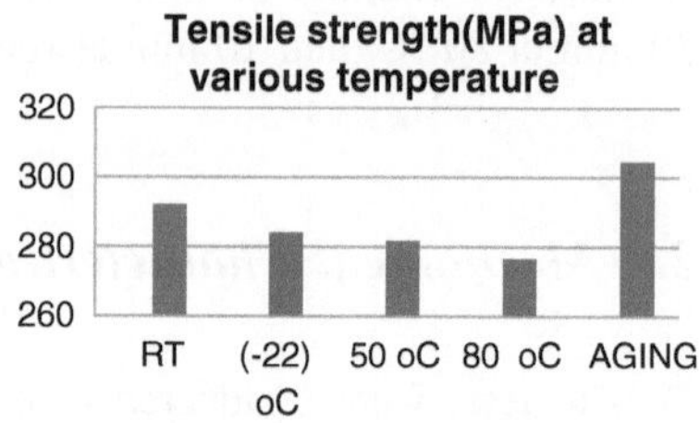

Fig. 1 Tensile strengths (in Mpa) at various temperature conditions

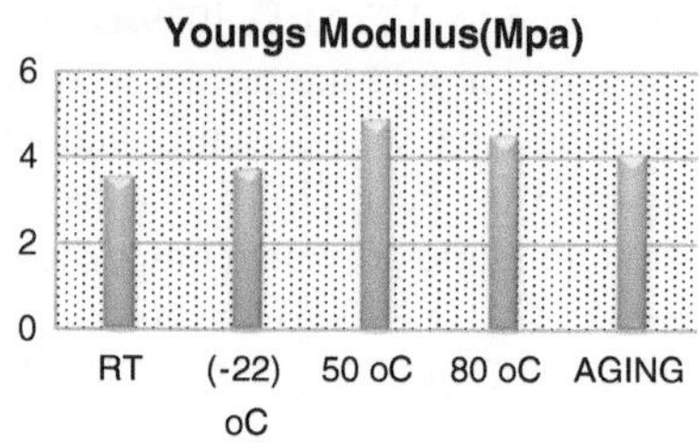

Fig. 2 Young's modulus (in Mpa) at various temperature conditions

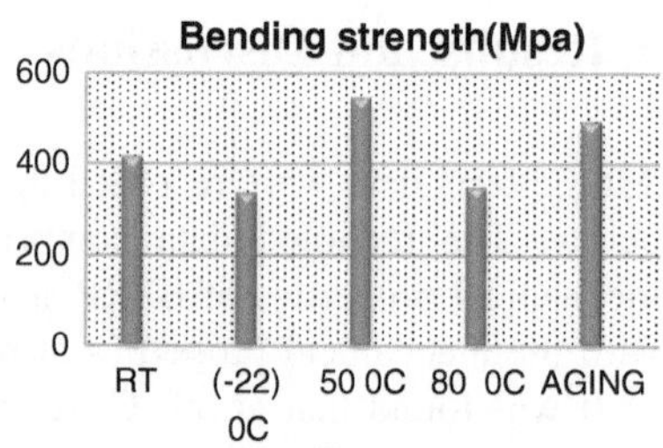

Fig. 3 Bending strengths (in Mpa) at various temperature conditions

the factors which is having influence with Inter laminar shear stress. By generating pure shear stress cause between the laminae, interlaminar shear failure can be induced. This is possible by using short beam shear (SBS) test [12]. This test is having many limitations [13]. ILSS is determined by the specimen prepared based on the standard ASTM D 790-03. Figure 4 depicts the interlaminar shear strengths of BFRP at various temperature conditions. BFRP bending stress and modulus are calculated for each test from the data obtained, and bar charts are plotted for various temperatures (Figs. 3 and 4). The bending strength and flexural moduli reach their maximum value at 50 °C but decrease just before the glass transition temperature, Tg of epoxy, while the composite loses its flexural strength at 80 °C

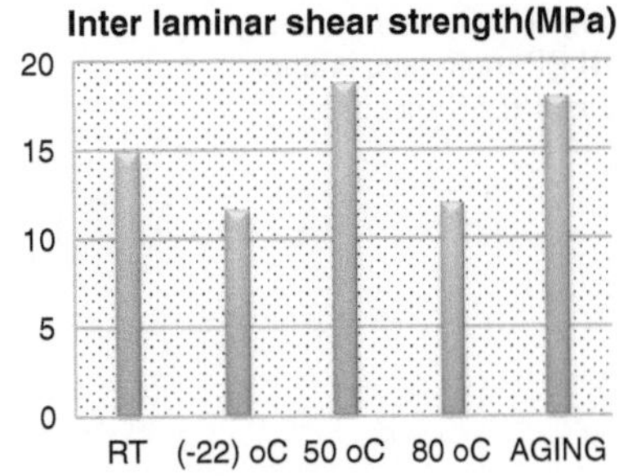

Fig. 4 ILSS (in Mpa) at various temperature conditions

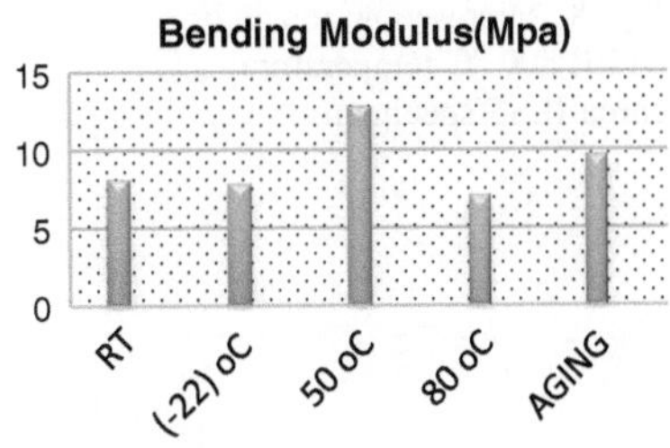

Fig. 5 Bending modulus (in Mpa) at various temperature conditions

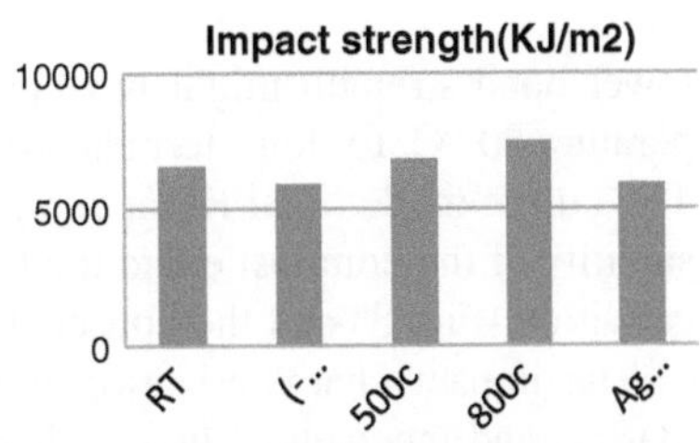

Fig. 6 Impact strength (in KJ/m^2) at various temperature conditions

(nearly Tg) rapidly. As a consequence, the ductile-to-brittle transition of epoxy is lower than −22 °C and this thermoplastic can be used until this temperature range (Figs. 5 and 6).

To have a better understanding of the break behavior, microscopic analyses are performed on the failure points for different temperatures (Figs. 7 and 8). At room temperature, at the tension side there is some delamination between the first few layers and buckling at the compression side. It can be seen in Fig. 8. During failure microcracks develop at different layers of resin near the breaking point alone which are induced due to failure of composites. During the production of laminate some amount of residual stresses are developed and are the root cause for the microcracks. When temperature increases to 50 °C (Fig. 8), the failure is clearly a tensile failure with crack propagation through the layers but not much interlaminar shear stress is recorded.

ILSS at 50 °C is higher than the room temperature condition. At 50 °C and the aging conditions the bond between the molecules of the epoxy resin is strengthened and correspondingly the adhesion factor of the epoxy and basalt is improved. The

Fig. 7 Microscopic image of BFRP room temperature cured ILSS specimen

Fig. 8 Microscopic image of BFRP 800 °C temperature cured ILSS specimen

lower bond strength might have promoted fiber-matrix debonding at higher temperature 80 °C, for long-term heating of BFRP and at freezing condition (−22 °C). The failure mode of BFRP at compressive load is buckling and delamination. The stability of the composite and the bonding strengths are increased due to the aging condition (Fig. 3) and the consecutive changes in temperature.

The impact characterization is essential for applications in the aeronautical, marine, and mechanical fields where crash aspects represent a key point. Most of the components in aircraft like landing gear struts, aircraft wings, fuselage are subject to impact loading. The impact energy absorbing character increases with increase in temperature (Fig. 6). At freezing condition the impact strength decreases. At freezing condition the material becomes brittle and microcracks developed are reduced. But at low temperatures other forms of failure mechanisms take place., i.e., internal damage area increases which are associated with ply delamination. This shows that interlaminar bonds are degraded due to low temperature. Another failure mechanism at low temperature is fiber breakage near notch of the specimen.

4 Conclusions

The mechanical behavior of BFRP is good when we compare the properties of GFRP. This material can be used in place of GFRP in aero applications. Tensile strength, bending strength, and ILSS are high at aging condition. One of the suggestions to improve mechanical properties of BFRP is aging (consequent heating and cooling) treatment can be given before it is used for some applications. Moreover, if this material is used for the construction of aircraft, the structure will be strengthened when the aircraft flies through various temperature regions. But the impact energy decreases at aging condition. Results also show that BFRP loses its mechanical characteristics at freezing condition due to high brittleness. Alternative treatment has to be given to improve the strengths for low temperature applications.

References

1. Wei B, Cao H, Song S (2010) Tensile behavior contrast of basalt and glass fibers after chemical treatment. Mater Des 31:4244–4250
2. Wei B, Cao H, Song S (2010) Environmental resistance and mechanical performance of basalt and glass fibers. Mater Sci Eng A 527:4708–4715
3. Pavlovski D, Mislavsky B, Antonov A (2007) CNG cylinder manufacturers test basalt fiber. Reinf Plast 51:37–39
4. Liu Q, Shaw MT, McDonnell AM, Parnas RS (2006) Investigation of basalt fibercomposites mechanical properties for application in transportation. Polym Compos 27:41–48
5. Liu Q, Giffard HS, Shaw MT, McDonnell AM, Parnas RS (2005) Preliminary investigation of basalt fiber composite properties for applications in transportation. The official newsletter of the international institute for FRP in construction, vol 2. p 6–8
6. Szabo JS, Czigany T (2003) Static fracture and failure behavior of aligned discontinuous mineral fiber reinforced polypropylene composites. Polym Test 22:711–719
7. Artemenko SE (2003) Polymer composites materials made from carbon, basalt and glass fibers, structures and properties. Fiber Chem 35:226–229
8. Militky, J, Kovacic V, Rubnerov J (2002) Influence of thermal treatment on tensile failure of basalt fibers. Eng Fract Mech 69:1025–1033
9. Lopresto, V, Leone C, De Iorio I (2011) Mechanical characterisation of basalt fibre reinforced plastic Compos: Part B 42:717–723
10. Lisakovski AN, Tsybulya YL, Medvedyev AA (2001) Yarns of basalt continuous fibers. In: Proceedings of the fiber society, Raleigh, May 2001
11. Dehkordi MT, Nosraty H, Shokrieh MM, Minak G, Ghelli D (2010) Low velocity impact properties of intra-ply hybrid composites based on basalt and nylon woven fabrics. Mater Des 31:3835–3844
12. Rosselli F, Santare MH (1997) Comparison of the short beam shear (SBS) and interlaminar shear device (ISD) tests. Compos A 28:587–594
13. Abali F, Pora A, Shivakumar K (2003) Modified short beam shear test for measurement of interlaminar shear strength of composites. J Compos Mater 37:453–464

Designing and Controlling the Performance of FMS: Application of Principle Component Analysis

M. Bragadish, N. Venkatachalapathi and A. Ramakrishna Rao

Abstract This paper attempts to establish a quantitative relationship between output indicators such as automated-guided vehicle (AGV) utilization and input variables like part arrival rate, processing times at machining centers, and various decision rules employed at AGV and Resources in a dynamic dedicated flexible manufacturing system (FMS). Dummy data are generated and the system is simulated to produce the output indicator for a given set of input variables and decision rules. Design and controlling of a FMS require much investment. So it is essential to make best decisions for selection of a FMS option. The proposed Principle Component Analysis is performed on a dataset that contains requirement of FMS alternatives.

Keywords Simulation · Resources · AGV · FMS · Principal component analysis

1 Introduction

A flexible manufacturing system (FMS) is one such system that meets the needs, when markets become surprisingly turbulent due to shortened product life cycles and power shifts toward the customer, the need for systems that rapidly and cost-

M. Bragadish (✉) · N. Venkatachalapathi
Mechanical Department, Velammal Engineering College, Chennai 600066, India
e-mail: brags92@gmail.com

N. Venkatachalapathi
e-mail: vecdr.venkat@gmail.com

A. R. Rao
Mechanical Department, Annamacharya Institute of Technology and Sciences, Tirupati, India

R. P. Bajpai et al. (eds.), *Innovative Design, Analysis and Development Practices in Aerospace and Automotive Engineering*, Lecture Notes in Mechanical Engineering, DOI: 10.1007/978-81-322-1871-5_27, © Springer India 2014

effectively develop products has become pressing [1]. An FMS is an embellished flow shop equipped with CNC machine tools and supporting workstations that are connected by an automated material handling system [2]. Such systems typically comprise process equipment, material handling equipment, a communications system, and a sophisticated computer control system. Modern FMS is a group of versatile machines that allows for the performance of more than one operation on a work piece without removal from the machine. This eliminates intermediate setups since the work piece does not have to leave the machine until all its operations are complete. This also reduces the need to move parts to various machines for their processing. The high investment cost of FMS justifies the use of computer simulation support [3]. In this paper, an attempt is made to look at the operational problems of FMS through simulation. Looking at the operational problems of FMS, such as scheduling and loading strategies, simulation methodology seems to be useful to address these issues. Many authors have used various criteria for the generation of optimum schedule, such as number of tardy jobs, number of completed jobs in process inventory, and machine utilization. Montazeri and Wassenhove [4] were stressed because the need for simulation prior to actually setting up the FMS. They use a user-oriented discrete event simulator to mimic the operation of a real life FMS. Stecke and Solberg [5] have carried out a detailed simulation of a real-life system. Gupta et al. suggest that two approaches have been used to solve FMS scheduling problems: The first approach involves decision making for prerelease, part-release and part-routing decisions, and the second approach involves formulating a loading model and evaluating the performance of such loading strategies. Simulation can be used to investigate the effect of scheduling rules on system performance. These models have been developed using general purpose programming languages (C, FORTRAN, PASCAL, etc.), general simulation languages (GPSS, SLAM, SIMSCRIPT, etc.; and special purpose simulation packages (WITNESS, SEE-WHY, ARENA, SIMFACTORY, etc.

The authors provide a critical review of scheduling rules in FMS. A set of important attributes while selecting simulation software for manufacturing is given by Law and McComas [6]. Generally, different authors give different statements of the functions of a simulation package tool, depending mainly on how detailed this statement is. In summary, the following advantages of simulation methodology for modeling FMS can be highlighted:

- Simulation can reduce the risk of installing an FMS which may not provide sufficient flexibility.
- A simulation model can represent important characteristics of an FMS more realistically. It may incorporate the complex interactions which may exist between various variables, for example, loading strategy at buffer and at workstations.
- Alternate FMS designs can be evaluated easily in a controlled environment.

2 Methodology

The objectives of the present simulation model include:

(a) To model a hypothetical FMS System in an easy-to-understand programming language and to develop a simulation model in a PC-based environment so that the model can be used as a pedagogical tool while teaching FMS.
(b) To study the effect of various loading rules at the buffer and the workstation.

The first step in any FMS simulation study is to establish the system configuration and then to develop an Arena simulation model. In this regard, keeping the objectives of the simulation model. ARENA 10.0 has been chosen as a programming package.

A short-time scheduling problem is attempted in this paper. Simulation methodology has been used to examine the effect of dummy independent variables used in the regression model. Various performance measures AGV Utilization, Parts total flow time, throughput etc., are examined. A simulation model is developed to address dummy independent variables in a hypothetical FMS environment such as depicted in Fig. 1. This is a hypothetical FMS with two part types and two work stations with four intersections.

The purpose of the present simulation is to analyze various combinations of dummy decision rules in the FMS system, i.e., given a set of dummy independent rules applied at the each station in the AGV request and effect of different rules at the resource set 1 are studied. In this particular simulation model, a rough cut design of the FMS system was made after having assumed a set of parts to be processed. As a first level approximation, certain assumptions were made which are given in the next section. Once the model was ready and validated, the assumptions were relaxed one by one and the model validated at each stage.

2.1 Assumptions

- There is a strictly ordered operation sequence for parts.
- There is only one operation at a time on a machine.
- Due date is not considered.
- Setup times are constant.
- The machines are continuously available for production.
- The same pallets are used for different parts.
- Operations once started cannot be interrupted before completion.
- Machine waiting time due to nonavailability of tools, jigs, fixtures, etc., is negligible.
- The Part A is processed as per predetermined sequence 1-2-3-4 and Part B is processed 1-3-2-4 network.
- The inter-arrival and processing times of the parts vary with exponential distribution of the mean value.

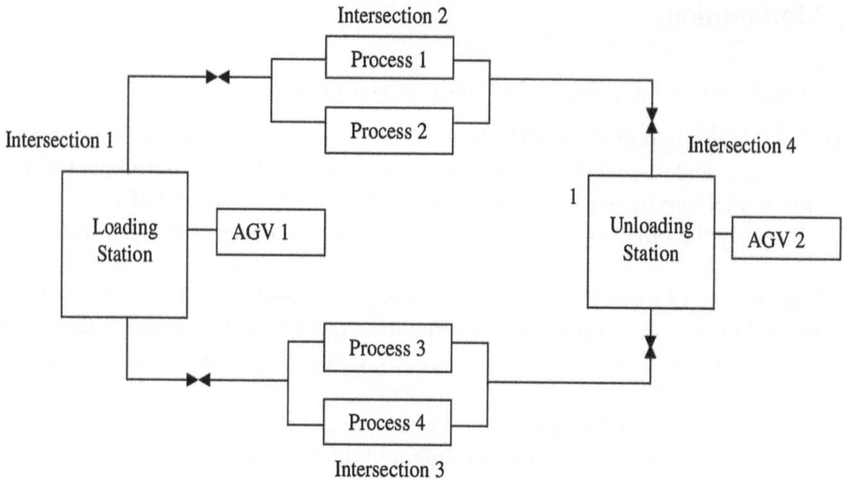

Fig. 1 Hypothetical flexible manufacturing system

3 Simulation Model with Arena

Arena provides an intuitive, flowchart-style environment for building an as-is model of the FMS operations process. Arena model in the Fig. 2 was created in this paper. The Parts arrival will generate into the system based on an Exponential distribution with a mean of 5 units. As soon as the parts are received at loading station, they are loaded into number of pallets and thereafter dispatched for processing. The part along with the pallet is stored in the empty buffer space of the concerned machine otherwise returned back to the loading/unloading station. Two varieties of parts have been considered in this work, Part A and Part B. Part A requires two operations in process 1 and process 3, based on an exponential distribution with a mean of 30 units. In this example, the Process 1 Operators set contains three resources, R_1, $R2$, and R_3. These members may contain efficiencies of 0.9, 0.85 and 0.75. The decision rules for selecting among the available set members are specified within the Process module. Part B requires another two operations in process 2 and process 4 based on an exponential distribution with a mean of 30 units of single resource in each. The AGV request module has to be used for transport, here we uses different types of decision rules. Parts are differentiated by decide module as a batch two parts are allowed to be transported for unloading station through AGV. Finally, in the Unloading Station all the parts are disposed off. We have pallets they are sent to pallet station. Parts A and B are shipped.

In this paper an attempt is made to consider multiple objectives for FMS scheduling. The results obtained in the earlier chapters, namely, the work on Simulation of FMS and the subsequent Regression models developed are utilized for developing multiple objective optimization models. Initially, a single objective

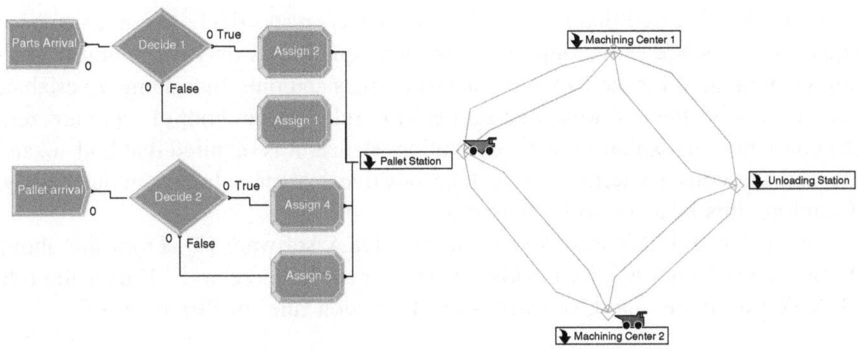

Fig. 2 Arena model for the flexible manufacturing system

Table 1 AGV utilization for decision variables

Objective/ decision rules	X_1 Constant	X_2 CYC	X_3 RAN	X_4 POR	X_5 SDS	X_6 LDS	X_7 CYC	X_8 RAN	X_9 POR	X_{10} LRC	X_{11} SNB	Z
AGV utilization (max.)	1	0	1	0	0	0	1	0	0	0	0	0.89

Maximize (Z) = $0.778 + 0.28833\ x_2 + 0.33166\ x_3 + 0.33083\ x_4 + 0.22916\ x_5 - 0.21666\ x_7 - 0.27866\ x_8 - 0.21666\ x_9 - 0.21666\ x_{10}$

integer linear programming model has been developed and then modified to incorporate multiple objectives. The method of Global criterion is applied for the solution of the above multiple objectives model.

All the objectives functions considered are all linear only and hence the model considered is a multiple objective linear integer programming problem of the 0–1 type. The regression equation for AGV utilization which was obtained from maximization of AGV utilization by using linear integer programming technique is shown in Table 1.

Subject to

i. $X_1 = 1$
ii. It is necessary that only one of the five decision rules for AGV transporter has to be selected. Hence, we require that
$X_2 + x_3 + x_4 + x_5 + x_6 = 1$
iii. Similarly only one of the five decision rules for Resource set can be selected at any given decision point. Therefore
$X_7 + x_8 + x_9 + x_{10} + x_{11} = 1$
iv. Since there are two AGV's their combined utilization should be less than or equal to two. Therefore we require
$0.778\ x_1 + 0.28833\ x_2 + 0.33166\ x_3 + 0.33083\ x_4 + 0.22916\ x_5 - 0.21666\ x_7 - 0.27866\ x_8 - 0.21666\ x_9 - 0.21666\ x_{10} \leq 2.$

It may be observed that the variables x_6 and x_{11} respectively are given a value equal to zero while estimating the regression equation as only four dummy variables can be used for the five AGV decision rules and only four dummy variables can be used for the five Resource set decision rules. When both of them are zero the constant term explains their contribution. Hence, it is required that both x_6 and x_{11} and the constant terms cannot take positive integer values, simultaneously. Therefore, this is achieved by letting $X_6 – x_{11} = 0$.

The above 0–1 ILP is solved by using TORA software [7]. From the above Table the maximum AGV utilization is 0.89 for the Random ($x_3 = 1$) decision rule of AGV transporter and Cyclical ($x_7 = 1$) decision rule for Resource set.

4 Principal Components

The principal components of a set of process variables x_1, x_2, $x_3 \ldots xp$ are just a particular set of linear combinations [8] of these variables say,

$$\begin{aligned} Z_1 &= c_{11}x_1 + c_{12}x_2 + \ldots\ldots\ldots \ldots + c_{1p}x_p \\ Z_2 &= c_{21}x_1 + c_{22}x_2 + \ldots\ldots\ldots \ldots + c_{2p}x_p \, . \\ Z_p &= c_{p1}x_1 + c_{p2}x_2 + \ldots\ldots\ldots \ldots + c_{pp}x_p \end{aligned} \tag{1}$$

Where the c_{ij}'s are constants to be determined. Geometrically, the principal component variables $Z_1, Z_2 \ldots Z_p$ are the axes of a new coordinate system obtained by rotating the axes of the original system (the x's). The new axes represent the directions of maximum variability.

Using MATLAB software package eigen values and eigen vectors are computed and principal components analysis is performed. Once the principal components have been calculated and subset of them selected the new principal component observations Z_{i1} and Z_{i2} are obtained by substituting the original observations x_{ij} into the set of retained principal components. Using FMS simulated data on the $p = 5$ variables, x_1, x_2, x_3, x_4, x_5 in Table 2 which are output performance variables. The first 10 observations in the upper panel of this table are first plotted against each other in a pairwise manner. This display is usually called a matrix of scatter plots, and it indicates that the third and fourth variables are highly correlated, where as the other three variables exhibit only moderate correlation. The principal component scores are shown in Fig. 3.

$\mu_1 = -2023.7$, $\mu_2 = 1024.8$, $\sigma_1^2 = 14198.31$, $\sigma_2^2 = 53.29683$, $\sigma_{12} = 0.623173$. Therefore, the statistics used for process control purposes is:

$$T^2 = \frac{n}{\sigma_1^2\sigma_2^2 - \sigma_{12}^2}\left[\sigma_2^2(\bar{x}_1 - \mu_1)^2 + \sigma_1^2(\bar{x}_2 - \mu_2)^2 - 2\sigma_{12}(\bar{x}_1 - \mu_1)(\bar{x}_2 - \mu_2)\right] \tag{2}$$

$T^2 = 2.60549$ for 1st sample.

Table 2 Sample means for performance measurements

Observations	AGV utilization X_1	Resource utilization X_2	Total flow time X_3	Parts throughput X_4	System number out X_5	Z_{i1}	Z_{i2}
1	0.8301	0.1876	163.0	101.6	5.2	−164.528	98.298
2	0.7604	0.2401	210.8	97.4	5.4	−212.236	93.699
3	0.6528	0.3946	325.6	95.4	14.4	−327.254	89.395
4	0.8958	0.1304	71.8	88.0	2.6	−73.107	86.025
5	0.8166	0.2334	156.2	95.8	7.2	−157.725	92.326
6	0.8294	0.2020	131.4	93.2	6.0	−132.862	90.147
7	0.8260	0.2206	164.0	97.8	7.8	−165.570	94.148
8	0.8778	0.1532	90.8	100.8	4.2	−92.336	98.283
9	0.7228	0.3092	249.4	105.4	7.4	−250.996	100.980
10	0.6982	0.3404	261.6	93.2	10.6	−263.131	88.337

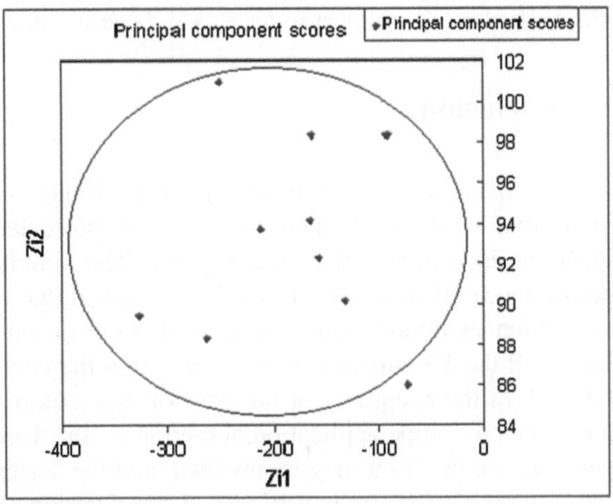

Fig. 3 Principal component score

As long as scores remain inside the ellipse there is no evidence that the process mean has shifted. If subsequent plot outside the ellipse then there is some evidence that the process is out of control. Figure 4 represents the Hotelling T^2 control chart for multi variable z_1 and z_2. Considering this, establishing statistical control [9] in the preliminary samples, and the Upper control limit is calculated from the Equation given below [10].

$$\text{UCL} = \frac{p(m-1)(n-1)}{mn - m - p + 1} F_{\alpha,\, p,\, mn-m-p+1} \tag{3}$$

UCL $= 9.9563$.
LCL $= 0$.

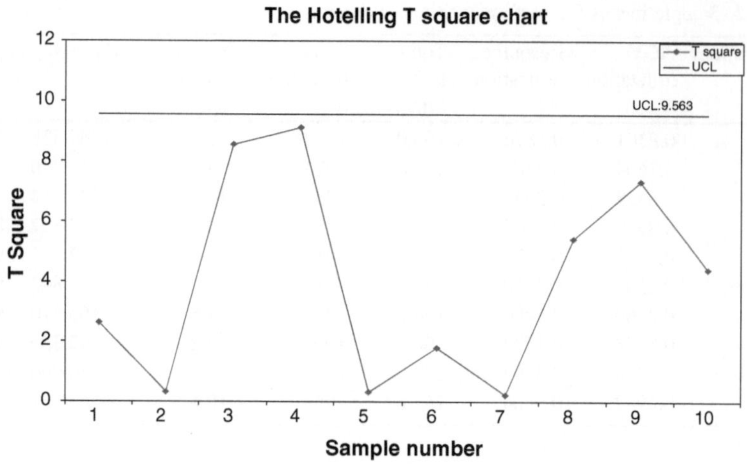

Fig. 4 Hotelling T^2 control chart for Z_{i1} and Z_{i2} table with 95 % confidence ellipse

5 Conclusion

In this paper, an attempt has been made to design control charts for performance monitoring of the FMS. The control charts presented consider all possible combinations in a multivariate frame work. The simultaneous effects of the entire performance criterion are shown. It is observed that from the data generated from simulation exhibited statistical control. Control chart for Z_{i1} and Z_{i2} in Fig. 4 shows all the 10 sample numbers are within the control limit at 95 % confidence interval. In the event any of the performance criterions is performing unsatisfactorily, in a real-time application, temporarily, the decision variables related to that performance criterion may be invoked, and the system brought under control. As soon as the system returns to normal the decision rules corresponding to global criterion optimization may be once again implemented. This switching helps to monitor and control a real application. The design of control charts help to this end.

References

1. Kidd T (1994) Agile manufacturing: forging new frontiers, Reading, MA, Addison-Wesley
2. Askin RG, Standridge CR (1993) Modelling and analysis of manufacturing systems. Wiley & Sons Inc., New York
3. Cho H (1992) An intelligent workstation controller for computer integrated manufacturing. PhD dissertation, industrial engineering, Texas A&M University, College Station, TX
4. Montazeri, M. and Wassenhove, L.N., "Analysis of Scheduling rules for an FMS", International Journal of Production Research, Vol. 28 No. 4, 1990, pp. 785-802

5. Stecke KE, Solberg JJ (1981) Loading and control policies for a flexible manufacturing system. Int J Prod Res 19(5):650–655
6. Law AM, McComas MG (1992) How to select simulation software for manufacturing applications. Ind Eng 7(7):29–33
7. Taha AH (2003) Operational research, an introduction, Prentice-Hall of India
8. Hatcher L, Stepanski E (1994) A step-by-step approach to using the SAS system for univariate and multi variate statistics. SAS Institute Inc, Cary
9. Ryan TP(2001) Statistical methods for quality improvement, 2nd edn. John Wiley and sons, Inc, New York
10. Montgomery DC, Lowry CA (1995) A review of multi variate control charts. IIE transactions 27:800–810

Improved Methods for Thermal and Compression Testing of Carbon Foams

S. Gowthaman, J. M. Babu, P. Mathiyalagan and K. Shivakumar

Abstract Although much interesting research on carbon foams has been published, their thermal conductivity and compression properties are not fully understood because of complexity of porous and brittle structure. These properties are critical for applying carbon foams in thermal management, composite tooling, and shock resistant applications. Special test methods are needed for accurate characterization of thermal and compression properties of carbon foams. This paper describes improved methods for measuring the thermal conductivity and compression properties of carbon foams. The thermal conductivity was measured and validated using a heat flow meter technique that was built in house and the compression properties were measured using end-potting technique. The results from improved techniques were compared with that of standard techniques. Compared to standard techniques, the improved techniques resulted in accurate measurement of thermal conductivity and compression properties with consistent and reproducible test data. Thus, these techniques are recommended for measuring the thermal conductivity and compression properties of carbon foams and similar open-cell brittle foams.

Keywords Cellular foams · Thermal conductivity · Compression properties · Failure mechanisms

---

S. Gowthaman (✉) · J. M. Babu · P. Mathiyalagan
Department of Mechanical Engineering, VelTech University, Chennai, India
e-mail: gswami100@gmail.com

J. M. Babu
e-mail: babumtech@nitc.ac.in

P. Mathiyalagan
e-mail: mathis09051970@yahoo.co.in

K. Shivakumar
Center for Composite Materials Research, Department of Mechanical Engineering, North Carolina A&T State University, Greensboro, NC, USA
e-mail: kunigal@ncat.edu

R. P. Bajpai et al. (eds.), *Innovative Design, Analysis and Development Practices in Aerospace and Automotive Engineering*, Lecture Notes in Mechanical Engineering,
DOI: 10.1007/978-81-322-1871-5_28,

1 Introduction

Carbon foam is a porous material made of interconnected open cell structure. Because of this structure, the foam can have controlled variable density ranging from 0.05 to 0.7 g/cc. Carbon foams have excellent thermal stability and conductivity, mechanical properties, and ability to absorb impact energy at high temperatures, which no other foams have. Because of these superior properties, carbon foams are used in thermal management and shock resistant structures. Although much interesting research on carbon foams has been published, their thermal conductivity and compression properties are still not fully understood. This is because of difficulty in accurate measurement of these properties due to complexity of porous and brittle structure.

Flash diffusivity method based on ASTM E1461 standard has been commonly used for evaluating the heat transfer properties of carbon foams [1–3]. In this method, the thermal diffusivity of the material is determined directly and then the thermal conductivity is determined from the knowledge of specific heat capacity and the density of the material. Although this method has various advantages like independent measurement of thermal diffusivity, minimal heal losses, fast data acquisition, and small specimens, the method has certain disadvantages when using porous materials like carbon foams. The porous structure of carbon foam specimen could cause deep penetration of absorbed heat into the specimens and this could change the characteristic of heat wave on the other face of the specimen. Klett et al. [2] discussed this issue and suggested that as long as the depth of heat penetration is less than 6 % of the specimen thickness, the flash diffusivity technique is reliable for determining the thermal diffusivity of carbon foams. The thickness of the specimen normally used in these experiments is only around 6 mm, and achieving the above mentioned criteria in a $\sim$6 mm thick specimen of low-density carbon foams is not always possible. Hence flash diffusivity method may not be suitable for measuring the thermal conductivity of low-density carbon foams and there is a need for the development of more reliable techniques. This paper describes the development and application of heat flow meter that was built in house for reliable measurement of thermal conductivity of various materials including carbon foams.

Similarly, there are also challenges in measuring the compression properties of carbon foams due to its low density and porous structure. There were three types of compression specimens used in the literature: cube [4–6], cylinder [7], and cylindrical dog-bone [4]. The cube and cylindrical specimens are based on the ASTM C365 Standard Test Method for Flatwise Compressive Properties of Sandwich Cores. The cube and cylindrical specimens are widely used because of their simplicity and the cylindrical dog-bone specimen is less widely used. When the porous, open cell and brittle foams are machined, the outer surface exposes partially cut cellular structure. A compression load between the foam specimen and the loading platen introduces high stresses at the contact edges of the cell wall and thus causing early failure (see Fig. 1). This failure progresses as the load is

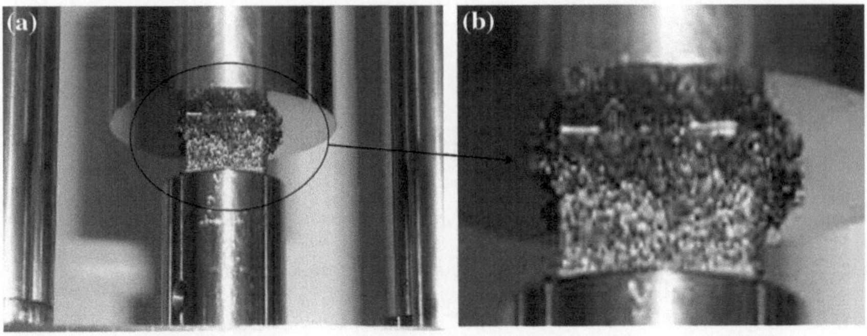

Fig. 1 Compression failure on-set in carbon foam specimen [8]

increased. Thus the measured compression modulus and strength tends to be much less than the real modulus and strength of the foam material. Researchers from AFRL [4] compared the compression results of cylindrical dog-bone specimen and cube specimen and observed significant differences, but did not provide a clear explanation. Min et al. [8] studied the compression of cube specimen (20 mm) and found the failure to be a layer-by-layer crushing of the material. Thus, there is a need to improve the compression testing of carbon foams. This paper proposes a simple modification to the ends of the specimen so that the premature edge compression could be avoided or mitigated. The modified test specimen was used to measure compression modulus and strength of carbon foams and the results are compared with that of the standard (not end-potted) test specimen.

2 Experiments

2.1 Materials and Methods for Thermal Conductivity Measurement

Figure 2 shows the carbon foam specimens used for the measurement of thermal conductivity. These specimens were cut from the carbon foams supplied by WMR Inc, USA. As shown in the figure, the structure of foams was not uniform throughout. There were some large pores in the specimens. The same specimens were later used for graphitization followed by thermal conductivity measurement.

The graphitization of carbon foam specimens was performed by heating at the rate of 1.8 °C/min from room temperature to 2,750 °C and holding for about 1 h. A slower heating rate was used in order to have sufficient time for stress relieving as the materials is heated up. After 1 h of heat soaking, the furnace was switched off and allowed to cool down to RT. The effect of this graphitization on the structure of carbon foam was studied using XRD analysis. The thermal

Fig. 2 Carbon foam specimens for thermal conductivity measurement

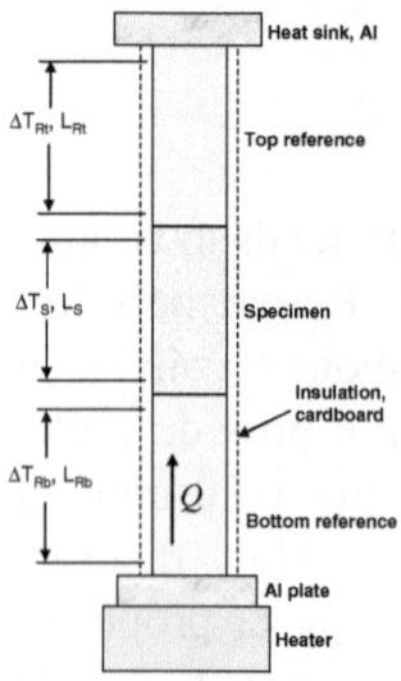

Thermal conductivity equation:

$$\lambda_S = \frac{L_S}{\Delta T_S} \cdot \frac{\lambda_R}{2} \cdot \left[\frac{\Delta T_{Rt}}{L_{Rt}} + \frac{\Delta T_{Rb}}{L_{Rb}} \right]$$

λ_S = Thermal conductivity of sample

λ_R = Thermal conductivity of reference

Test #	Specimen	Thermal conductivity W/mK	Reported data W/mK
1	Copper	398.42	390
2	Aluminium	169.38	167
3	Aluminium	173.6	167
4	Steel	16.32	16.3
5	Epoxy	1.25	1.2
6	Epoxy	1.31	1.2

Fig. 3 Schematic of heat flow meter

conductivity of carbon and graphite foams ware measured using the heat flow meter that was built in-house. The schematic of heat flow meter (based on ASTM E1225 standard) is shown in Fig. 3. The readings were taken using thermocouples after attaining the steady state of heat condition. Steel specimens were used as the reference materials. The test setup was validated by measuring the thermal conductivity of different types of materials ranging from metals to polymers as shown in Fig. 3.

2.2 Materials and Methods for Compression Testing

Carbon foam (CFoam 20) supplied by Touchstone Research Laboratory (USA) was used in the evaluation of the proposed method for measuring the compression properties. The material designation follows the following specification. CFoam represents carbon foam; the number represents the density—"20" for 0.23 g/cc.

Cylindrical specimens used for compression testing of carbon foams are shown in Fig. 4. Two types of specimens—standard (no end-potting) and end-potted specimens were prepared by coating the two end surfaces of the specimens using Epoxy DP460 adhesive. The resin coating was then cured for 2 h at 70 °C.

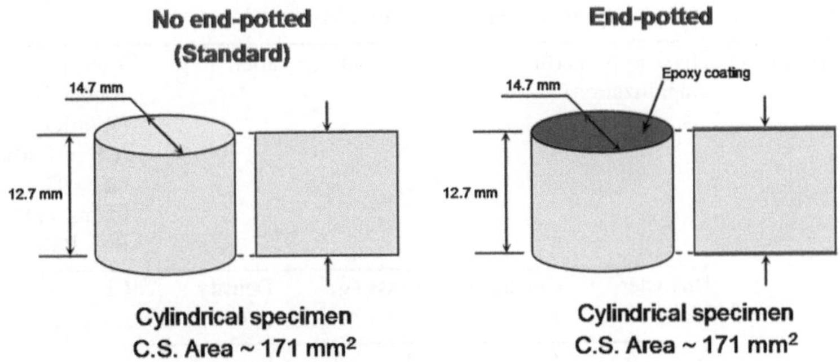

Fig. 4 Compression test specimen configurations

The resin filled the open cells and potted the cell walls. The depth of the potting zone was measured by studying the cross section of end-potted specimen under Scanning Electron Microscope. The depth of the potting zone was about one-cell thick from the end surfaces. The potting strengthened the cell walls and also formed a smooth flat area for uniform load transfer between the sample and loading platen. Compression test was performed using MTS machine at the displacement rate of 0.51 mm/min according to ASTM C365 test method. To understand the failure onset and propagation, the test was continuously monitored using Photron FASTCAM SA4 high-speed camera. The final failure mode was also photographed. After the failure, the fracture surface of the specimens was studied using scanning electron microscope (SEM) to understand the fracture morphology and the stress state.

3 Results and Discussion

3.1 Thermal Conductivity

In XRD analysis, the 002, 101, and 004 peaks, which are characteristic peaks of graphite, were narrow and intense in graphitized foam than in carbon foam. This was indicative of ordered graphite structure in graphitized foam. For comparison purposes, the XRD analysis of Pocofoam was also performed. Pocofoam was graphite foam supplied by Pocographite Inc., USA with a highly ordered graphitic structure, as could be observed from very narrow and intense peaks in XRD analysis. The thermal conductivity of carbon and graphitized foams measured using the in-built heat flow meter as shown in Table 1. The thermal conductivity of graphitized foams was about 15 to 19 times that of carbon foams. Sample 1103.MP29 was broken during graphitization and sample 1103.MP30 was disintegrated during graphitization. An estimated thermal conductivity ($\sim$15 times that of carbon foam) of those samples are reported in Table 1. The measured thermal

Table 1 Thermal conductivity of carbon and graphite foams

WMR Specimen#	Physical properties (B = before and A = after graphitization)								Thermal conductivity, λ_s (W/mK) (CF = Carbon, GF = Graphite foams and R = Ratio)		
	Diameter (mm)		Thickness (mm)		Mass (g)		Density (g/cc)		CF	GF	R
	B	A	B	A	B	A	B	A			
1103.MP17	12.20	12.00	19.63	19.33	0.505	0.498	0.220	0.228	1.28	22.62	18
1103.MP28	12.43	12.10	25.41	25.35	0.896	0.948	0.291	0.325	2.18	33.25	15
1103.MP29	12.45	–	16.35	–	0.729	–	0.366	–	1.72	25.80*	15
1103.MP30	12.33	–	15.35	–	0.481	–	0.262	–	1.14	17.10**	15
1103.MP32	12.34	12.02	21.51	21.14	1.014	1.003	0.394	0.418	1.76	25.01	14
0807.MP07	12.53	12.05	13.03	12.75	1.632	1.117	1.06	1.117	2.03	38.63	19

*Specimen broken during graphitization (estimated $\lambda_s = \sim 15$ times of CF)
**Specimen lost in graphitization (estimated $\lambda_s = \sim 15$ times of CF)

conductivities were accurate and as expected. The proposed method eliminated the restriction on the thickness of the specimen and thus can be used for the accurate measurement of thermal conductivity of low-density porous structures.

3.2 Compression Properties

The standard carbon foam specimens failed by local crushing of cell walls on the top loading surface. The specimens did not become dense and the failure was observed to be a progressive crushing of cells. The photographs of this type of failure are shown in Fig. 5. Such a failure is due to the buckling of open-cell structure at the loading end and this resulted only in the end-compression of the specimen.

The end-potted carbon foam specimens avoided layer-by-layer crushing of cells and failed in a pure shear mode. The failure occurred very fast and sudden load drop was observed during the testing. For better understanding of this sudden failure in end-potted specimens, the compression test was continuously monitored using high-speed camera. Typical failure initiation and propagation is shown in Fig. 6. The initial failure occurred along a shear plane and away from the loading end, and further propagated predominantly along shear planes. The local microscopic failure analyses showed shear-like deformation of cells along the shear plane and no deformation of cells away from the shear plane as shown in Fig. 7.

The average modulus (440.2 MPa) and strength (4.72 MPa) of end-potted specimens measured from the experiments were significantly higher than that of

Fig. 5 Layer-by-layer failure progress in a standard carbon foam specimen

Fig. 6 Typical failure initiation and propagation in end-potted carbon foam specimen

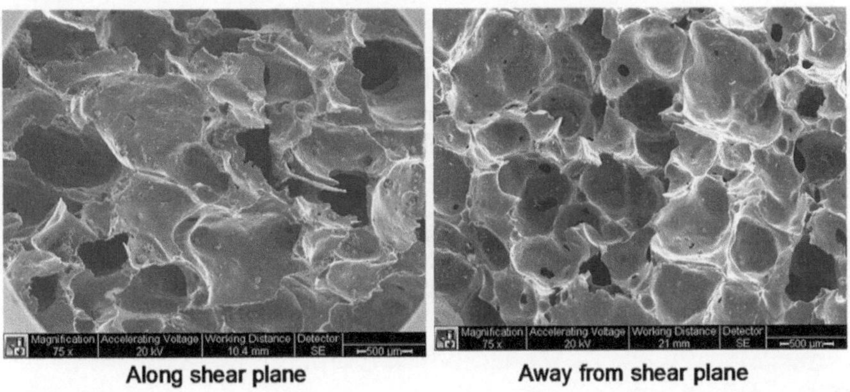

Fig. 7 SEM images of the fractured surfaces on the end-potted carbon foam specimen

the standard specimens. The increments in modulus and strength were about 5.3 and 3.5 times, respectively. In addition, the standard deviation of the test data was much smaller for end-potted specimens than the standard specimens. The increase in the modulus and strength of the end-potted specimens is purely due to the uniform compression of foam specimens and is not affected by the depth of the potting zone which was about one-cell thick. The measured data with end-potting can be used as the true property of foams. Thus the end-potted specimen is recommended for all open cell foams because of simplicity of specimen preparation, accuracy of the test results and limited data scatter.

4 Conclusions

Measurement of thermal conductivity and compression properties of carbon foams is challenging because of their low density porous structure. Various test methods published in the literature were not adequate enough to accurately characterize and describe these properties. This paper described improved methods for accurately measuring the thermal conductivity and compression properties of carbon foams. A heat flow meter was built in house and validated to measure the thermal conductivity. Compared to flash diffusivity method that was widely used in the literature, the proposed method eliminated the restriction on the thickness of the specimen and resulted in accurate and correct measurement of thermal conductivity of carbon foams as well as graphite foams. The paper also presented a simple method of minimizing the stress concentration between the cell walls and the loading platen thereby improving the measurability of compression modulus and strength of carbon foam. The method is called end-potting of the foam samples. The end potting involved coating the top and bottom contact surfaces of the specimen by a thin layer of resin and curing. The coated and noncoated samples were represented as the end-potted and standard samples, respectively. Under compression test, the standard samples failed by end crushing of cell walls followed by progressive crushing of layers from one end of the sample. On the other hand, the end-potted specimens failed suddenly in a pure shear mode, typically like a brittle material. The proposed compression test method showed accurate failure modes and test data with limited data scatter. Thus, the improved testing techniques proposed in this paper are recommended for measuring the thermal conductivity and compression properties of carbon foams and similar open-cell brittle foams.

Acknowledgments The authors acknowledge the financial support by Wright Materials Research Co. Grant # 0II-0620353 and NASA Grant # NNX09AV08A. The authors also thank Touchstone Research Laboratory for providing the carbon foam materials.

References

1. Gaies D, Faber KT (2002) Thermal properties of pitch-derived graphite foam. Lett Ed/Carbon 40:1131–1150
2. Klett JW, Hardy R, Romine E, Walls C, Burchell T (2000) High-thermal-conductivity, mesophase-pitch-derived carbon foams: effect of precursor on structure and properties. Carbon 38:953–973
3. Klett JW, Mcmillan AD, Gallego NC, Walls CA (2004) The role of structure on the thermal properties of graphitic foams. J Mater Sci 39:3659–3676
4. Moore EH (2005) Carbon foam round robin test method development. AFRL-ML-WP-TR-2006-4118
5. Bunning TJ, Jeon HG, Roy AK, Kearns KM, Farmer BL, Adams WW (2003) Polyurethane-infiltrated carbon foams: a coupling of thermal and mechanical properties. J Appl Polym Sci 87:2348–2355

6. Klett JW (2013) High thermal conductivity, mesophase pitch-derived carbon foams. http://Physics.lbl.gov/~gilg/ATLASUpgradeRandD/HighKFoam/Graphite_Foams.pdf. Accessed 25 May 2013
7. Chen C, Kennel EB, Stiller AH, Stansberry PG, Zondlo JW (2006) Carbon foam derived from various precursors. Carbon 44:1535–1543
8. Min G, Zengmin S, Weidong C, Hui L (2007) Anisotropy of mesophase pitch-derived carbon foams. Carbon 45:141–145

Integration of eLCAr Guidelines into Vehicle Design

Patricia Egede, Frank Nehuis, Christoph Herrmann and Thomas Vietor

Abstract Road transportation contributes considerably to environmental concerns like the increase of greenhouse gases in the atmosphere. As the world population grows and urbanization increases, outdoor air pollution becomes another environmental concern. In this context, it is a great challenge to meet mobility needs in an ecologically sound manner. Electric vehicles (EVs) are one solution to achieve this goal depending on the electricity mix in use. To support decision makers minimizing the environmental impacts of EVs, a detailed analysis and assessment over the entire life cycle is necessary. A suited method is Life Cycle Assessment (LCA). This approach accounts for all relevant flows and estimates the environmental impact of a product. Calculating the LCA of an EV encompasses challenges. For example, defining the correct system boundaries particularly for the comparison with other technologies is very significant to achieve fair and comparable results. Changes made to a single component can have multiple impacts on other components of the vehicle leading to a high complexity and the necessity to handle these interdependencies. The E-Mobility Life Cycle Assessment Recommendations (eLCAr) guidelines have been developed to help the LCA practitioner to overcome these challenges when assessing the environmental impact of EVs. In order to improve the performance of EVs regarding environmental criteria, the environmental impacts must be considered during the design phase of the vehicle. Against this background, an integrated approach is presented which includes the LCA method and the eLCAr guidelines in the design phase.

Keywords Life cycle assessment · Electric vehicles · Vehicle design

P. Egede · F. Nehuis · C. Herrmann (✉) · T. Vietor
Technische Universität Braunschweig, Braunschweig, Germany
e-mail: C.Herrmann@tu-bs.de

R. P. Bajpai et al. (eds.), *Innovative Design, Analysis and Development Practices in Aerospace and Automotive Engineering*, Lecture Notes in Mechanical Engineering,
DOI: 10.1007/978-81-322-1871-5_29,

1 Introduction

Road transportation contributes to different challenges around the world. It burdens cities with traffic congestions, contributes to the increase of greenhouse gases in the atmosphere as well as the depletion of natural resources and creates noise and air pollution. Lim et al. [1] estimate that 2.3 million people died of outdoor air pollution in 2010 with motor transportation being a major source of air pollution [2]. Significant effort is made to reduce the impact of the use phase, which originally was dominating the environmental impact of vehicles heavily over the entire life cycle. The weight of the vehicle, its long life time and the proportional effect of the fuel combustion causes this effect. However, as the first oil crisis of the 1970s unsettled the market, energy efficiency was demanded by customers and focused on by car manufacturers [3]. Since the beginning of the 1990s, the exhaust emissions in Europe have been regulated according to the European emissions standard. More and more rigorous standards have been implemented. The next stage is reached in 2014 when the EURO 6 standard becomes mandatory for newly sold light passenger and commercial vehicles. Even though CO_2 emissions are not affected by the EURO standard, reductions have been achieved due to voluntary agreements between the EU and car manufacturers. Using lightweight materials have the potential to reduce the vehicle weight and therefore the fuel consumption. However, this comes with a trade-off in the raw material extraction and the manufacturing phase. Materials like magnesium and aluminium have a much higher environmental impact than the production of steel [4]. Fuels like compressed natural gas are used because they cause fewer emissions of certain gases than petrol. New technologies like electric vehicles (EVs) can reduce the impact of the use phase significantly depending on the source of the electricity. The materials used for the battery and the electric engine intensify the importance of the raw material extraction. It can be concluded that within the last four decades the importance of the use phase has shrunk and the upstream processes have become more important. This development is depicted qualitatively in Fig. 1.

The impact of the use phase of EV depends highly on the source of energy. When energy sources with low environmental impacts are used, the focus to decrease the environmental impact shifts toward the materials used and the recyclability. Material selection takes place during the design stage. But also the recyclability is mainly determined by design decisions. When making decisions in the product design phase, problem shifting from one life cycle phase to another or from one product component to another has to be avoided. An evaluation of the environmental advantageousness of a product or a design solution with alternatives, in most cases, cannot be made by qualitative data evaluation. Life Cycle Assessment (LCA) is a well-established method supporting a quantitative evaluation.

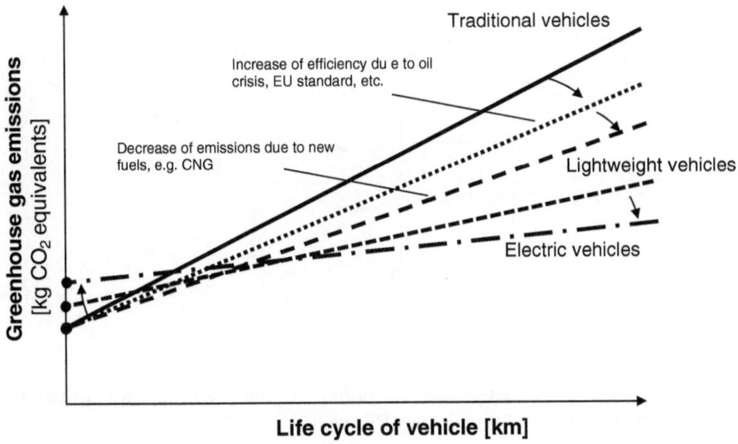

Fig. 1 Development of greenhouse gas emissions of vehicles

2 Life Cycle Assessment of Electric Vehicles

The LCA method assesses the environmental impact of a product or process over the entire life cycle from cradle to grave by considering all input and output flows of the system. The method is an iterative approach of four steps. First, the goal and scope of the study are determined. Most importantly, a functional unit is defined which represents the value of the system and is the foundation of the calculation. Furthermore, the system boundaries are defined which decides what is considered in the assessment and what is left out. Second, all exchanges across the system boundaries, meaning all input and output flows are collected in the inventory analysis. In the third step of the impact assessment, the elementary flows collected in the previous step are transferred into impact categories allowing the LCA practitioner to interpret the environmental impact in the final step of the LCA. The process of LCA is shown in the left part of Fig. 2.

Calculating the LCA of an EV is a challenging task as EVs are complex products with a large number of different, interdependent components. Changes of the battery can, e.g., affect the capacity, the impact of the raw material extraction, the construction of the glider, and the requirements for the electric engine. Therefore, assessments become challenging to carry out and LCAs are often not comparable because different assumptions are made for each analysis. The *E-Mobility Life Cycle Assessment Recommendations* (eLCAr) guidelines [6] have been designed with the purpose to assist LCA practitioners in calculating LCAs of EVs and to increase the comparability of these LCAs. The guidelines are based on the *International Reference Life Cycle Data System* (ILCD) Handbook which was developed by the Joint Research Centre of the European Commission and complies with the ISO 14040 *Environmental management—Life cycle assessment—Principles and framework* standard. The eLCAr guidelines provide information on all phases of the EV life cycle. It also contains average data for the most important parameters of

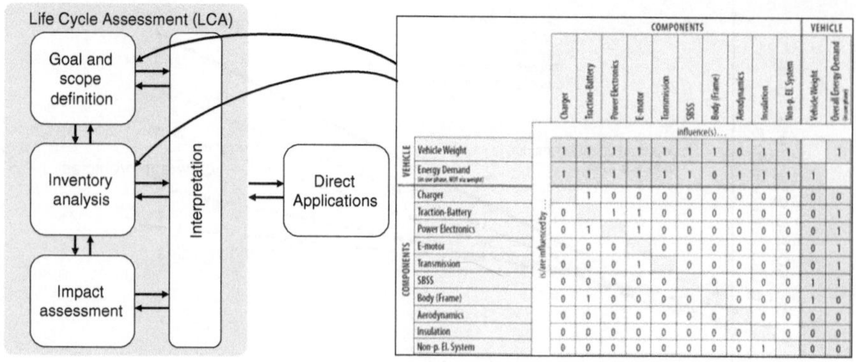

Fig. 2 Method of life cycle assessment according to the ISO standard 14040 [5] (*left*), Interdependency matrix of eLCAr guidelines [6] (*right*)

the EV as well as an interdependency matrix (IM). The IM links the different components of the EV and highlights which changes occur to other components if one component is altered. The right side of Fig. 2 shows the IM. It can be used during the goal and scope definition to define the system boundaries as well as during the inventory analysis to define different values of the components. This allows the LCA practitioner to calculate LCAs of single components whilst including significant changes of the remaining vehicle into the assessment. This is essential, e.g., for comparative LCAs of components to ensure a fair and comprehensive comparison. An example is the comparison of two different batteries made of different materials. Of the two batteries with the same capacity, one might be lighter than the other affecting the construction of the glider and the energy consumption in the use phase. These changes of the vehicle have to be taken into account in order to make a well-informed decision, which reflects the overall optimum of the entire vehicle and not only the local optimum of the battery component.

As stated above, the significance of the use phase of EVs diminishes as electricity from renewable resources with fewer emissions becomes available. This increases the significance of other factors, e.g., material choices and design choices affecting the recyclability which can hardly be altered after the production. Subsequently, this type of information should be included into the design phase. This integration requires the availability of the necessary knowledge on the technical interdependencies influencing the environmental impact as given in the eLCAr guidelines as well as experience with the method of LCA.

3 Integrated Approach

In order to satisfy this demand an integrated approach is required which combines vehicle design, LCA and the eLCAr guidelines. This approach should enable the designer of an entire vehicle or single component to incorporate the assessment of

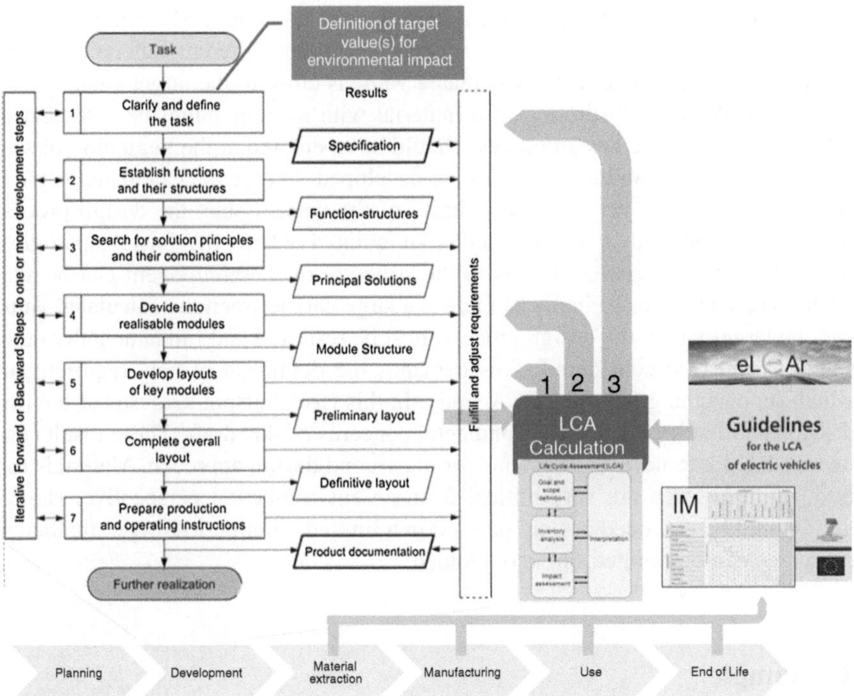

Fig. 3 Integrated approach of environmental assessment into vehicle design

the environmental impact of the entire EV into the design phase. Such an approach is presented in Fig. 3. It contains the design process as described in the VDI standard 2221 *Systematic approach to the development and design of technical systems and products* (Verein Deutscher Ingenieure, engl.: The Association of German Engineers) [7], the product life cycle, the LCA method, and the eLCAr guidelines. The VDI 2221 can be used for the development and design process of vehicles. Hence, it is located in the second step of the life cycle following the planning and previous to the production, use and end of life phase. The eLCAr guidelines provide information on the production, use and end of life phase of EVs.

When starting the design process according to the VDI 2221 standard, a target value for one or several impact categories must be fixed. In the following, the design steps should be completed according to the standard. When a preliminary layout and a first draft of the list of parts have been created (outcome of step 5), a LCA can be calculated. The list of material should contain information on the material and the according weight of the parts. The information of the eLCAr guidelines can be made available via a continuously updated database. It should be pointed out that particularly the interdependency matrix is crucial when designing vehicle components in order to include the entire life cycle of the vehicle into the LCA calculation. The results of the LCA must then be compared to the threshold values, which have been specified in the beginning of the design process. If the target value is reached, the design process can be continued with no further adjustments. If the target value is not reached, three different outcomes of the comparison are possible. If the calculated value is close to the target value, minor alterations like the substitution of a material with another might be sufficient to achieve the goal. The design process should be re-entered at the beginning of step 5 where the layouts and key modules are developed. In case the calculated value is close but not within very close proximity of the target value, the design process should be re-entered at the beginning of point 4 where the basic topological solution is created. Here, changes to the structure of the component can be adequate to reach the goal. Finally, if there is a large gap between the calculated value and the target value, the design process must be restarted and fundamentally other solutions must be considered. This demands the definition of new requirements which support the goal of the LCA study. In this case, hotspot analyses of the first design can identify critical environmental concerns of this draft. These results can be translated into new requirements for the second design approach. Materials can be identified which are very critical from an environmental perspective. Hence, they should be excluded or only be used in a limited amount. Likewise, inefficient processes can be pointed out and avoided.

4 Example

The use of the integrated approach is illustrated with an example of designing a battery. A target value of 1.6 kg CO_2-eq is defined with a functional unit of 50 MJ charged and used to power the vehicle. With a nickel metal hydride battery a value of 3.1 kg CO_2-eq [8] is reached. As this is almost the double of the target value a new design attempt is necessary to reach the goal. The analysis shows that the material nickel contributes significantly to the environmental impact [8]. Therefore, a solution should be sought which requires less nickel. The design process is restarted from the top and a lithium-ion battery is chosen, which has a longer life time and contains significantly less nickel and hardly any rare earth metals (see also [9]). A value of 1.9 kg CO_2-eq [8] is reached with a nickel cobalt manganese lithium-ion battery. As this is rather close to the target value, the design process is re-entered at step 5 and some minor changes are done to the battery design. An iron phosphate lithium-ion battery gives an output of 1.4 kg CO_2-eq [8], which complies with the target value of 1.6 kg CO_2-eq.

5 Conclusion

With an increasing pressure resulting from environmental concerns such as global warming and outdoor air pollution, increasingly strict regulations of exhaust emissions and the demand of customers, automobile manufacturers have to consider the environmental performance of their products. With regard to the complexity of assessing the environmental impact of EVs, an integrated approach to consider the environmental performance of EVs in an early design phase is necessary. Such an integrated approach was proposed combining the VDI 2221 standard, the LCA method and the eLCAr guidelines. A joint design and assessment lab can contribute to the integration of the assessment of the environmental impact into the early design phase of a product. By bringing together vehicle designers and LCA practitioners as well as the software programmes they use, an exchange can be stimulated. This also allows including other software tools with additional features such as the evaluation of the end of life phase [10]. Further research is necessary in order to define the target value as well as the thresholds for the comparison of the target value and the calculated value. Also, the software implementation of the method has to be defined in detail.

References

1. Lim SS, Vos T, Flaxman AD et al (2012) A comparative risk assessment of burden of disease and injury attributable to 67 risk factors and risk factor clusters in 21 regions, 1990–2010: a systematic analysis for the Global Burden of Disease Study 2010. Lancet 380:2224–2260
2. Dybing E, Fletcher T, Greenbaum DS et al (2013) Air pollution and cancer. In: Straif K, Cohen A, Samet J (eds) IARC Scientific Publications No. 161, p 161
3. Hohensee J (1996) Der erste Ölpreisschock 1973/74: die politischen und gesellschaftlichen Auswirkungen der arabischen Erdölpolitik auf die Bundesrepublik Deutschland und Westeuropa. Steiner Verlag Wiesbaden GmbH, Stuttgart
4. Krinke S (2011) Implementing life cycle engineering efficiently into automotive industry processes. In: Hesselbach J, Herrmann C (eds) Glocalized solutions for sustainability in manufacturing. 18th CIRP international conference on life cycle engineering, Braunschweig, 2–4 May 2011, Springer, Heidelberg, p 11
5. ISO 14040:2009-11 (2009) Environmental management—life cycle assessment—principles and framework
6. Althaus HJ, Del Duce A, Egede P et al (2013) Guidelines for the LCA of electric vehicles. http://www.elcar-project.eu/fileadmin/dokumente/Guideline_versions/eLCAr_guidelines.pdf. Accessed 10 Sept 2013
7. VDI Standard 2221 (1993) Systematic approach to the development and design of technical systems and products, Düsseldorf
8. Majeau-Bettez G, Hawkins TR, Strømman AH (2011) Life cycle environmental assessment of lithium-ion and nickel metal hydride batteries for plug-in hybrid and battery electric vehicles Guillaume. Environ Sci Technol 45:4548–4554
9. Notter DA, Gauch M, Widmer R et al (2010) Contribution of Li-ion batteries to the environmental impact of electric vehicles. Environ Sci Technol 44:6550–6556
10. Krause FL, Herrmann C, Frad A, Raga Z, Luger T (2006) Integrating end-of-life evaluation in conceptual design. In: IEEE Computer Society, International Association of Electronics Recyclers (eds) Proceedings of the 2006 IEEE international symposium on electronics and the environment, ISEE 2006—and the 7th electronics recycling summit, San Francisco, California, 8–11 May 2006, Piscataway, NJ, p 245

Design of Magnetic Wheel Rotor

S. Yuvaraj and M. Anantha Kumar

Abstract The aim of this paper is to use magnetic force of attraction as an alternate fuel to drive the shafts. Here we have designed a system which makes a shaft to rotate automatically with the use of the magnetic force. The efficiency of the system is improved by utilizing less energy for higher loads. This can be applied to the rotating machines like IC engines and Gas Turbine Engines.

Keywords IC–internal combustion · NS–North South poles · HP–horse power · Rpm-rotation per minute

1 Introduction

A shaft is an element used to transmit power and torque. It is subjected to bending and torsional loads [1]. In this chapter, a flat plate is coated with magnetic powder which is fixed below the shaft with a gap between the shaft and the plate. A restrictor is placed in the air gap, which covers the shaft at the four ends. The flat plate coated with magnetic powder acts as an permanent magnet. If one side of the shaft is attracted by the magnetic field, it moves down. When it reaches bottom, the field is restricted by the restrictor. Similarly the next portion is attracted and moved down. This continuous attraction and restriction makes the wheel to rotate continuously (Fig. 1).

S. Yuvaraj (✉)
Department of Aeronautical Engineering, Karpagam University, Coimbatore 642120, India
e-mail: syuva91@gmail.com

M. Anantha Kumar
Department of Aeronautical Engineering, Sri Ramakrishna Engineering College, Coimbatore 641022, India

R. P. Bajpai et al. (eds.), *Innovative Design, Analysis and Development Practices in Aerospace and Automotive Engineering*, Lecture Notes in Mechanical Engineering, DOI: 10.1007/978-81-322-1871-5_30, © Springer India 2014

Fig. 1 Shaft and magnetic field arrangement

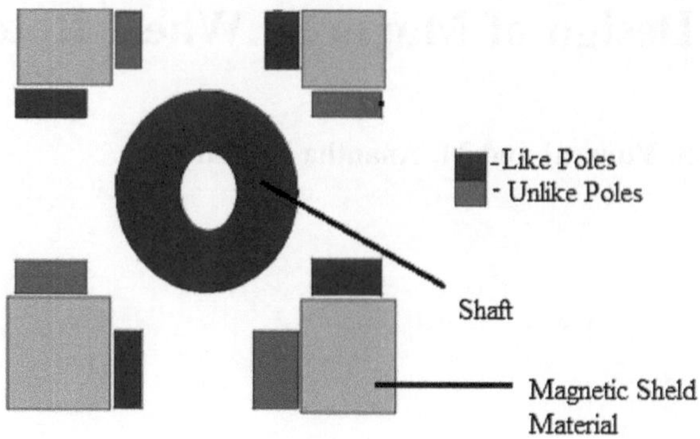

2 Construction

In magnetic rotator wheel, we use two types of magnet arrangement namely hollow circular magnet and rectangular block magnet.

2.1 Circular Hollow Magnet

In magnetic rotator wheel, shaft is inserted tightly to the circular hollow magnets without any air gap for proper rotational operation occurs by magnetic attraction and the shaft ends are connected to the wheel. Maximum numbers of circular hollow magnets are present on the shaft, not only at the same dimensions but also arranged in opposite poles like NS, NS, etc. Pictorial representation for a circular hollow magnet arrangement around the shaft is shown in Fig. 2.

We choose power of the circular hollow magnet that is only less than 1,000 Gauss for increasing the efficiency of the pull force for rectangle block magnet.

2.2 Rectangular Block Magnet

Rectangular block magnet is placed at each corner (four sides) of the circular hollow magnet. In each corner of the circular magnet, a pair of rectangular block magnet is placed perpendicular to each other in an opposite poles. It resembles the symbols "L." We think, if the two rectangular block magnets are arranged in perpendicular with opposite polesit will be attracted by one another. For this reason, some material is placed under the magnet like "" structure and then magnet is tightly covered with stell of less thickness (5 mm). In one corner, 20 pieces of block magnet are placed. If magnetic rotator wheel has four corner, then 4*20 = 80 magnet pieces are used. In 80 block magnetic pieces, 40 magnet pieces for attraction and remaining 40 magnet pieces are for repulsion.

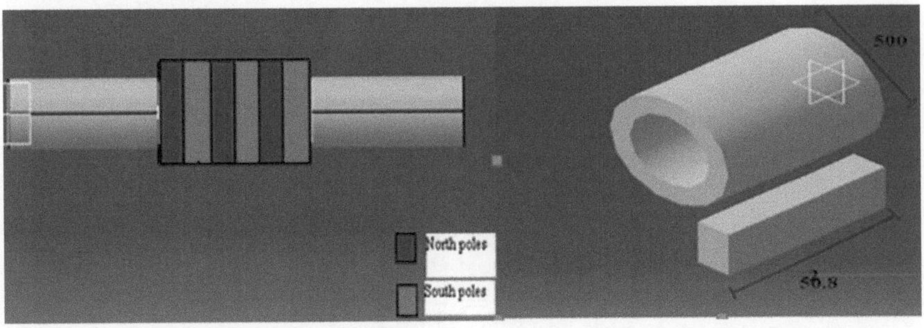

Fig. 2 Circular hollow magnet and its polarity

In shaft, circular hollow magnet is placed three by fourth of the area and remaining area is arranged by circular hollow disk with a certain gap, the support is given to the disk by the base of the automobiles, which is used for disk brake. Maximum two disk brakes are used at the end of the hollow magnets, for our safety.

3 Working

Take one corner of the block magnets, it has a pair of magnet with opposite polarity. In pairs one magnet is placed vertical, it is denoted as "A," and the other magnet is placed horizontal, it is denoted as "B." If the hollow magnet and magnet "A" are arranged in same poles means, magnet "B" is arranged in opposite poles compare to magnet "A." Due to this arrangement, hollow magnet is attracted by the magnet "B" and repelled by the magnet "A." By this continuous operation, hollow magnet containing shaft is rotated in one direction. Direction of the shaft rotation changed is achieved by using stepper motor. The above-mentioned operation occurs in each corner of the circular hollow magnet.

4 How to Select the Magnetic Dimensions

In magnetic wheel rotator, both the circular hollow magnet outer radius and rectangular block magnet length dimension are must important for rotational operation when compared to other dimensions of both magnet. Because, both the hollow magnet outer radius and block magnet length determine the rotational operation. In two magnets, if any one of the dimension is not proper then the remaining magnet will not be ready to rotate. Magnetic dimension selection for both the hollow magnet outer radius and block magnet length is related to each other is given below:

$$\text{i.e. } \mathrm{L} = \frac{r}{5} + 1\,\mathrm{mm} \tag{1}$$

where,

L Length of the block magnet in mm,
r Outer radius of the hollow magnet in mm.

Here, r = 250 mm.

$$\text{Length of the magnet} = \frac{r}{5} + 1\ \text{mm}, = 51\ \text{mm}.$$

5 Pull Force Concept

"When two magnets are involved in attraction, one magnet must have less power than the other then only gets the maximum pull force." The same concept is followed in this project. In this project, circular hollow magnet has less gauss power than the rectangular block magnet.

5.1 Pull Force Calculation

We already know about pull force formula for two identical rectangular block and disk magnet to a certain distance. In this project, we choose two different types of magnet (circular hollow and rectangular block magnet) in different size. Pull force formula between two different magnets in a different size is not available. For this reason, we choose one as a powerful magnet (rectangular block magnet) and the other one as not a powerful magnet (circular hollow magnet) , which is less than 1000 Gauss. This time, we need not calculate the pull force between two magnets, but just calculate pull force for a powerful magnet at a certain distance which is distanced between two magnets (Table 1).

Pull force calculation for a block magnet at a certain distance is calculated into two steps:

i. Calculate the magnetic flux density at a required distance,
ii. Calculate pull force for a calculated magnetic flux density value.

5.2 Magnetic Flux Density Calculation

Magnetic flux density formula for a block magnet at a certain distance is:

Table 1 Relation between distance in cm and magnetic flux density in gauss are given below by using magnet (N48), with length 2 inch, width 2 inch, and thickness 2 inch given below:

Distance in (cm)	1	2	3	4	5	6	7	8	9	10
Magnetic flux Value in (Gauss)	3,943	2,420	1,496	961	645	451	326	243	186	145

$$\text{Magnetic flux density} = \frac{Br}{\pi}\left(\tan^{-1}\left(\frac{LW}{2Z\sqrt{(4Z^2+L^2+W^2)}}\right) - \tan^{-1}\left(\frac{LW}{2(D+Z)\sqrt{4(D+Z)^2+L^2+W^2}}\right)\right)\text{Gauss} \tag{2}$$

where,

Br Remanence field,
W Width of the block magnet,
L Length of the block magnet,
D Thickness of the block magnet,
Z Distance from the block magnet.

$$\text{Magnetic flux density} \propto \frac{1}{\text{distance from the magnet}} \tag{3}$$

5.3 Pull Force Calculation

$$\text{Pull force} = 0.576 * (Br)^2 * \text{TH} * \sqrt{\text{Area}}\ \text{in lbs}, \tag{4}$$

where,

Br Magnetic flux density in Kilo gauss,
TH Thickness of the magnet in inches,

$$\sqrt{\text{Area}} = \text{Area of the magnet in inches } (\text{L} * \text{W}).$$

i.e., **Pull force** $\propto$ **thickness of the magnet**.

6 Theoretical Proof

Rectangular block magnet dimensions:

$$\textbf{Neodymium magnet (N52)} : \textbf{14,800}\ \text{Gauss},$$
$$\text{Length} : 2\ \text{inches},$$
$$\text{Width} : 2\ \text{inches},$$
$$\text{Thickness} : 2\ \text{inches}$$

The value of magnetic flux density at a 1 inch distance is calculated by using online calculator (magnetic flux density vs. distance), is 1,994.78 Gauss.

By using the calculated magnetic flux density, pull force for a block magnet is found.

$$\begin{aligned}\text{Pull force} &= 0.576 * \text{Br}^2 * \text{TH} * \sqrt{\text{Area}}\\ &= 9.1608\ \text{lbs.}\end{aligned}$$

7 Torque and Horse Power Calculation

Torque calculation for attraction force in block magnet:

Torque = (sum of the pull force of the block magnet in lbs. and weight of the hollow magnet in lbs.)*

Radius of the circular hollow magnet in meter.

- Radius of the circular magnet is calculated by using formula,

$$\textbf{Length of the block magnet} = \left(\frac{\text{Radius of the hollow magnet}}{5}\right) + 1\ \text{mm}$$

$$\begin{aligned}\text{Radius of the hollow magnet} &= (\text{length of the block magnet} - 1) * 5\ \text{mm},\\ 2\ \text{inch} &= 50.8\ \text{mm} = 49.8 * 5 = 249\ \text{mm}.\end{aligned}$$

And then weight of the hollow magnet is also used to calculate torque. Magnetic wheel required size of the circular hollow magnet is not available there, so we just consider weight of the circular hollow magnet is 10 kg. We required magnet value in pounds, so we convert hollow magnet from kg to pounds. One kilogram is equal to £2.205 and then 10 kg is equal to £22.05.

$$\begin{aligned}\text{Therefore; Torque} &= \text{pull force for a block magnet}\\ &\quad + \text{weight of the hollow magnet}\\ &= 138.8256\ \text{N}.\end{aligned}$$

$$\begin{aligned}\text{Torque radius} &= \text{Torque in N} * \text{Radius of the hollow magnet in meter}\\ &= 34.568 \text{ Nm.}\end{aligned}$$

Assume speed of the train is 70 Km/h.

$$\text{Speed} = \frac{\left(\text{speed in} \frac{\text{km}}{\text{h}} * 1000\right)}{(2 * \pi * r * 60)} \text{ in rpm,}$$

where,
R = radius of the train wheel in meter = 0.5 m

$$\begin{aligned}\text{Speed} &= \frac{(7o * 1000)}{(2 * \pi * 0.5 * 60)} \text{ rpm}\\ &= 371.36 \text{ rpm.}\end{aligned}$$

w.k.t

$$\begin{aligned}\text{Horse power} &= \frac{(\text{torque in Nm} * \text{speed in rpm})}{5252}\\ &= 2.44 \text{ hp.}\end{aligned}$$

Finally, we concluded one block magnet is produced 2.44 hp by pull force. 40 magnets are used in pull force in magnetic rotator wheel,

Therefore, horse power produced by pull force magnet = 97.6 hp.

We know that, pull force is nearly equal to repulsive force. Therefore, we consider repulsive force is 2 hp.

Therefore, horsepower produced by repulsive force magnet,

$$\begin{aligned}&= 40 * 2\\ &= 80 \text{ hp.}\end{aligned}$$

$$\begin{aligned}\text{Total horse power} &= (\text{HP produced by pull force} + \text{HP produced by Repulsive force})\\ &= 177.6 \text{ hp.}\end{aligned}$$

Theoretical calculated, one magnetic wheel rotator is achieved 177.6 hp.

7.1 Magnetic Shielding Material

The function of the magnetic shielding material is redirecting the power to the magnet (or) saturate within them. Rectangular block magnet is fully covered with magnetic shielding material expect the operational side. If the magnet is not fully covered, all the attached materials are attracted by the block magnet.

$$\text{Magnetic shielding property} \propto \frac{1}{\text{permeability value of a material}}$$

7.2 Magnetic Shield Material Selection

Attenuation is used to select the magnetic shielding materials. In magnet, Attenuation means "weakening the strength of the magnetic flux density." By using attenuation, we also calculate how much magnetic flux density shielded by the material.

$$\text{Attenuation (A)} = \frac{(\text{permeability value of material} * \text{diameter of the material})}{\text{Thickness of the material}} \text{ in dB}$$

By using Attenuation value, calculate how much magnetic flux density presents inside the material,

$$\text{Br} = \frac{H_O}{A} \text{ Gauss,}$$

where,

Br Magnetic flux density presents inside the shielded material,
H_O Magnetic flux density presents outside the shielded material,
A Attenuation.

In my project, we use the magnetic shielding material is **ultra-low carbon steel**, because it has permeability ranges from 1,000–4,000 and its saturation level is 22,000 Gauss.

Shield material	Saturation value in gauss	Permeability value	
		Maximum	Minimum
Ultra-low carbon steel	22,000	4,000	1,000

7.3 Theoretical Calculation for Magnetic Shielded Material

First, we calculate the attenuation for a selected magnetic shielded material using formula:

$$\text{Attenuation} = \frac{(\text{permeability of the material} * \text{thickness of the material in inch})}{\text{diameter of the material in inch}}$$

$$\text{Permeability of ultra} - \text{low carbon steel} = 1,000,$$
$$\text{Thickness} = 0.016,$$
$$\text{Diameter} = 2.$$
$$\text{Attenuation} = 8 \text{ dB}$$

By using the calculated attenuation value, how much magnetic flux density is present in the ultra-low carbon steel is found.

Magnetic flux density is present inside the material in gauss (Br)

$$= \frac{\text{magnetic flux density is present outside the material } (H_0) \text{ in gauss}}{\text{Attenuation } (A) \text{ in dB}}$$

Magnetic flux density present outside the material is calculated by using magnetic flux density versus distance using online calculator, is 1,994 Gauss, at a distance 1 inch. Using magnet for calculating magnetic shield is magnet N52 with a dimension is 2*2*2.

Therefore,

$$\begin{aligned}\text{Br} &= \frac{H_0}{A} \text{ in Gauss}\\ &= 249.25 \text{ Gauss.}\end{aligned}$$

Final magnetic flux density is calculated by using formula:

$$\begin{aligned}&\text{Final magnetic flux density}\\ &\quad = (\text{Normal magnetic flux density at a required distance}\\ &\qquad - \text{Magnetic flux density is present inside the material}) \text{ in Gauss.}\\ &\quad = 610.27 \text{ Gauss.}\end{aligned}$$

8 Conclusion

The magneto rotator wheel shows that there is a way to obtain the required horse power for automobiles to fly with different speed that do not need any fuel. So that the cost of transporting is reduced as there is no need of fuel. This chapter shows the main issue of fuel efficient automobile can be satisfied using magnet. The main fact depends on the arrangement of magnet in such a way that the repulsive and attractive force is properly used for the continuous transportation.

References

1. Beer FP, Johnston ER et al. (2001) Mechanics of materials, 3rd edn. McGraw-Hill, New York, p 150
2. Furlani EP (2001) Permanent magnet and electromechanical devices: materials, analysis and applications, Academic Press Series in Electromagnetism, London
3. McCaig M (1977) Permanent magnets in theory and practice. Halsted Press (Division of Wiley and Sons), New York
4. Moskowitz L (1976) Permanent magnet design and application hafldbook. Cahners Book International, Boston
5. Cullity BD (1972) Introduction to magnetic materials. Addison-Wesley Publishing Company, Reading, MA
6. Schuler K, Brinkman K (1970) Dauermagnete–Werkstoffe and Anwendungen. Springer, Berlin
7. Parker RJ, Studders RJ (1962) Permanent magnets and their applications. Wiley, New York
8. Hadfield D (ed) (1962) Permanent magnets and magnetism. Iliffe Books Ltd, London, Wiley, New York

The Effect of Injection Pressure on Engine Performance While the Engine is Running on Electrolytically Generated Hydrogen Oxygen Mixture

J. M. Babu, B. Harish Babu, M. Rajesh Kumar, S. Gowthaman, R. Mariappan and P. Mathiyalagan

Abstract Day by day fossil fuel energy reserves are depleting and the cost of the fuel was increasing. This is because of rapid usage of fossil fuel and it also causes increase of pollution, global warming and fuel scarcity. So this is the time to search for alternative fuels. Some of the alternative fuels are Ethanol, methanol, natural gas, bio diesel blends and hydrogen etc. Hydrogen was one of the best alternative fuel among all these because it having so many advantages like (high heating value, renewable energy source, absolutely no pollution and available at cheap price), It has the potential to supplement and if possible replace the fossil fuels for the production of energy in the near future, applicable to fuel for vehicle and rocket, internal combustion engine using hydrogen storage, direct combustion for heat and so on. Water electrolysis is a key technology to produce hydrogen and is one of the best method. In diesel engine injection pressure plays an important role. In this experimental study, effects of injection pressure on engine performance and exhaust emissions have been investigated. Experiments have been conducted on kirloskar constant speed single cylinder 4 stroke water cooled diesel engine while the engine is running on electrolytically generated hydrogen oxygen mixture. The experiments was conducted by changing the injection pressure from 200 to 300 bar with an interval of 50 bar by applying the minimum load to maximum load of 17.8 kg on rope brake dynamometer. Hydrogen oxygen mixture is continuously supplied to the engine at a constant rate of 500 ml/min. According to the results maximum performance has been obtained at 300 bar and at high injection pressure these values are O_2, SO_2 and CO_2 minimum. At lower injection pressure NO_x, emissions and smoke levels is very low. As the injection pressure increases efficiency will increase, and exhaust emissions will decrease.

J. M. Babu (✉) · B. H. Babu · M. R. Kumar · S. Gowthaman · R. Mariappan · P. Mathiyalagan
Department of Automobile Engineering, Vel Tech Dr. RR and Dr. SR Technical University, Avadi, Chennai, India
e-mail: babumtech@nitc.ac.in

B. H. Babu
e-mail: bachina.harish@gmail.com

R. P. Bajpai et al. (eds.), *Innovative Design, Analysis and Development Practices in Aerospace and Automotive Engineering*, Lecture Notes in Mechanical Engineering, DOI: 10.1007/978-81-322-1871-5_31,

Keywords Electrolysis process · Hydrogen · Injection pressure · Engine performance · Exhaust emissions

1 Introduction

In recent years the usage of diesel engines was more concentrated because of high efficiency and less fuel consumption compared to all other engines. Diesel engines were the most trusted power sources in transportation industry. The growing energy needs, depleting fossil fuel reserves and stringent emission norms have increased the focus on research into use of cleaner alternate fuels. Alternative fuels for internal combustion engines, such as LPG, natural gas, ethanol, hydrogen and bio diesel etc. have been the subject of extended research. Considering the pollutant emissions [1] by hydrocarbon fuels, the green house effect and the uncertainty of energy supply, hydrogen is considered as the best fuel for the future. Hydrogen production via electrolysis of alkaline aqueous electrolytes can be used to make electrolysers, which can supply hydrogen and oxygen gas which can be stored and used for various purposes. Commercial stainless steel or nickel based alloys in alkaline solutions can be utilized for making the electrolysers [2] Research is going on to reduce the overvoltage by realizing water electrolysis under high temperature and pressure or developing new electrode materials so as to reduce the overall energy requirement for the process [3].

The first hydrogen engine was developed by Cecil in 1820, which operated on vacuum principle. In this engine atmospheric pressure drives a piston back against a vacuum, created by burning hydrogen air mixture allowing it to expand and then cool, to produce power. Because of its high energy to weight ratio, it is extensively used in space program. The energy crisis, with the rapidly dwindling resources of gasoline and diesel, has necessitated the requirement to develop rapidly, alternative sources of energy to power the automobiles. Research is on for utilizing hydrogen as an automotive propellant on a commercial scale. Researchers and industries have directed their efforts to numerous aspects of hydrogen energy related to generation, storage and utilization of hydrogen [4].

Coming to experimental set up experiments were conducted on single cylinder 4-stroke water cooled vertical direct injection diesel engine. The load is applied on the rope brake dynamometer from minimum load to maximum load of 17.8 kg where speed is kept constant through out the experiment. The hydrogen supplied at a constant rate of 500 ml/min. By using Quintox exhaust gas analyzer exhaust gas analysis was done.

Table 1 General details single cylinder, four stroke, compression, ignition, constant speed, vertical, water cooled, direct injection

Bore	80 mm
Stroke	110 mm
Swept volume	553 cm^3
Clearance volume	36.87 cm^3
Compression ratio	16.5:1
Rated output	3.7 kW at 1500 rpm
Rated speed	1500 rpm
Injection pressure	240 bar

2 Experimental Set up

The engine used for experimental set up is Kirloskar single cylinder 4 stroke water cooled direct injection (DI) diesel engine, rated power is 3.78 Kw at rated speed of 1,500 rpm. The specifications of the engine is shown in Table 1. Hydrogen is supplied from the electrolysis cell at a constant rate of 500 ml/min. The pressure of the hydrogen will be around 1 bar. Hydrogen is first supplied to the flame trap by regulating the valve and then passed to the rotometer where it is measured. The rotometer gives the flow rate of hydrogen in l/min. Here we are using a flame trap to arrest the flames which is coming from the engine. The flame trap is used as a safety device and acts as a non return valve (NRV) then hydrogen is passed to the engine and suction process will start. Before going to the inlet manifold hydrogen will mix with air and both hydrogen oxygen mixture and air will combinedly will go inside the engine. Mixing of air and hydrogen oxygen mixture is called enrichment process. Normally the injection pressure of the engine is 200 bars. Experiments were conducted by changing the injection pressure from 200 to 300 bars with an interval of 50 bars with a constant flow rate of hydrogen of 500 ml/min.

The fuel injector is shown in Fig. 1. By adding or by removing shims can change the fuel injection pressure. This fuel injector is used for single cylinder four stroke direct injection system (Figs. 2, 3, 4).

3 Specification of the Engine

3.1 Results and Graphs

According to the results the engine performance is higher at 300 bar injection pressure. The highest brake thermal efficiency [5] is obtained at 300 bar is 40.01 % while the engine is running with hydrogen oxygen mixture where as at 200 bar it is 33.7 % is only with diesel. This means that the the maximum

Fig. 1 Photographic view of experimental set up

Fig. 2 Modification made to the engine for hydrogen inlet

efficiency improved is 6.31 %. By seeing the graph we can say that as the injection pressure increases the fuel consumption was decreasing, coming to mass of air (Graph 1, 2, 3, 4).

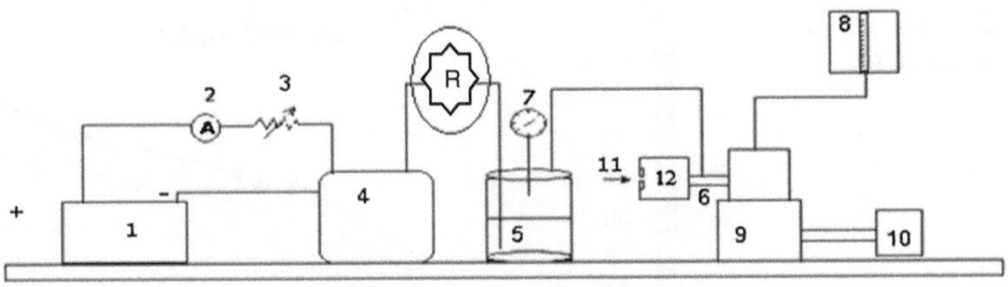

Fig. 3 Schematic arrangement of experimental setup, **1**-battery, **2**-Ammeter, **3**-rheostat, **4**-electrolyser cell, **5**-flame trap, **6**-intake manifold, **7**-pressure gauge, **8**-diesel tank with burette, **9**-engine, **10**-rope brake dynamometer, **11**-air inlet, **12**-air box with orifice meter **R** Rotometer

Fig. 4 Hydrogen checking

Graph 1 Brake power versus brake thermal efficiency

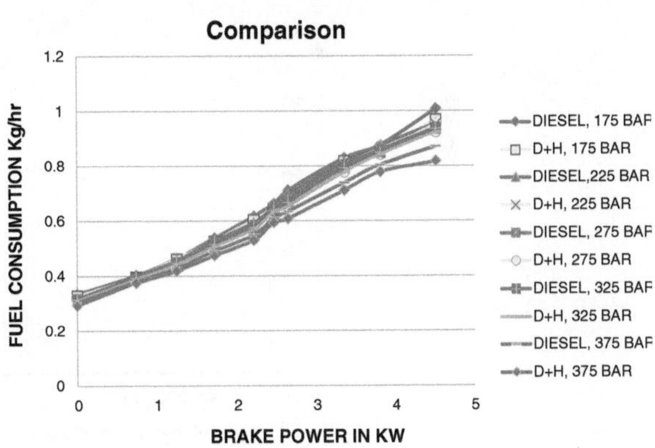

Graph 2 Brake power versus air fuel ratio

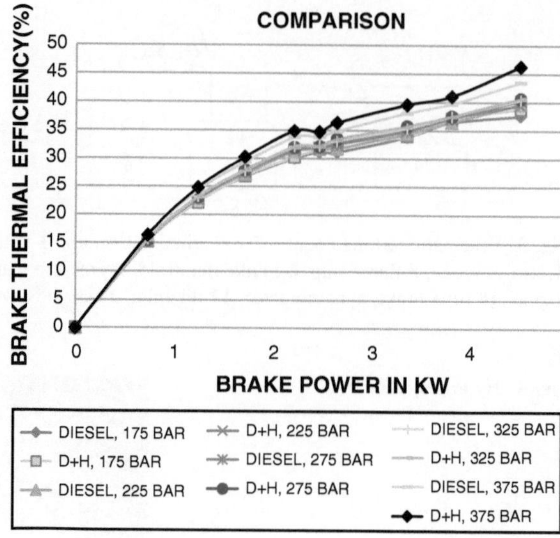

Graph 3 Brake power versus fuel consumption

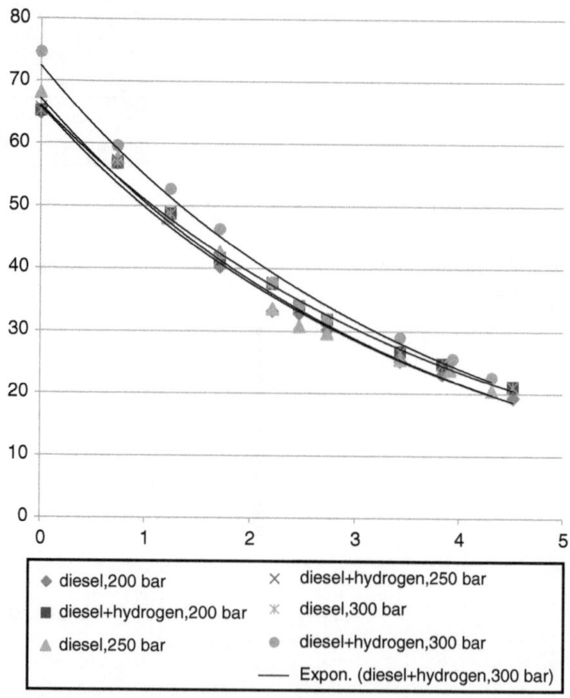

Graph 4 Brake power versus mass of air comparison

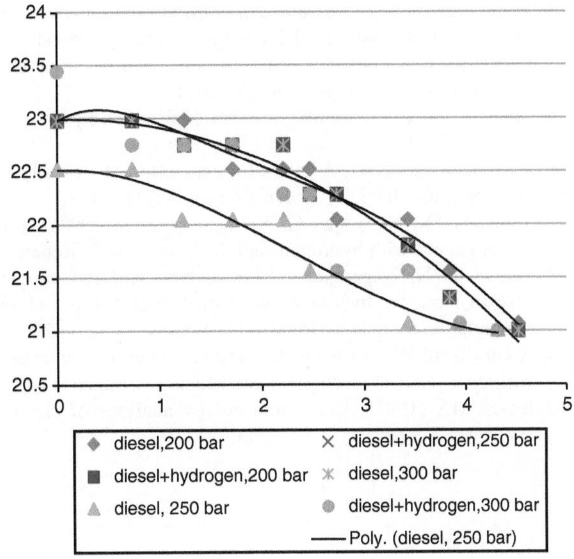

4 Conclusion

Electrolysis cell was designed and fabricated. Experiments have been conducted on the kirloskar single cylinder 4 stroke water cooled direct injection diesel engine and performance characteristic graphs has been plotted. Results showing that fuel consumption was reduced as the injection pressure decreased, brake thermal efficiency was increased 6.31 % at 300 bar injection pressure running with diesel-hydrogen compared with 200 bar injection pressure with only diesel. As the injection pressure increases the air fuel ratio was good compared with low injection pressure. Finally conclusion is as the injection pressure increases efficiency, air fuel ratio was increasing and fuel consumption, mass of air, volumetric efficiency, specific fuel consumption was decreasing. Important thing is when the engine is running with diesel with increase in injection pressure giving high efficiency but if it running with diesel + hydrogen is giving highest efficiency at the highest injection pressure.

References

1. Zeng K, Zhang D (2009) Combustion science. In: Recent progress in alkaline water electrolysis for hydrogen production and applications progress in energy (in press) Corrected proof. Available online 1 Dec 2009
2. Lj Stojic D et al (2003) Hydrogen generation from water electrolysis—possibilities of energy saving. J Power Sources 118:315–319

3. Nagai N et al (2003) Existence of optimum space between electrodes on hydrogen production by water electrolysis. Int J Hydrogen Energy 28:35–41
4. Ulleberg O (2003) Modelling of advanced alkaline electrolyzers: a system simulation approach, Int J Hydrogen Energy 28:21–33
5. Marcelo D et al (2008) Economical electrolyser solution. Int J Hydrogen Energy 33:3041–3044
6. Viswanath RP (2004) A patent for generation of electrolytic hydrogen by a cost effective and cheaper route. Int J Hydrogen Energy 29:1191–1194
7. Saravanan N, Nagarajan G, Narayanasamy S (2008) An experimental investigation on DI diesel engine with hydrogen fuel 16 March 2007. Renewable Energy 33:415–421
8. Saravanan N, Nagarajan G, Kalaiselvan KM, Dhanasekaran C (2008) An experimental investigation on hydrogen as a dual fuel for diesel engine system with exhaust gas recirculation technique. Renewable Energy 33:422–427
9. Abdul-Resul M, Al-Baghdadi S The safe operation zone of the spark ignition engine working with dual renewable supplemented fuels (hydrogen + ethyl alcohol)
10. Rosen MA (1995) Energy and energy analyses of electrolytic hydrogen production. Int J Hydrogen Energy 20(7):547–553

Design and Fabrication of Air Preheater for Diesel Engine

Vikas Sharma, J. M. Babu, R. Naresh, S. Gowthaman and R. Mariappan

Abstract In a diesel engine, the fuel-air mixture is introduced at room temperature and burned inside the cylinder. After the fuel is burned, the combustion products are discharged at atmospheric pressure and high temperature. The heat generated during the combustion of fuel is partially converted into work to drive the car (or any load) and remaining is wasted to the atmosphere through exhaust gas and coolant. The heat generated by combustion of fuel is converted into the work because of the pressure created by the combustion process. The efficiency of the engines is between 20 and 30 %. This paper highlights provision for incorporating a heating vessel for preheating air before admitting into the cylinder of a diesel engine. The heating vessel comprises heating elements for heating the air flowing past said heating element, and a body for mechanically holding and electrically contacting said heating element. The effect of preheated air on standard diesel fuel engine indicated a good result on emission control. NOx and CO emissions at intake air temperature of 55 °C were less when compared at intake air temperature of 32 °C. Higher inlet air temperature causes lower ignition delay, which is responsible for lower NOx formation. Uniform or better combustion is occurred due to pre-heating of inlet air, which also causes lower engine noise.

V. Sharma (✉)
Department of Mechanical Engineering, Gojan School of Business and Technology, Chennai, India
e-mail: vikas6844@gmail.com

J. M. Babu · S. Gowthaman · R. Mariappan
Department of Mechanical Engineering, Vel Tech Dr. RR and Dr. SR Technical University, Avadi, Chennai, India

R. Naresh
Department of Mechanical Engineering, Gitam University, Hyderabad, Andhra Pradesh, India

R. P. Bajpai et al. (eds.), *Innovative Design, Analysis and Development Practices in Aerospace and Automotive Engineering*, Lecture Notes in Mechanical Engineering, DOI: 10.1007/978-81-322-1871-5_32, © Springer India 2014

Easy vaporization and better mixing of air and fuel occur due to warm up of inlet air, which causes lower CO emission.

Keywords Air pre-heater · Diesel engine · Performance

1 Introduction

Diesel engine efficiency depends on multiple complex parameters like heat losses during cooling of engine, heat losses in exhaust gases, friction loss, transmission efficiency losses, etc. However, intake air temperature plays a predominant role in achieving better efficiency. Air Intake preheaters (also known as grid and manifold heaters) are used in diesel applications needing quick, reliable, and environmentally friendly starts. Air Intake heaters are installed in the intake manifold and preheat the combustion air to the required temperature for ignition of fuel. Powered by the vehicle battery, air intake heaters provide an on-board, unplugged, cold weather starting aid. This type of engine preheating is cost-effective and good for the environment. In addition to producing fast starts, air intake heaters reduce white smoke ("cold emissions"), engine wear, battery consumption, and fuel consumption during start up. Lower temperature intake air leads to inadequate final compression temperature, increase in emission delay, longer time between the injection of the fuel to ignition, local over-enrichment, incomplete combustion, and high pressure gradients due to abrupt mixture conversion in the cylinder. These factors lead to knocking of the diesel engine, increase in emission of hydrocarbons in the exhaust gas leading to severe loading of the environment. In order to avoid these, a prefilter is introduced to heat intake air during start and/or warm running [1] of the engine with the help of a electrical heating element. Heating period is selected on the basis of the temperature of the unheated air taken in by the internal combustion engine. Temperature controller is provided to control the outlet temperature at the down stream of heater. Based on air outlet temperature, the relay supplies current to the heater to attain desired temperature [2].

2 Overview

This project relates to incorporating of a heating vessel for preheating air in an intake line leading to an internal combustion engine. Air is drawn from the upstream side of an intake manifold using an air supply unit. The air is heated using an intake heater and is supplied (injected) from the fuel injection side of a fuel injection valve [2]. In this way, vaporization of the fuel is enhanced. One feature of the present disclosure provides an intake air heating apparatus having improved cold startability. By heating air in an intake passage and a circulation passage to a sufficiently high temperature, high-temperature air flows into a combustion chamber during cranking and, therefore, vaporization of fuel is promoted. A warm air intake (WAI) is better than a cold air intake (CAI) on a diesel engine. A WAI improves fuel economy on a gas engine for three reasons:

(1) The warm air is less dense, so the throttle opens up more to get the same air, therefore throttling losses are reduced.
(2) The warm air improves the flame speed
(3) The warm air improves the vaporization of the fuel.

In cold ambient air temperature environments [3], especially below zero degrees Fahrenheit, or when a cold engine is started, the coolant rarely becomes hot enough to boil. Thus, the coolant does not need to flow through the radiator. Nor is it desirable to dissipate the heat energy in the coolant in such environments since internal combustion engines operate most efficiently and pollute the least when they are running relatively hot. A cold running engine will have significantly greater sliding friction between the pistons and respective cylinder walls than a hot running engine because oil viscosity decreases with temperature. A cold running engine will also have less complete combustion in the engine combustion chamber and will build up sludge more rapidly than a hot running engine. In an attempt to increase the combustion when the engine is cold, a richer fuel is provided. All of these factors lower fuel economy and increase levels of hydrocarbon exhaust emissions.

In a typical internal combustion engine, it is ideal to heat the air entering the intake manifold to about 120 °C Fahrenheit. Heating the intake air to temperatures higher than about 130 °C Fahrenheit reduces combustion efficiency. This is because that air expands as it is heated [1]. Consequently, as the air volume expands, the number of oxygen molecules per unit volume decreases. Since combustion requires oxygen, reducing the amount of oxygen molecules in a given volume decreases combustion efficiency. Although preheating intake combustion air is not beneficial in all environments, preheating the air in relatively cold ambient temperature environments (e.g., below 20 °F.) provides many benefits, including improved fuel economy, reduced emissions and the creation of a supercharging effect.

A cold internal combustion engine emits exhaust gases containing a large amount of noxious ingredients such as hydrocarbons (HC's) and carbon monoxide (CO), which contaminate the atmosphere until the engine has warmed up. Accordingly, reduction in the period required of the engine to warm up results in reduction or elimination of the total amount of air pollutants in engine exhaust gases.

It is, therefore, to provide a warming-up system for use in an internal combustion engine which is adapted to rapidly warm up a cold engine.

It is necessary to provide a warming-up system for use in an internal combustion engine which is adapted to maximize combustion efficiency of a combustible mixture during the engine warming up period.

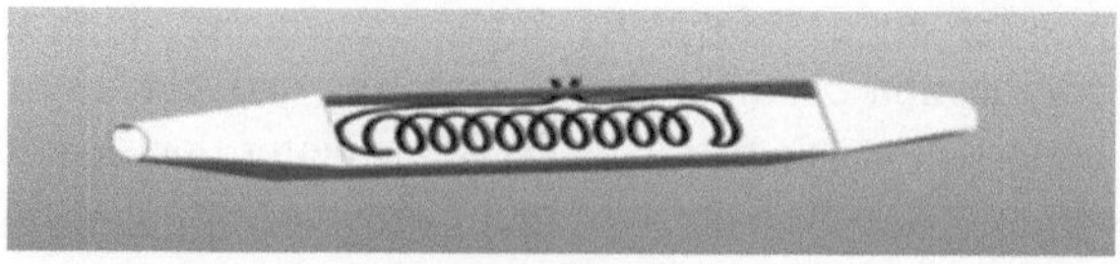

Fig. 1 Heater coil

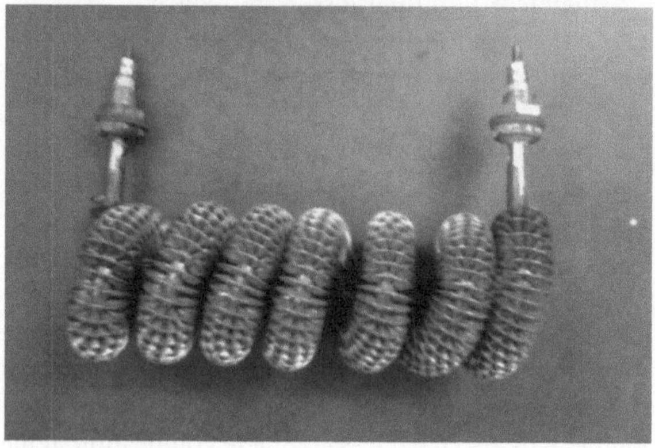

Fig. 2 Heater vessel

3 Design and Fabrication of Preheater

The heating vessel for the preheater is fabricated using 2 mm mild steel plate. Six plates of required size was cut and fabricated as a cuboid by welding its edges. Proper type welding rod was used for welding. Healthiness of welding was ensured by dry penetrate test. Two nozzles were incorporated for facilitating air to flow inside and exit out. A filter has been provided at the inlet of heater for removing the dirt present in the air. Thermocouples have been incorporated at the inlet and outlet lines of the heater. In addition, one more thermocouple is fixed to measure the exhaust gas temperature. The heater is installed in the heating vessel and electrically isolated from the heating vessel. The outlet of heater is connected to the engine (Figs. 1, 2).

A temperature controller is used to control the out let temperature of the heater.

The signal from the heater out let temperature is connected to this controller. Based on the set point, heater is switched on and after attaining the required temperature, power is cut off by the temperature controller.

3.1 Experimental Procedure

After connecting the test set up, diesel engine was run without switching on the electrical heater and parameters were logged. First, engine ran and datas were collected without switching on the heater when the ambient temperature is 35 °C.

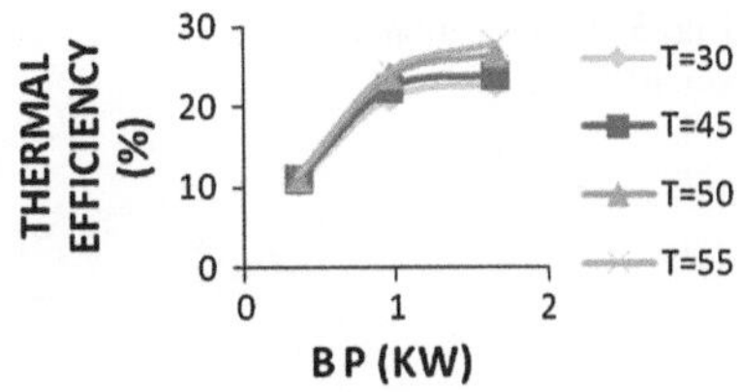

Fig. 3 Thermal efficiency versus Brake horse power

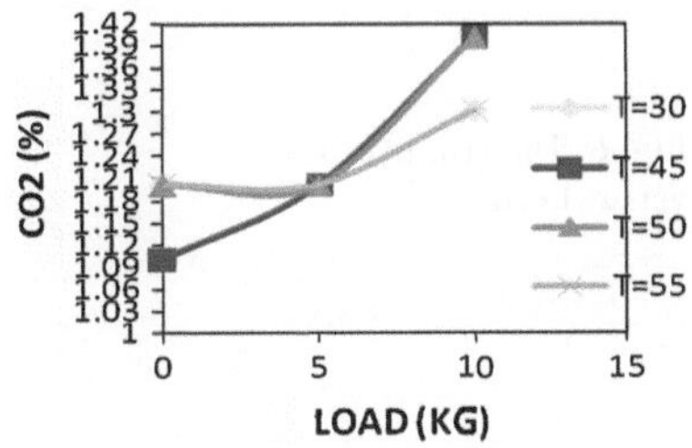

Fig. 4 Carbon dioxide versus Load

Then heater was switched on and heater outlet temperature was maintained at 45, 50, and 55 °C at two different loads viz. 53.955 N and 93.195. At each stabilization step, parameters CO, UBHC, CO_2, O_2 and NOx were logged. In each load, the heat input was calculated based on the fuel consumption. Using these data, brake thermal efficiency was calculated and plotted as shown.

4 Test Results

The heat input required for the engine reduces with increase in intake air temperature. This was observed in both loads. Since heat input reduces with increase in intake air temperature for the same load, brake thermal efficiency increases. The graph shows that increase in thermal efficiency with respect to brake power at different temperatures (Fig. 3).

CO content in the exhaust gas slightly reduces with increase in intake air temperature. The graph shows that increase in CO_2 with respect to brake power at different temperatures, CO content in the exhaust gas slightly reduces with increase in intake air temperature. The graph shows that decrease in CO_2 with respect to brake power at different temperatures (Figs. 4, 5).

Un burnt hydro carbon (UBHC) reduces with temperature. The graph shows increase UBHC in with respect to break power at different temperatures (Fig. 6).

NOx content in the exhaust gas slightly increases with increase in intake air temperature. The graph shows that increase in NOx with respect to brake power at different temperatures (Fig. 7).

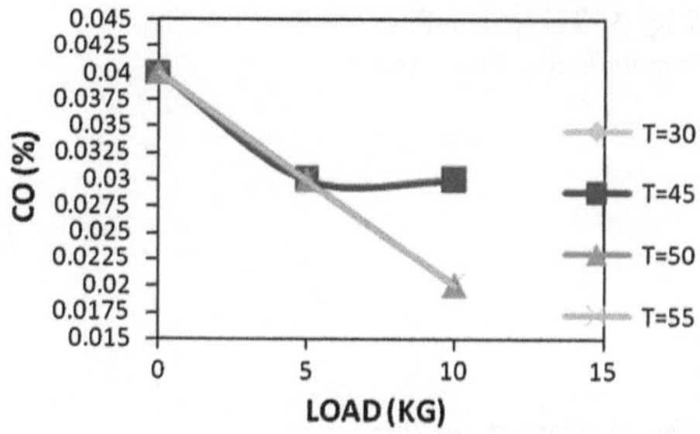

Fig. 5 Carbon monoxide versus Load

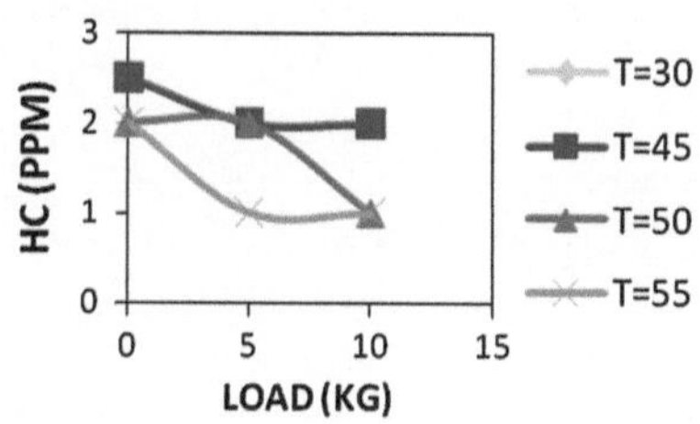

Fig. 6 Unburnt hydrocarbon versus Load

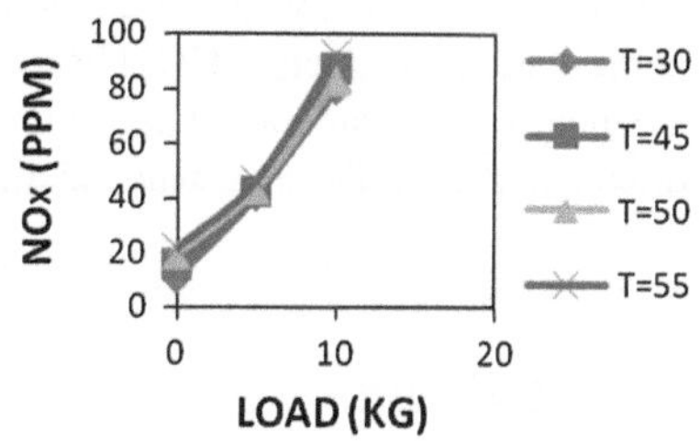

Fig. 7 NOx versus Load

5 Discussion

The following observations were made from the test.

- The heat input required for the engine reduces with increase in intake air temperature. This was observed in both loads.
- Since heat input reduces with in intake air temperature for the same load, brake thermal efficiency increases.
- CO content in the exhaust gas slightly reduces with increase in intake air temperature.
- CO_2 and O_2 content in the exhaust gas remains unaltered in the exhaust with increase in intake air temperature.
- Where as NOx content in the exhaust gas slightly increases with increase in intake air temperature.

From the test it is clear that the fuel consumption reduces and brake thermal efficiency increase with increase intake air temperature. The CO_2 and O_2 content remains unaltered.

6 Conclusion

From the test it is clear that the fuel consumption reduces and brake thermal efficiency increases with increase in intake air temperature. The carbon dioxide and oxygen content remain unaltered in the exhaust gas. Carbon monoxide content in the exhaust gas slightly reduces with increase in intake air temperature. Whereas NOx content in the exhaust gas slightly increases with increase in intake air temperature. Hence, the test results indicated that the advantages gained are more with increase in intake air temperature though there are few disadvantages.

References

1. Tamilvendhan D (2011) A performance, emission and combustion investigation on hot air assisted eucalyptus oil direct injected compression ignition engine. Mod Appl Sci 5(4) 243–248
2. Shahadat MMZ, Nabi MN, Shamim Akhter M (2005) Diesel NOx reduction by preheating inlet air. In: Proceedings of the international
3. Yilmaz N (2012) Effects of intake air preheat and fuel blend ratio on a diesel engine operating. Fuel 94:444–447

Studies on Tropical Conditions in Hot Sunny Days in India and Their Effects on Interior Temperature Rise in Truck Cabin

Suraj Mani Chaurasiya, Swanand Kirpekar and Avinash R. Arankalle

Abstract Most of the tropical countries experience heat suffocations inside the commercial vehicles when soaked in sunny hot days. The solar heat radiation pass inside the vehicle cabin by convection, conduction, and radiation through roof, side panels, glass windows and windshield when all windows are kept closed. Even the heat dissipation to inside of cabin takes place from the engine compartment. The solar radiations are composed of IR, UV and visible rays which generate heat and is absorbed by the interiors to give rise of truck cabin temperature to as high as to 70–80 °C as stated in the review articles. This heat suffocation is injurious to human health and in particular to growing kids. There are many reported cases of children and pets dying due to suffocation of heat. In addition it also deteriorates durable life of interior components like seats, upholstery and dashboard. In India, the similar study, as done elsewhere in the world is required to understand the effects of hot weather conditions in different parts of the country and thus design for on i-cool system that is affordable for a common man to control interior cabin temperatures, other than use of air conditioning system. With this background to design an innovative i-cool system, the project was carried out to study temperature and humidity conditions in major metropolitan and hottest 15 cities in India to determine maximum temperatures during period of March to June interval. The mathematical modeling was used to determine peak temperature and total heat content that gets accumulated during 3 h from 10 am to

S. M. Chaurasiya (✉) · A. R. Arankalle
Vel Tech Dr. RR and Dr. SR Technical University, Avadi, Chennai
e-mail: chaurasiya.suraj@gmail.com

A. R. Arankalle
ARAI-FID, New Ajantha Avenue, Building 3C, Wing A2, Flat No. 24, Paud Road, Kothrud, Pune 411038, India
e-mail: avinash.arankalle@gmail.com

S. Kirpekar
Padmabhooshan Vasantdada Patil Institute of Technology, Budhgaon, India
e-mail: swanand21@yahoo.com

R. P. Bajpai et al. (eds.), *Innovative Design, Analysis and Development Practices in Aerospace and Automotive Engineering*, Lecture Notes in Mechanical Engineering, DOI: 10.1007/978-81-322-1871-5_33, © Springer India 2014

4 pm at intervals of 10, 20, 30, 60, 90, 120, 180 min of soaking of vehicle on the road respectively. CFD FLUENT software package was further used to determine temperature distribution inside the cabin at various locations such as dashboard, seats, roof and Windshield for soaking of vehicle for the duration of 30 min at peak temperature for the selected Metropolitan city, viz. Chennai because of the reasons that it is the second largest hottest and humid metropolitan city in India. Moreover, after Rajasthan and Gujarat, Tamil Nadu receives more solar radiations and also truck population is one of the highest in the country. The mathematical analysis was conducted and it show that the interiors of truck cabin reaches as high as to 56–78°C within 30 min of soaking of vehicle under the solar radiations during the peak summer days from ambient temperature of 40 and 47 °C respectively. The results are alarming and to design of system to control interior temperature to comfort zone, various i-Cool materials like—Phase Change Materials (PCM), Thermoelectric Materials (TEM), and Organic Dye Cell Materials (OCM) are required to be studied as a future work for their critical properties, availability and cost along with an innovative integration system.

Keywords Tropical countries · I-cool system · Interior temperature distribution · Ambient temperature · Phase Change Materials (PCM) · Thermoelectric Materials (TEM) · And Organic Dye Cell Materials (OCM)

1 Solar Heat and Its Effect on Truck cabin Temperature

Sun is the abundant source of energy and existence of Mankind. When we measure the midday intensity of sunlight at the Earth's surface, we find that about 136.7 mW falls on every square centimeter. We call this number "The Solar Constant" and designate it by the Greek letter sigma (σ). Thus, 1 A.U.: $\sigma = 136.7$ mW/cm^2. The spectrum of solar radiance consists of 5 % UV, 46 % Visible and 49 % Infra-red. These electromagnetic waves from the sun enter the vehicle through the glass windows, metallic body structure and are then absorbed by the seats and other interior plastic components as shown in Fig. 1. These components, then, re-emit electromagnetic radiation of a different wavelength that gets trapped inside the car, causing the temperature to rise. Car becomes an Oven Trap in an hour when soaked in hot and sunny day leading to—risk of heat cramps, Exhaustion, and heat stroke, in addition to damage of interior components and upholstery. This is particularly true in most of tropical countries like India creating discomforts, suffocations to the vehicle passengers, till the sufficient ventilation and or air conditioning system is energized. There are many reported accidental deaths of infants /pets after being left unattended in vehicles and is alarmingly as high as to 34 % fatality [1, 2].

To study tropical conditions in India and their effect on truck interior temperatures, it is essential to study both weather and climate which are characterized by the certain variables known as climatic factors. They are—Solar radiation,

Fig. 1 Solar rays and spectrum of radiance, heat dissipation inside Truck Cabin

ambient temperature, air humidity, precipitation, wind and sky conditions. For a good evaluation of thermal environment within truck driver cabin, the knowledge of these factors are essential in mathematical modeling and CFD analysis to design of i-Cool system. The most important part is solar radiation and it comes in many forms, such as visible light, radio waves, heat (infrared), x-rays, and ultraviolet rays. Earth receives only one-half of one-billionth of the Sun's total energy output through different layers like troposphere, stratosphere, mesosphere, thermosphere, and exosphere. The temperature in troposphere 16 km above the earth is important in design of i-cool system for truck cabin because of the temperature in it is high and keep earth warmest near the ground level.

The solar radiation incident from sun travels through all these layers. Only 55 % of the solar radiation reaches surface in that only 51 % of solar radiation absorbed by surface and 4 % gets reflected from surface. The 19 % of solar radiation absorbed by clouds and atmosphere, 20 % is reflected from clouds and only 6 % reflected from atmosphere. All high energy X-rays are absorbed by the atmosphere. These rays not come to ground level. Likewise most of the UV rays are blocked by thermosphere, mesosphere, and stratosphere. The relatively low energy, long wavelength portion of UV spectrum reaches the ground with some of the shorter wavelength IR frequencies.

The solar constant, a measure of flux density, is the amount of incoming solar electromagnetic radiation per unit area that would be incident on a plane perpendicular to the rays, at a distance of one astronomical unit (AU). The "solar constant" includes all types of solar radiation, not just the visible light. Its average value is approximately 1.366 kW/m^2. The radiation incident on a surface varies from moment to moment depending on its geographic location (latitude and longitude of the place), orientation, and season, time of day and atmospheric conditions. Solar radiation is the most important weather variable that determines whether a place experiences high temperatures or is predominantly cold. Ignoring clouds, the daily average irradiance for the Earth is approximately 250 W/m^2 (i.e., a daily irradiation of 6 kWh/m^2), taking into account the lower radiation intensity in early morning and evening, and its near-absence at night.

Other important factor that will govern design of i-Cool system is Sky condition which generally refers to the extent of cloud cover in the sky or the duration of sunshine. Under clear sky conditions, the intensity of solar radiation increases; whereas it reduces in monsoon due to cloud cover.

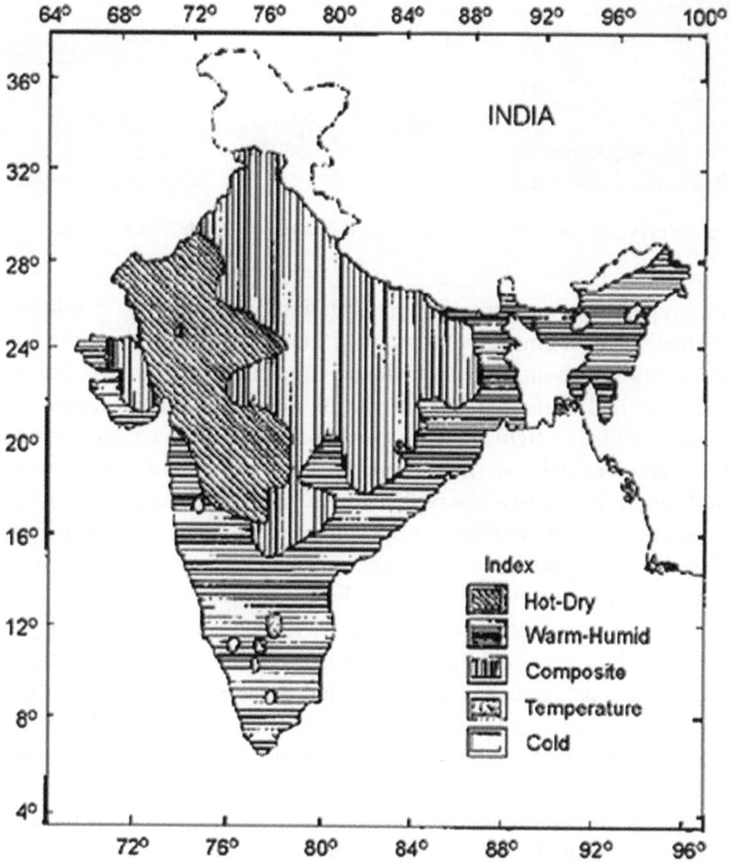

Fig. 2 Five major climatic zones in India

2 Climatic Zones in India

According to recent code of Bureau of Indian Standards, the climatic conditions in India are divided into five zones as shown in Fig. 2 viz. Hot-Dry, Warm-Humid, Composite, Temperate and Cold. Majority of Indian map falls in first four zones which play the important role in design of i-Cool materials for truck cabin interior. In the project work major 15 hottest cities in these four zones were considered to study interior temperature and heat contents in the truck cabin [3]. The data of temperatures in these cities at the interval of 10 am to 2 pm was collected for month of March to June for 2007–2012, and average max temperature and Relative Humidity (RH) were identified per year as summarized in Table 1 [4]. These values are further used for the calculations of Zenith (altitude) angle, azimuth angle, and Sky temperature as summarized in Table 2.

Table 1 Summary of max temperatures and RH in major 15 hot cities

City	2007	2008	2009	2010	2011	2012	Average Temperature	Average RH	Heat index
New Delhi	46	46	47	46	47	45	46	54	82.7
Calcutta	40	40	41	42	41	40	41	71	78
Chennai	41	41	42	40	42	41	41	70	77
Mumbai	39	40	39	39	41	39	40	75	77
Ahmedabad	42	42	42	40	41	41	41	55	62.3
Hyderabad	42	42	42	42	42	42	42	56	67.2
Pune	39	39	40	39	40	41	40	59	61.8
Jaipur	43	42	42	43	43	43	43	64	80.4
Akola	45	45	46	44	45	45	45	49	72.2
Hisar	44	44	45	44	46	44	45	60	85.6
Jabalpur	46	47	45	46	45	45	46	76	116
Kota	40	40	41	40	40	41	40	44	50.7
Bhuj	41	40	42	41	40	41	41	63	69.7
Solapur	40	40	40	42	45	44	42	49	60.8
Nagpur	47	47	47	46	46	47	47	53	86.1

Table 2 Solar Zenith and Azimuth angle and sky temperatures for selected hot cities

City	Solar zenith angle	Solar azimuth angle	Dew point temperature	Sky temperature
New Delhi	75.3	172.7	36	44.73
Calcutta	77.8	228.4	34	39.73
Chennai	88.4	311.3	37	41.82
Mumbai	82.1	128.5	33	40.72
Ahmedabad	79.1	144.4	36	45.68
Hyderabad	86.7	169.6	34	43.67
Pune	83.3	128.9	34	40.75
Jaipur	77.1	165	38	44.77
Akola	83.4	164.2	37	45.71
Hisar	74.6	167.3	39	45.78
Jabalpur	80.9	184.1	39	44.81
Kota	78.5	163.3	35	46.61
Bhuj	77.2	133.8	38	44.77
Solapur	85.3	135.5	34	43.67
Nagpur	83.0	181.1	38	45.75

Further in the mathematical modeling and CFD analysis, heat transfer mechanism and equations for conduction, convention and radiation through vehicle body and window glasses is important to determine interior truck cabin temperature, distribution and heat. The calculations demonstrate rise in truck cabin temperatures for all these selected cities that develops after soaking a vehicle in sunny days for 10, 20, 30, 60, 90, 120 and 180 min in Table 3.

Table 3 Truck cabin temperature after soaking up to 180 min in hot sunny days

City	Average Temperature (°C)	10 min	20 min	30 min	60 min	90 min	120 min	180 min
New delhi	46	62	70	75	78	80	81	82
Calcutta	41	50	56	60	62	63	64	65
Chennai	41	52	58	62	64	65	66	67
Mumbai	40	48	53	56	58	59	60	61
Ahmedabad	41	54	61	65	67	68	69	70
Hyderabad	42	54	61	65	67	68	69	70
Pune	40	48	52	56	58	59	60	61
Jaipur	43	56	63	67	69	70	71	72
Akola	45	60	68	73	76	78	79	80
Hisar	45	58	60	71	74	76	77	78
Jablapur	46	62	70	75	78	80	81	82
Kota	40	50	56	60	62	63	64	65
Bhuj	41	52	58	62	64	65	66	67
Solapur	42	50	56	60	62	63	64	65
Nagpur	47	64	73	78	81	83	84	85

3 CFD Analysis

In CFD analysis for distribution of heat inside truck cabin is conducted for Chennai city is based on the value of solar irradiation and also due to the fact that it is one of the hottest and relatively humid city in India, highly populated with light to commercial trucks in the country. The CFD analysis was conducted at critical locations inside the HCV at—Dashboard, Seat, Roof and Windshield for distribution of heat in first 30 min of soaking time and solar rays penetrating inside through door panel, front and back panel, roof panel, floor panel, windshield, window glasses, seat, dashboard surfaces. The CFD analysis is shown in Fig. 3.

The outcome of CFD results are as follows:

- The maximum temperature is found on the dashboard as well as on the seat area.
- The glasses are contributing maximum to the interior temperature rise.
- The mid plane vector view shows the direction of air flow after temperature rise is in the upward direction.
- The concentration of more hot air is found to be in upper zone.

4 Conclusions and Future Work

- Indian tropical conditions indicate that majority of states and cities fall in Hot-Dry, Warm-Humid, Composite and Temperate zones as defined by Bureau of Indian standards code.
- The study of Indian tropical conditions were conducted by narrowing down cities from 93 to 15 Nos. divided in four zones and defined by BIS code.

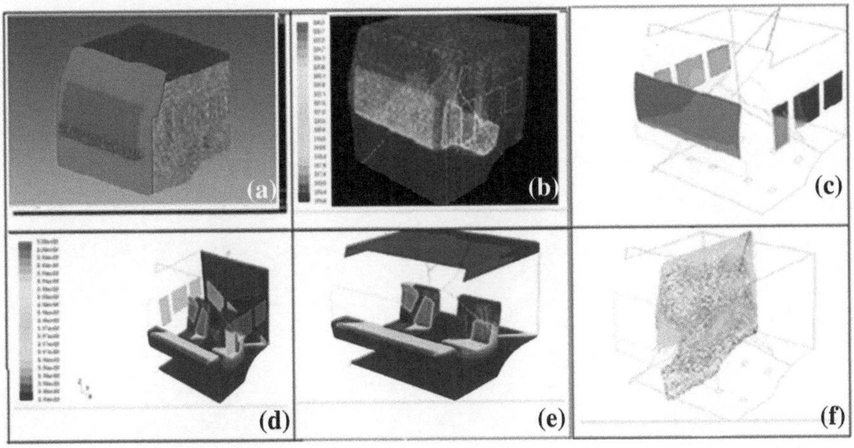

Fig. 3 CFD analysis for truck cabin

- It is seen that in hot summer days the interior of truck cabin temperature rises to minimum 61 °C and maximum 85 °C when soaked in hot sunny days for 180 min when the ambient temperature is 40 and 47 °C respectively.
- CFD analysis indicate that maximum heat rise is on dashboard and more heat is absorbed by seats and re-emitted back in the truck cabin. More heat is radiated inside the cabin by the windshield and window glasses. Due to air flow heat moves up the cabinet.
- The rise of temperature inside the cabin is alarmingly high and more rise is in the first 20–30 min of soaking of vehicle in the sunny days.
- A future work is proposed to study various i-Cool materials like—Phase Change Materials (PCM), Thermoelectric Materials (TEM), and Organic Dye Cell Materials (OCM) to design truck cabin and an integrated system that shall take care to keep interior temperature to comfort zone when vehicle is parked in sunny days and is stand-alone in off mode. This will help to reduce heat suffocations, dis-comforts of drivers and reduce use of air-conditioning systems that average truck owner cannot afford.

References

1. Saidur R, Masjuki HH, Hasanuzzaman M (2009) Performance of an improved solar car ventilators. Int J Mech Mater Eng 4(1):24–34 (Department of Mechanical Engineering University of Malaya, 50603 Kuala Lumpur, Malaysia)
2. S. K. U. In and around vehicle safety, 8 12 2011. Available at http://www.safekids.org/assets/docs/ourwork/research/in-and-around-cars.pdf
3. Bansal NK, Minke G (1988) Climatic zones and rural housing in India. Kernforschungsanlage, Juelich
4. Indian Meteorology Department, Government of India. www.imd.gov.in

Fabrication of Bamboo Fiber Reinforced Polylactide Biocomposites

Wen-Bin Young, Bo-Yuan Tian and Yu-Chi Tsau

Abstract Bamboo fiber (BF) is a good candidate as the reinforcement for green composites. The applications of bamboo fibers as fabric or structural reinforcements are being widely investigated. This study went through the process from the bamboo strips to the fibers and compounded with the PLA to form the composite pallets for the following injection molding process. The alkali-treated BF at 100 °C for 12 h could properly delignified the bamboo fiber strips, resulting in the required bamboo fiber bundles. After compounding with PLA using a banbury mixer, the resulting composite after injection molded composites can improve the mechanical properties including the tensile strength, tensile modulus, flexural strength, flexural modulus, and impact strength. The reinforced effect of the BF on PLA was quite apparent in our experiments. A simple rule of mixture was used to approximate the tensile modulus and strength of the PLA/BF composite.

Keywords Polylactide · Bamboo fiber · Biocomposite · Injection molding

1 Introduction

Bamboo fiber (BF) is a good candidate as the reinforcement for green composites. The applications of bamboo fibers as fabric or structural reinforcements are being widely investigated [1–5]. The bamboo culm consists of vascular bundles with the fiber strands and sheaths that confines the conducting elements. An ultra hierarchical structure of BF wall model has been proposed by Parameswaran and Liese [6]. The lamellation consisted of alternating layers of broad and narrow lamellae with the longitudinal and transverse microfibril angles, respectively.

W.-B. Young (✉) · B.-Y. Tian · Y.-C. Tsau
Department of Aeronautics and Astronautics, National Cheng-Kung University, 70101 Tainan, Taiwan, ROC
e-mail: youngwb@mail.ncku.edu.tw

R. P. Bajpai et al. (eds.), *Innovative Design, Analysis and Development Practices in Aerospace and Automotive Engineering*, Lecture Notes in Mechanical Engineering, DOI: 10.1007/978-81-322-1871-5_34, © Springer India 2014

Bamboo fiber can be processed into different form for applications [7]. The node and epidermis portions of the bamboo are often removed with the remaining culm used for the following extracting process. The bamboo strips, by cutting the bamboo culm, can be directly used as reinforcements and combined with thermosetting polymer to form composites. BF can be extracted from the bamboo strips by either the chemical or mechanical methods [7–9]. Both the raw and delignified BF could be used.

The PLA-based green composites have been studied extensively recently. Different natural fibers reinforced PLA reports were reviewed by Bajpai et al. [10]. Different processes were used to fabricate the green composites as hot pressing, injection molding, compression molding, prepreg, film stacking, etc. Tokoro et al. studied the mechanical properties of PLA/BF green composites for short and medium length BF [9]. A twin screw extruder was used to compound the PLA/BF pallet followed by the injection molding process. For the medium length, a hot pressing method was applied to form the composite in order to preserve the fiber length of about 3–6 mm. Large improvement for the impact strength was observed for the PLA and medium length BF composite. Improving toughness of PLA/BF composite was also reported by Okubo et al. by employing the microfibrillated cellulose (MFC) technology [11]. The MFC was introduced to the polymer matrix as a second reinforcement phase to suppress the crack propagation. Gonzalez et al. [12] used auto hydrolysis process to extract the BF and form the PLA/BF composites. Alkali pretreatment and coupling agent were applied to the BF before forming of the PLA/BF composite to enhance the thermal and mechanical properties [13]. The PLA composite containing delignified BF by alkali treatment showed higher tensile and flexural modulus. Kumar et al. reported that the nanoclay could be introduced to the PLA/BF composite to improve the dielectric and mechanical properties [14].

In this study, the alkali treatment was used to delignify the bamboo fiber. Instead of the twin screw extruder, a banbury mixer was used to compound the PLA/BF pallet to reduce the breakage of the BF during the process. A longer BF can be derived in the compounded pallet. The mechanical properties of the PLA/BF composite were studied after an injection molding process.

2 Experimental

2.1 Materials

Polylactic acid, one of the green plastic materials, was thermoplastic aliphatic polyester made from renewable resources, such as corn starch, tapioca roots, or sugarcane. The injection molding grade PLA polymer (Ingeo^{TM} Biopolymer 3001D from Nature Works LLC) was used in this study.

The thorny bamboo used in this study was first cut into short strips in a length about 5–7 cm and thickness 0.5–1 mm. The bamboo strips were treated in a sodium

Fig. 1 Microstructure of bamboo fibers treated in the sodium hydroxide (NaOH) solution of **a** 0.1 N, **b** 0.2 N, **c** 0.3 N for 12 h at the temperature of 100 °C

hydroxide solution (0.3 N) under a specified temperature. In order to find a proper concentration of sodium hydroxide solution and suitable treating time, the bamboo strips were put in the solution of 0.1, 0.2, and 0.3 N up to 12 h at the temperature of 100 °C. The alkali-treated bamboo strips were washed in water, and rubbed and split with hand to get the BF bundles. The BF bundles were dried before compounding with PLA. Figure 1 shows the microstructure of the resulting BF after alkali treated and drying for different concentrations of sodium hydroxide solution. Higher concentration of alkali solutions were also tried in our experiments, but the resulting bamboo fibers tended to break easily. The BF used in the following study was treated in a 0.3 N alkali solutions for 12 h at a temperature of 100 °C.

2.2 Compounding and Fabrication

PLA/BF composite pellets were compounded by using a banbury mixer (YKI-3, Goldspring Enterprise, Inc., Taiwan). Three different composite pellets with the BF contents of 10, 20, and 30 % by weight, were fabricated. The composite from the banbury mixer was a chunk block that was granulated to the composite pellets by a plastic shredder machine. In this way, the BF was avoided to pass through the spider of the extruder and the breakage of the fiber could be reduced.

An injection machine (*KM50-160C2*, KRAUSS MAFFEI) was used to fabricate the PLA/BF composite. Test specimens were molded at a melt temperature of 180 °C under a package pressure of 30 bar for 20 s. The mold temperature was 30 °C and the injection ram speed was 50 mm/s with a screw diameter of 25 mm.

2.3 Mechanical Characterizations

A tensile testing machine (CY-6102, Chun Yen Testing Machines Co., Taiwan) was used to perform the tensile and flexural tests of the PLA/BF composites. Dog bone-shaped specimens of the composites were fabricated by the injection molding process. Three specimens of each sample were tested to failure under a tension force, and the stress and strain data were recorded to derive the corresponding strength and modulus.

Three-point bending test was performed to measure the flexural properties of the composites. The dimension of the bend test specimen was 126×12.6 mm^2 with a thickness of 3.2 mm. The support separation for the three-point bending was 50 mm. For the notched Izod impact test, the samples had a dimension of 63×12.6 mm^2 with a thickness of 3.2 mm. The impact test was conducted using a pendulum weight of 0.82 Kg. All the specimens for testing were directly molded by the injection molding process without any other machining.

3 Results and Discussion

3.1 Tensile Test

Figure 2 shows the tensile strength and elongation at breakage from the tensile test for the PLA/BF composites with different BF contents. With a low BF content of 10 %, the tensile strength remains the same as the PLA polymer. As the fiber content increases up to 30 %, the tensile strength increases correspondingly with about 12 % improvement. The BF reinforced PLA composite can increase the tensile strength while the ductility is reduced. The elongation at breakage shown in Fig. 2b indicates the decrease with the increase of BF content. Brittle fracture of the PLA/BF composite was observed in the tensile testing, and there was little necking behavior and low elongation at breakage as well.

3.2 Bending Test

The flexural strength and modulus with different BF content are shown in Fig. 3. As compared to the results reported by Tokoro et al. [9], much improvement of the

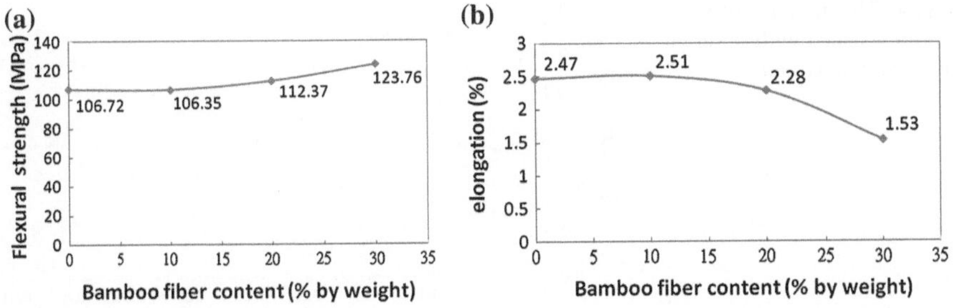

Fig. 2 **a** The tensile strength and **b** elongation at breakage of the PLA/BF composite with different BF contents

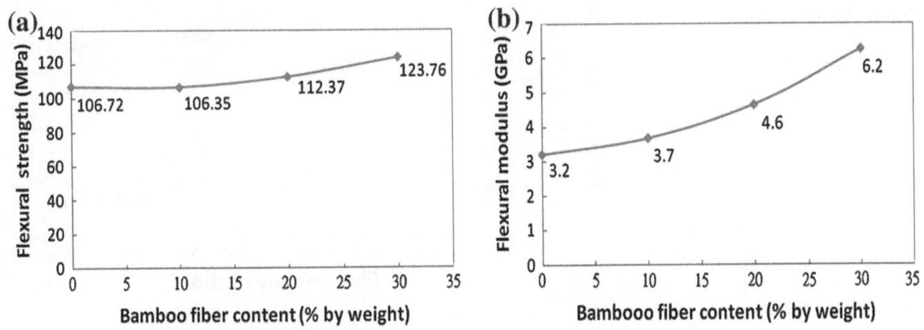

Fig. 3 The bending **a** strength and **b** modulus of the PLA/BF composites with different BF contents

flexural strength (about 15 % for the 30 % BF content) is observed for the alkali-treated PLA/BF composite. The tensile and flexural strength for the PLA/BF composites increase with the BF content that is different from the results reported by Kumar et al. [14] and Kang et al. [13]. Their tests showed that the flexural strength decreased for the PLA/BF composites. On the other hand, the increase of flexural modulus with the adding of BF fiber shows the same improvement.

3.3 Modulus and Strength

The modulus of the PLA/BF composite for different fiber contents is shown in Fig. 4. The fiber volume content was calculated from the weight fraction with the density of the PLA about 1240 Kg/m^3 and that of the BF around 800 Kg/m^3 from literatures [7, 13]. A simple rule of mixture is used to fit the experimental data and the equation is

$$E_c = KE_f V_f + E_m V_m$$

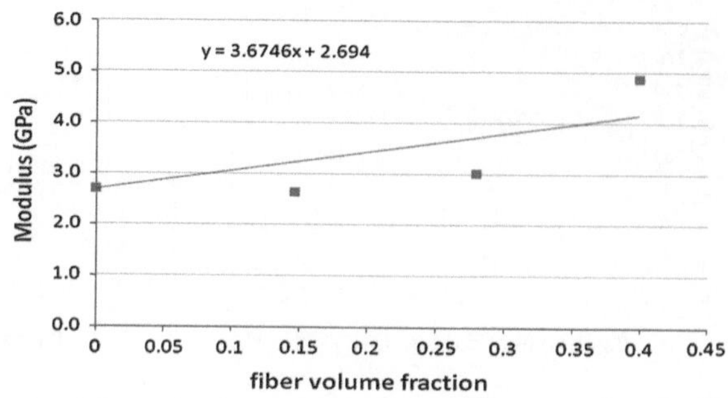

Fig. 4 The modulus of PLA/BF composites with different fiber volume content

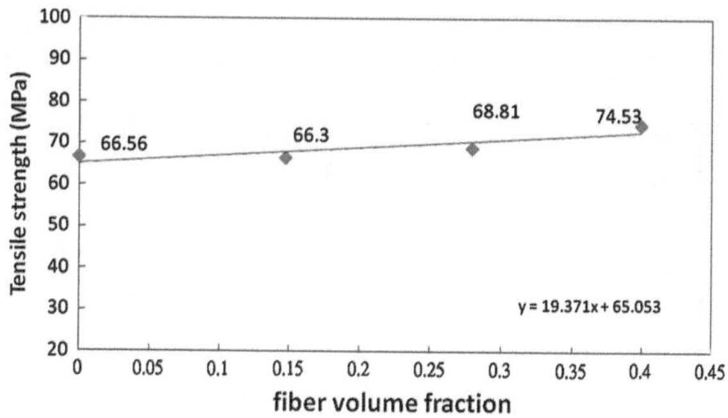

Fig. 5 The strength of PLA/BF composites with different fiber volume content

where E_c, E_f, and E_m are the modulus of composite, fiber, and matrix. V_f and V_m are the fiber volume content and matrix volume content. K is the reinforcement efficiency factor for the random short bamboo fibers. From the curve in Fig. 4, the modulus of the composite can be expressed as:

$$E_c = 6.4V_f + 2.7V_m$$

The resulting modulus of the BF and PLA are about 17 GPa and 2.7 GPa if a factor 0.375 is used for K due to the random orientation of the short BF fibers. The modulus of BF from literature studies is quite scattering because of the fiber shape and dimension. However, the derived modulus of BF in this study is comparable to the reported data from literatures.

The tensile strength can also be approximated by the rule of mixture as:

$$\sigma_c^* = K\sigma_f^* V_f + \sigma_m' V_m$$

where σ_c^* and σ_f^* are the tensile strength of the composite and BF respectively and σ_m' is the stress of the matrix upon failure of the fiber. Form the data in Fig. 5 of tensile strength with respect fiber volume fraction, the calculated strength of the BF is about 225 MPa and σ_m' is equal to 65 MPa with a factor 0.375 for K. This is the average tensile strength of the short BF after the compounding and the following injection molding in this study.

4 Conclusions

Bamboo fiber-reinforced PLA composite is one of the green materials that can reduce the environmental impact. This study went through the process from the bamboo strips to the fibers and compounded with the PLA to form the composite pallets for the following injection molding process. The alkali-treated BF at 100 °C for 12 h could properly delignified the BF strips, resulting in the required BF bundles. As compounding with PLA using a banbury mixer, the resulting composite after injection molding can increase the mechanical properties including the tensile strength, tensile modulus, flexural strength, flexural modulus, and impact strength. A simple rule of mixture was used to approximate the tensile modulus and strength of the PLA/BF composite. The resulting mechanical properties of the PLA/BF composite all had the reinforced effect in this study. This was different from some of the reports in literatures. It was believed that this was caused by the resulting larger fiber length in the composite fabricated in this study.

References

1. Lee SH, Wang SQ (2006) Biodegradable polymers/Bamboo Fiber biocomposite with bio-based coupling agent. Compos Part a: Appl S 37:80–91
2. Takagi H, Ichihara Y (2004) Effect of fiber length on mechanical properties of "green" composites using a starch-based resin and short bamboo fibers. Jsme Int J a: Solid M 47:551–555
3. Defoirdt N, Biswas S, De Vriese L, Tran LQN, Van Acker J, Ahsan Q, Gorbatikh L, Van Vuure A, Verpoest I (2010) Assessment of the tensile properties of coir, bamboo and jute fibre. Compos Part a: Appl S 41:588–595
4. Chen XY, Guo QP, Mi YL (1998) Bamboo fiber-reinforced polypropylene composites: a study of the mechanical properties. J Appl Polym Sci 69:1891–1899
5. Wang YP, Wang G, Cheng HT, Tian GL, Liu Z, Xiao QF, Zhou XQ, Han XJ, Gao XS (2010) Structures of Bamboo Fiber for Textiles. Text Res J 80:334–343
6. Parameswaran N, Liese W (1976) On the Fine Structure of Bamboo Fibres. Wood Sci Technol 10:231–246
7. Liu DG, Song JW, Anderson DP, Chang PR, Hua Y (2012) Bamboo fiber and its reinforced composites: structure and properties. Cellulose 19:1449–1480
8. Ogawa K, Hirogaki T, Aoyama E, Imamura H (2008) Bamboo Fiber Extraction Method Using a Machining Center. J Adv Mech Des Syst 2:550–559
9. Tokoro R, Vu DM, Okubo K, Tanaka T, Fujii T, Fujiura T (2008) How to improve mechanical properties of polylactic acid with bamboo fibers. J Mater Sci 43:775–787
10. Bajpai PK, Singh I, Madaan J (2014) Development and characterization of PLA-based green composites: a review. J Thermoplast Compos 27(1):52–81
11. Okubo K, Fujii T, Thostenson ET (2009) Multi-scale hybrid biocomposite: Processing and mechanical characterization of bamboo fiber reinforced PLA with microfibrillated cellulose. Compos Part a: Appl S 40:469–475

12. Gonzalez D, Santos V, Parajo JC (2011) Manufacture of fibrous reinforcements for biocomposites and hemicellulosic oligomers from bamboo. Chem Eng J 167:278–287
13. Kang JT, Kim SH (2011) Improvement in the Mechanical Properties of Polylactide and Bamboo Fiber Biocomposites by Fiber Surface Modification. Macromol Res 19:789–796
14. Kumar V, Sharma NK, Kumar R (2013) Dielectric, mechanical, and thermal properties of bamboo-polylactic acid bionanocomposites. J Reinf Plast Comp 32:42–51

Emission Control by Selective Exhaust Gas Recirculation Scavenging System in Two-Stroke Engine

V. Ramakrishnan, K. Purushothaman and S. Mohammed Ghouse

Abstract For small two-stroke engines, it is important that the newer concepts/systems should not adversely affect their existing advantages, particularly those of simplicity and low production and maintenance cost. With this in mind, we decided to do our project for improvement of the conventional engine. The project also aims to reduce short circuiting of fresh charge by admitting cooled exhaust gas to pass through reed valves fitted at the upper end of the transfer passage, in a crank case scavenged two-stroke engine. This results in appreciable decrease in specific fuel consumption and in HC/CO emissions. Major works involved in this project are designing and machining two aluminium flanges to accommodate the reed valves, and also providing an arrangement at the upper end of the transfer passage in the engine cylinder to fasten the flange and fabrication of heat exchanger in order to cool the exhaust gas which is at a temperature of about 400 °C to avoid pre ignition inside the engine cylinder. The performance of the engine is tested using eddy current dynamometer.

Keywords SI Engine · Emission · EGR · Reed block transfer duct · Engine speed

---

V. Ramakrishnan (✉)
SP University, Avadi, Chennai 600054, India
e-mail: rukmini_ramki@hotmail.com

K. Purushothaman
SPCET, Avadi, Chennai 600054, India
e-mail: purukrishnan@yahoo.co.in

S. Mohammed Ghouse
SVCET, Tirupachur, Chennai 631203, India
e-mail: ghouse967@gmail.com

R. P. Bajpai et al. (eds.), *Innovative Design, Analysis and Development Practices in Aerospace and Automotive Engineering*, Lecture Notes in Mechanical Engineering,
DOI: 10.1007/978-81-322-1871-5_35,

1 Introduction

Two-stroke spark ignition engines are used mainly in two wheeler vehicles which are meeting the present day individual transportation needs to a great extent [1]. As there is no distinct exhaust stroke in these engines, the responsibility of driving out the combustion product from the cylinder rests with the fresh charge entering into the cylinder.

Two-stroke SI engines in spite of their advantages like high specific output, ease of maintenance suffer from serious drawbacks like high HC/CO emissions and low brake thermal efficiency. It is found that the fuel consumption and HC emissions of two-stroke SI engines are 1.5–2 times of equivalent four stroke SI engines. It is required to meet the upper exhaust norms prescribed by EU norms for two wheelers [2].

In the crank case scavenged two-stroke SI engines, considerable amount of fresh fuel–air mixture gets short circuited through the exhaust port during the scavenging process. Also a lot of exhaust gas gets retained inside the cylinder and dilute the fresh charge. The loss of fresh charge through the exhaust port varies from 25 to 40 % of what comes into the cylinder depending upon the operating conditions and method of scavenging.

The above factors are responsible for high specific fuel consumption and high HC/CO emissions. It is evident that any attempt to flush out the exhaust gas properly without losing a considerable amount of fresh charge will result in an improved performance of the two-stroke SI engine [3].

The intake mixture is generally rich because of the exhaust gas dilution resulting in high CO emissions. The NOX emissions are low because of low combustion peak temperature and fuel-rich mixture.

Experimental results show that it is possible to meet Euro IV emission norms with Pcci-Di combustion concept with economical after treatment solution [4].

The exhaust gas dilution of the fresh fuel and air mixture result in a decreased flame which is reflected as decrease in efficiency due to low rate of combustion in the two-stroke SI engines. This also puts a restriction on the range of air–fuel ratios that can be used and gives the designers a very little opportunity to employ methods which regulate load to a certain extent by various mixture qualities and in effect leaves them with the option of controlling the load mostly by throttling alone.

It is clear that any effort to flush out the exhaust gases properly without loss of any fresh fuel–air mixture will result in an improved performance of the two-stroke engines. In the present work, a new concept of two-stroke gasoline engine by which short circuiting of fresh fuel–air mixture though the exhaust port is considerably reduced, has been developed that design improvements made on an existing two-stroke engine incorporating the concept for trapping more fresh fuel–air mixture inside the cylinder and scavenging more of the exhaust products are described. The experimental results indicating better reduced CO and HC emissions obtained with the new concept are presented and explained. Table 1 gives the specifications of the cub engine used for our experimental purpose.

Table 1 Engine specifications

Engine specifications (bajaj cub 100 cc engine)	
Type	Two stroke
No of cylinders	One
Bore	50 mm
Stroke	50 mm
Displacement	98.17 cm^3
Maximum horse power	4.4 kw @ 620 C rpm
Ignition timing	Spark advance of 22 ± 10 BTDC
Intake system	Rotary valve type
Ignition system	Flywheel magneto system
Spark plug	MICO h2 175 zi
Carburetor	Down drought 17 mm venturi
Clutch	Wet, multi disc type
Transmission	3 speed gear box
Dimension (1*w*w)	1,740 mm × 103 mm × 670 mm
Wheel base	1,200 mm
Dry weight	95 kg
Dynamometer specifications	
A. Electrical	
1. Total alarm inputs	5
	a. High dyn case temp
	b. High dyn excitation current
	c. Low dyn water pressure
	d. Over speed
	e. Spare available or back panel
B. Mechanical	
Overall dimension	482 mm × 88 mm × 380 mm
Cut out dimension	480 × 89 mm

2 The Concept

In the new design, two reed valves are fitted at the upper ends of the transfer ducts connecting the crank case and cylinder. The experimental setup shown in Fig. 1 was used to determine the performance of the engine under normal conditions and with EGR arrangement using reed valves.

Figure 2 shows the schematic diagram for the experimental setup. The two reed valves permit exhaust gas to enter the transfer ducts during the upward movement of the piston due to partial vacuums created in the transfer ducts and the crank case. This results in a complete or at least a partial filling up of the transfer dusts by exhaust gas. F/A mixture is inducted into the crank case in a conventional way. During the downward stroke of the piston the pressure inside the crank case increases and transfer ports opens and the exhaust gas is trapped in the transfer ducts first enters the cylinder and participates in the scavenging. Thereby, it reduces the loss of the following fresh fuel–air mixture.

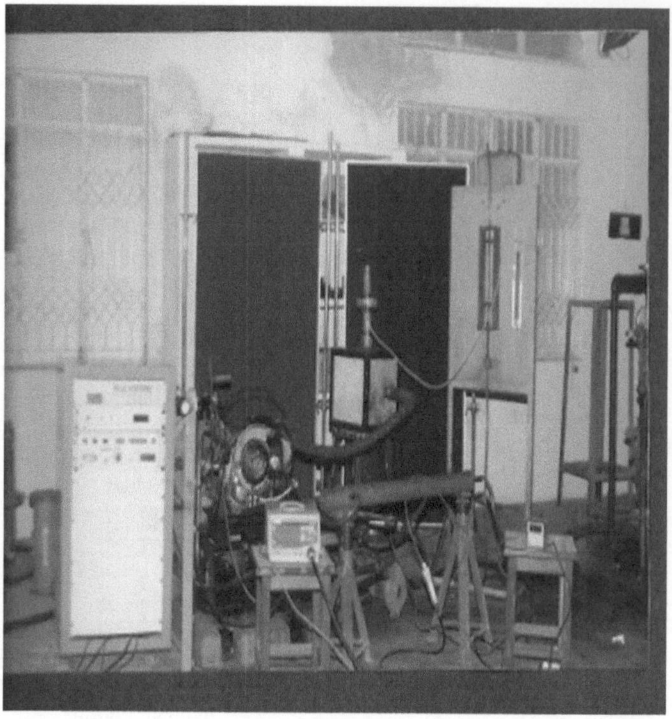

Fig. 1 Experimental setup

Fig. 2 Schematic diagram for the experimental set up

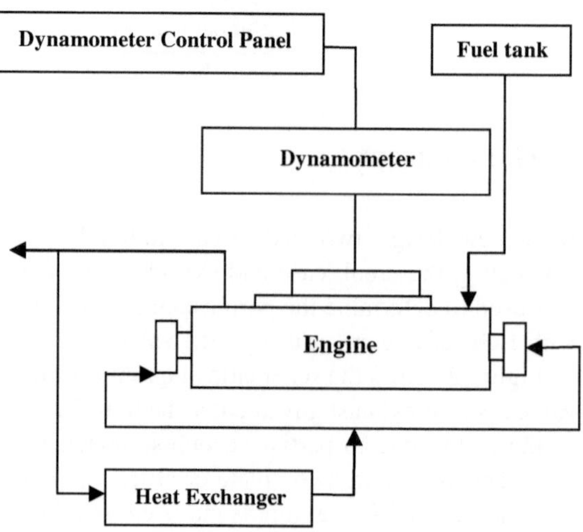

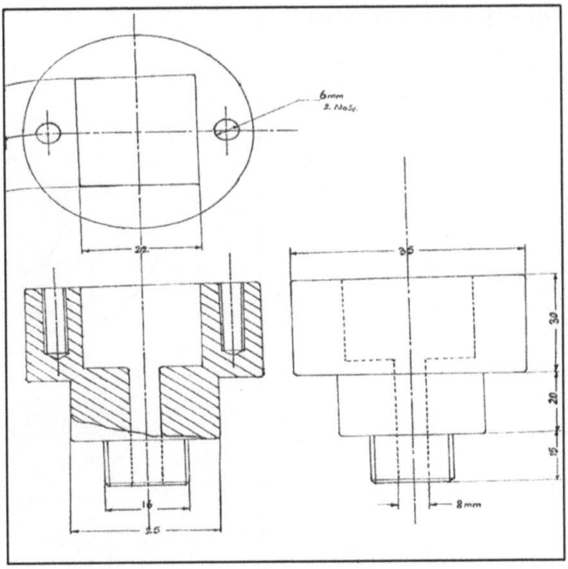

Fig. 3 Exploded view of reed block

3 Experimental Works Involved

3.1 Flange Design and Fabrication

The reed valves currently used in TVS50 moped engine are enclosed in separately fabricated aluminium alloy container. Gaskets are used in either side of the reed valve to prevent leakage. Care is taken to keep the volume of that portion of the container between the engine and the reed valve to a minimum. Due to this volume gets added to the crank case volume of the unmodified engine and may affect the scavenging pressure developed. The exploded view of the reed block is shown in Fig. 3.

The reed valve assembly is screwed to the top of the transfer ducts after the fins around that area are removed by milling and a hole was drilled and tapped suitably. The connections are made in such a way that the air through the reed valve perpendicular to the cylinder axis as shown in Fig. 4.

Figure 5 shows the way the reed valve assemblies are fixed. Care is taken to ensure that the reed valve connection does not extend over the inside surface of transfer duct wall and obstruct the flow.

3.2 Working Principles

In modified engine, air–fuel mixture and exhaust gas are supplied in the methods given below:

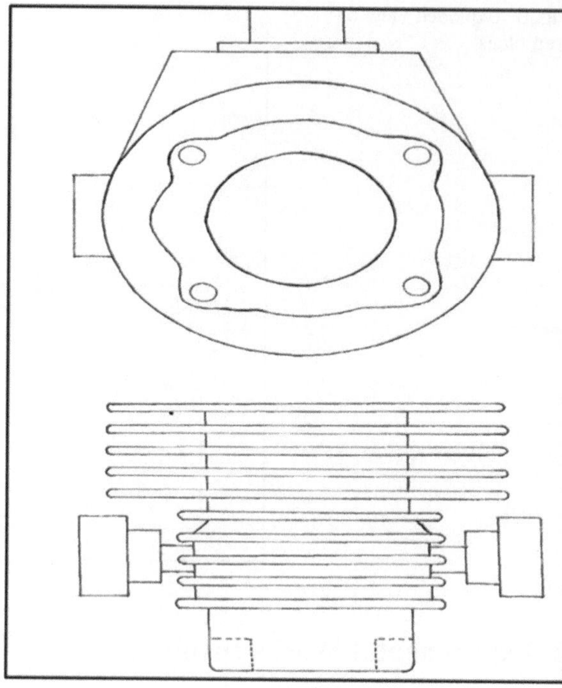

Fig. 4 Location reed valves

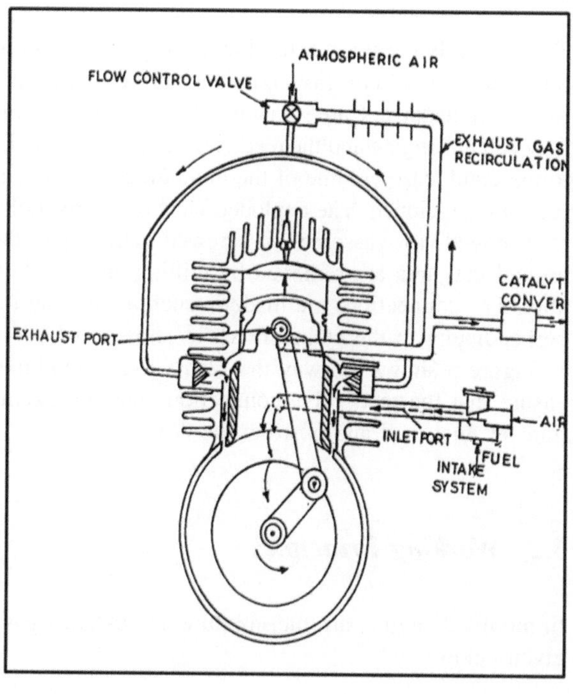

Fig. 5 Schematic assembly of the engine with EGR

Fig. 6 Full throttle—engine speed (rpm) versus HC%

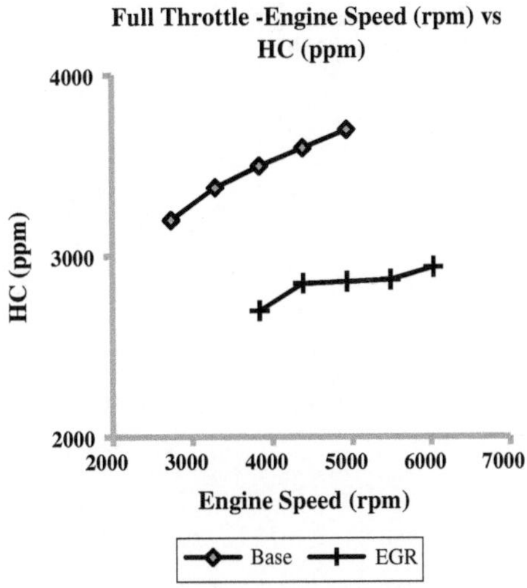

Air–fuel mixture through the carburetor in the conventional way.

Exhaust gas through the reed valves fitted at the transfer ports.

When the piston moves from BDC to TDC the pressure inside the crank case is below atmospheric pressure. This causes the air from atmosphere to flow into the crank case as mentioned earlier.

When the exhaust gas flows through the reed valves the air–fuel mixture present in the transfer duct as a result of the previous cycle is forced into the crank case and exhaust gas takes its place.

The flow of air through the carburetor per cycle for the same throttle positions sets reduced after the modifications. This is because the supplied exhaust gas will push the residual gas in the cylinder thereby fresh air–fuel short circuiting is reduced. During the downward stroke of the piston, pressure built up in the crank case and the transfer ducts as soon as the inlet port closes and hence the reed valve closes. When the piston descends further, the transfer ports open and the exhaust gas in the transfer ducts enter into the cylinder instead of air–fuel mixture as in the conventional engine. This exhaust gas pushes the burnt gases out, so that scavenging is done by exhaust gas. The air–fuel mixture which follows the exhaust gas is retained inside the cylinder to a large extent. Hence short circuiting of fresh air–fuel mixture can be reduced considerably.

4 Results

Figure 6 gives the comparison of HC emission for the base engine with and without EGR for full throttle conditions. Figure 7 gives the comparison of HC emission for the base engine with and without EGR for 3/4 throttle conditions. Figure 8 gives the comparison of HC emission for the base engine with and

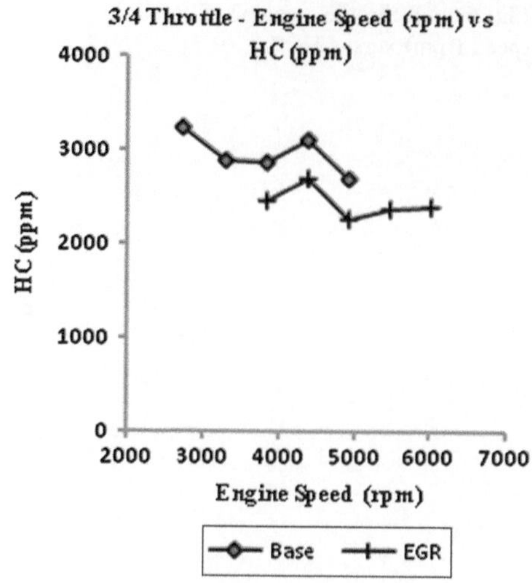

Fig. 7 ¾ throttle—engine speed (rpm) versus HC%

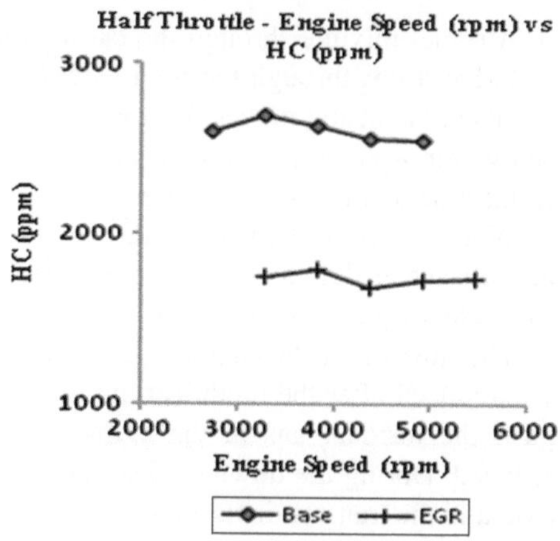

Fig. 8 Half throttle—engine speed (rpm) versus HC%

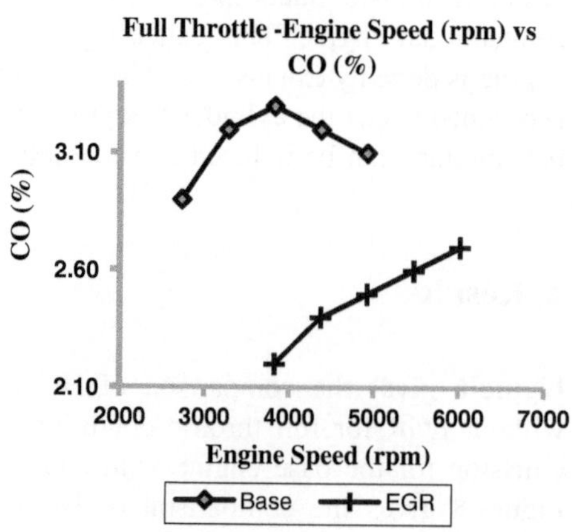

Fig. 9 Full throttle—engine speed (rpm) versus CO%

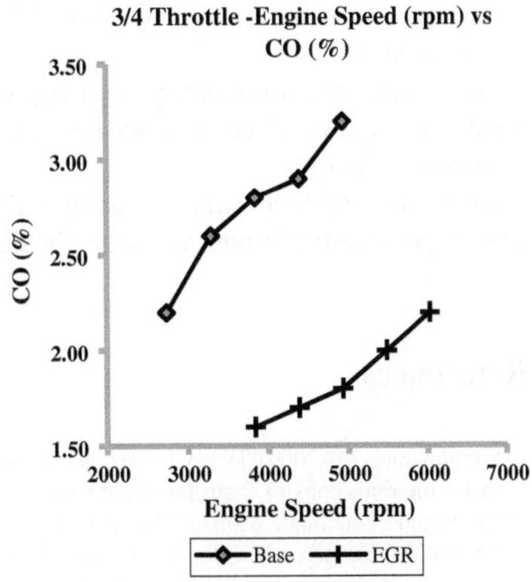

Fig. 10 3/4 throttle—engine speed (rpm) versus CO%

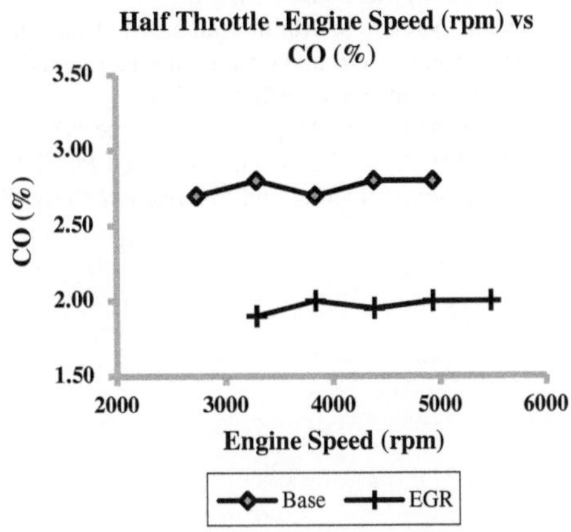

Fig. 11 Half throttle—engine speed (rpm) versus CO%

without EGR for 1/2 throttle conditions. Figure 9 gives the comparison of CO emission for the base engine with and without EGR for full throttle conditions. Figure 10 gives the comparison of CO emission for the base engine with and without EGR for 3/4 throttles conditions. Figure 11 gives the comparison of CO emission for the base engine with and without EGR for 1/2 throttle conditions.

5 Conclusions

The experimental results conclude that in half throttle conditions SFC is very low. Power output increases in all engine speed conditions since short circuiting of fresh air–fuel mixture is avoided and unburned hydrocarbons are reduced by means of improved scavenging technique.

HC emissions are reduced by about 20–25 % and CO emissions are reduced by about 30–40 %.

Serpentine heat exchangers with appropriate material selection can be incorporated in the two wheeler applications for improved performance and decreased pollutant emission.

However with increasing norms for stringent emission standards there is a long way to go investigate alternative methods.

References

1. Subramanian M (2009) Two wheeler technologies-An insight into measures taken in India to lower the emissions of 2-stroke two wheelers. SAE—EC JRC Joint International Conference Autodromo Nazionale Monza / Milano / Italy, 11–12 June
2. Aleonte M, Cosgărea R, Jelenschi L, Cofarul C (2011) Technical solutions for improving the efficiency of a two stroke SI engine. Bulletin of the Transilvania University of Braşov, Series I: Engineering Sciences 4(2):1–6
3. Swami Nathan S, Mallikarjuna JM, Ramesh A (2010) Effects of charge temperature and exhaust gas re-circulation on combustion and emission characteristics of an acetylene fuelled HCCI engine. Fuel 89(2):515–521
4. Juttu S, Mishra P, Thipse SS, Marathe NV (2011) The Automotive Research Association of India, MK Gajendra Babu. Combined PCCI-DI combustion to meet Euro-IV norms on LCV engine—experimental and visulisation study. SIAT, India, SAE Paper No.2011-26-0031

Design of Airfoil Using Backpropagation Training with Combined Approach

K. Thinakaran and R. Rajasekar

Abstract Here, we investigate a method for the inverse design of airfoil sections using artificial neural networks (ANNs). The aerodynamic force coefficients corresponding to series of airfoil are stored in a database along with the airfoil coordinates. A feedforward neural network is created with aerodynamic coefficient as input to produce the airfoil coordinates as output. In this paper, we explore different strategies for training this neural network. From our test, the most promising backpropagation strategy is to initially use steepest descent algorithm and then continue with linear and nonlinear constraint in the algorithm. Results indicate that our combined approach optimally trains artificial neural network and may accurately predict airfoil profile.

Keywords Linear error · Nonlinear error · Steepest descent method · Airfoil design · Neural networks

1 Introduction

In inverse design methods, the aim is to generate geometry for airfoil. There are many inverse techniques in use, for example, hodo-graph methods for two-dimensional flows [1–3] and other two-dimensional formulations using panel methods [4, 5]. The above methodologies have also been extended to the three-dimensional case [6–8]. The inverse method backpropagation takes long time to

K. Thinakaran (✉)
Sri Venkateswara College of Engineering and Technology, Thiruvallur, Tamil Nadu, India
e-mail: thina85@rediffmail.com

R. Rajasekar
Kalaignar Karunanidhi Institute of Technology, Coimbatore, Tamil Nadu, India
e-mail: rajasekar080564@yahoo.com

R. P. Bajpai et al. (eds.), *Innovative Design, Analysis and Development Practices in Aerospace and Automotive Engineering*, Lecture Notes in Mechanical Engineering, DOI: 10.1007/978-81-322-1871-5_36,

converge. Some focused on better function and suitable learning rate and momentum [9–11]. To design fast algorithm, Abid et al. proposed a new algorithm by minimizing sum of squares of linear and nonlinear errors for all output [12]. Jeong et al. proposed learning algorithm based on first- and second-order derivatives of neural activation at hidden layers [13]. Han et al. proposed two modified constrained learning algorithms: first new learning algorithm and second new learning algorithm to obtain faster convergence rate. The additional cost terms of the first algorithm are selected based on the first derivatives of the activation functions of the hidden neurons and second-order derivatives of the activation functions of the output neurons, while the additional cost terms of second one are selected based on the first-order derivatives of the activation functions of the output neurons and second-order derivatives of the activation functions of the hidden neurons [14].

The objective of this work is to show combined steepest descent and linear and nonlinear methods offer promising advantages in neural network training.

One of the issues when designing a particular neural network is to calculate proper weight for neural activities. These are obtained from the training process of neural network. The steepest descent method (First order) sometimes lack fast convergence [15–17]. Specifically, the process of obtaining appropriate weight in a neural network design utilizes two set of equations. First, the feedforward equation is used to calculate the error function. The feedback equation is next used to calculate the gradient vector. This gradient vector is used for defining search directions in order to calculate weight change. The gradient direction is the direction of steepest descent. But, the learning rate may be different in different directions. This makes the learning in steepest descent to be very slow and inefficient. The linear and nonlinear constraint introduces certain modification in directions in order to speed up the convergence. The combined approach of steepest descent and linear and nonlinear constraint resulted fast convergence.

2 Ann Training Method

Consider a neuron i with weights w_{i1}, w_{i2}, $w_{i3}, \ldots, w_{ij}$ as in Fig. 1.

Here, g() is a general nonlinear function called activation function [18]. The weight w_{ij} represents the strength of the connection between neurons i and j. μ_i is the threshold value for neuron i.

An ANN consists of a large number of simple highly interconnected such neurons. Neural networks offer a very powerful and general framework for representing nonlinear mappings from several input variables to several output variables. The general architecture of a two-layer neural network with feedforward connections and one hidden layer is shown in Fig. 2. The weighted sum of the inputs must reach the threshold value for the neuron to generate the output. It is normally very difficult to interpret the values of the connecting weights w_{ij} in terms of the task being implemented.

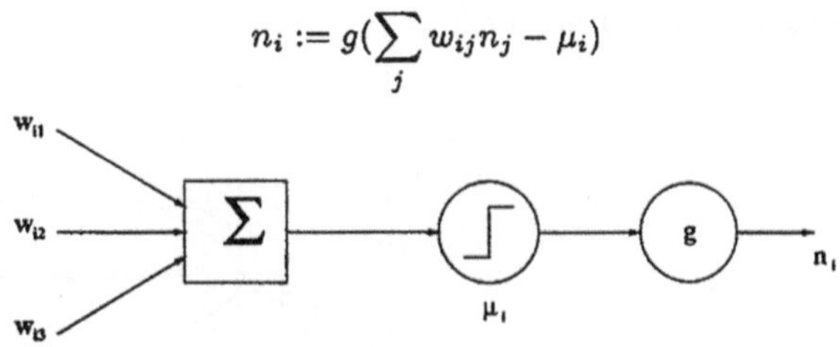

Fig. 1 Schematic diagram for a simple neuron

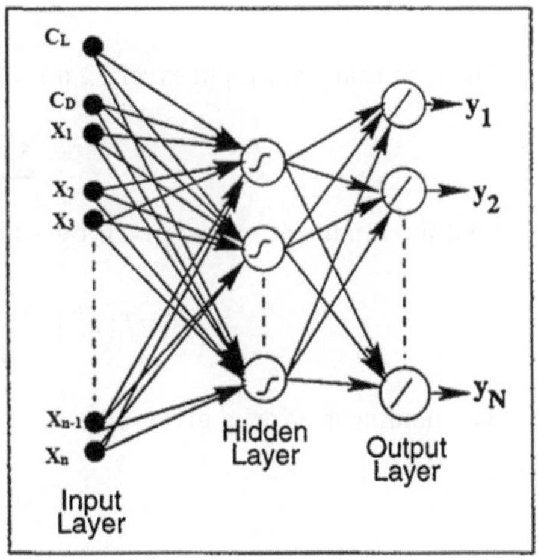

Fig. 2 The neural network model trained to predict surface *Y*-coordinates and C_l, C_d and *X*-coordinates are the inputs

2.1 ANN Architecture Used

An airfoil profile can be described by a set of *x*- and *y*-coordinates, as shown in Fig. 3. The aerodynamic force coefficients corresponding to series of airfoil are stored in a database along with the airfoil coordinates. A feedforward neural network is created with the inputs—aerodynamic coefficients C_l, C_d, and *X* coordinates to produce the outputs—the airfoil *Y*-coordinates.

Training generally involves minimization of error with respect to the training set. Learning algorithms such as the backpropagation algorithm for feedforward multilayer networks [19] help us to find set of weights by successive improvement from an arbitrary starting point.

We used the below sigmoidal activation function to generate the output.

$$f(x) = \frac{1}{1 + e^{-x}} \tag{1}$$

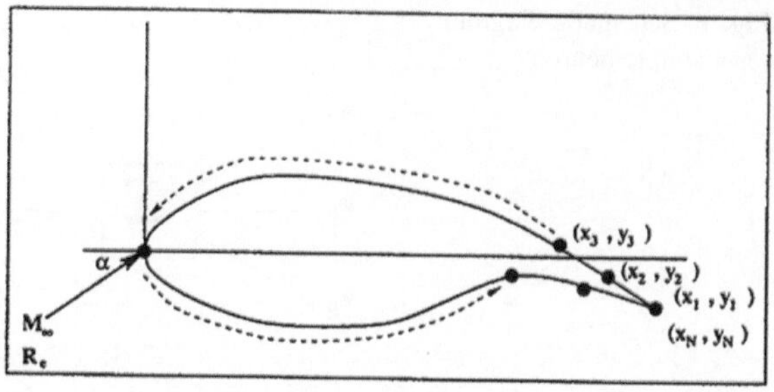

Fig. 3 Flow field and airfoil data

For a neuron j at output layer L, the linear outputs given by Abid et al. is:

$$u_j^L = \sum w_{ji} y_i^H \tag{2}$$

And the nonlinear output given by Abid et al. is:

$$f\left(u_j^L\right) = \frac{1}{1 + \mathrm{e}^{-\mathrm{u}_j^L}} \tag{3}$$

The nonlinear error is given by:

$$e_{1j}^L = d_j^L - y_j^L \tag{4}$$

where d_j^L and y_j^L respectively are desired and current output for jth unit in the Lth layer.

The linear error is given by:

$$e_{2j}^L = ld_j^L - u_j^L \tag{5}$$

where ld_j^L is defined as below $ld_j^L = f^{-1}(d_j^L)$ (Inversing the o/p layer).

The learning function depends on the instantaneous value of the total error. As per the modified backpropagation by Abid et al., the new cost term to increase the convergence speed is defined as:

$$E_p = \frac{1}{2}(e_{1j}^L)^2 + \lambda \frac{1}{2}(e_{2j}^L)^2 \tag{6}$$

where λ is Weighting coefficient

Now, this new cost term can be interpreted as below:

$$\mathrm{Ep} = (\text{nonlinear error} + \text{Linear error}).$$

2.2 Learning of the Output Layer

Now, the weight update rule is derived by applying the gradient descent method to Ep.

Hence we get weight update rule for output Layer L as:

$$\Delta w_{ji} = -\mu_L \frac{\partial E_p}{\partial w_{ji}}$$

where μ_L is the network learning parameter

$$\begin{aligned}\Delta w_{ji} &= -\ \mu_L e^L_{1j} \frac{\partial y^L_j}{\partial w_{ji}} + \mu_L e^L_{2j} \frac{\partial u^L_j}{\partial w_{ji}} \\ \Delta w_{ji} &= -\ \mu_L e^L_{1j} \frac{\partial y^L_j}{\partial u_j} \frac{\partial u^L_j}{\partial w_{ji}} + \mu_L e^L_{2j} y^H_i \\ \Delta w_{ji} &= -\ \mu_L e^L_{1j} f^{'}(u^L_J) y^H_i + \mu_L e^L_{2j} y^H_i\end{aligned} \tag{7}$$

.

2.3 Learning of the Hidden Layer

And we get the weight update rule for the hidden layer H as:

$$\begin{aligned}\Delta w_{ji} &= -\ \mu_H \frac{\partial E_p}{\partial w_{ji}} \\ \Delta w_{ji} &= -\ \mu_H e^H_{1j} \frac{\partial y^H_j}{\partial w_{ji}} + \mu_H e^H_{2j} \frac{\partial u^H_j}{\partial w_{ji}} \\ \Delta w_{ji} &= -\ \mu_H e^H_{1j} \frac{\partial y^H_j}{\partial u_j} \frac{\partial u^H_j}{\partial w_{ji}} + \mu_H e^H_{2j} y^F_i \\ \Delta w_{ji} &= -\ \mu_L e^H_{1j} f^{'}(u^H_J) y^F_i + \mu_L e^H_{2j} y^F_i\end{aligned} \tag{8}$$

The network learning parameter μ is initialized, which plays an important role in minimizing the error. Then the network is trained with corresponding change of weight for both hidden and output layer. And the weight is determined by calculating linear and nonlinear error.

2.4 Learning Algorithm: Steepest Descent

The learning algorithm [3] is the repeated application of the chain rule to compute the influence of each weight in the network with respect to an arbitrary error function E.

The weight change is as below:

$$\Delta W = \eta \frac{\partial E_K}{\partial W_{ik}}$$

The Error function is as follows:

$$E_k = \frac{1}{2}(T_K - O_{\text{ok}})^2 \tag{9}$$

Now, we compute $\frac{\partial E_k}{\partial W_{\text{ik}}}$ by applying the chain rule of differentiation as:

$$\frac{\partial E_k}{\partial W_{\text{ik}}} = \left(\frac{\partial E_k}{\partial o_{\text{ok}}}\right)\left(\frac{\partial o_{\text{ok}}}{\partial I_{\text{ok}}}\right)\left(\frac{\partial I_{\text{ok}}}{\partial W_{\text{ik}}}\right)$$

where,

$$\frac{\partial E_k}{\partial O_{\text{ok}}} = -\,(T_k - o_{\text{ok}})$$

$$\frac{\partial I_{\text{ok}}}{\partial W_{\text{ik}}} = O_{\text{Hi}}$$

To get the term $\frac{\partial o_{\text{ok}}}{\partial I_{\text{ok}}}$, we first calculate the output of kth output neuron o_{ok} using below equation

$$o_{\text{ok}} = \frac{1}{1 + e^{-\lambda(I_{\text{ok}} - \theta_{\text{ok}})}}$$

$$=_{\text{ok}} = (1 + e^{-\lambda I_{\text{ok}}} e^{\lambda \theta_{\text{ok}}})^{-1}$$

Now, differentiate o_{ok} w.r.to I_{ok} gives,

$$\frac{\partial o_{\text{ok}}}{\partial I_{\text{ok}}} = \lambda e^{-_{\text{ok}}} e^{\lambda \theta_{\text{ok}}} (1 + e^{-\lambda I_{\text{ok}}} e^{\lambda \theta_{\text{ok}}})^{-2}$$

$$=\gg \frac{\partial o_{\text{ok}}}{\partial I_{\text{ok}}} = \lambda o_{\text{ok}}(1 - o_{\text{ok}})$$

Hence,

$$\frac{\partial E_k}{\partial W_{\text{ok}}} = -(T_k - o_{\text{ok}}) \cdot \lambda o_{\text{ok}}(1 - o_{\text{ok}}) \cdot O_{\text{Hi}} \tag{10}$$

The problems in steepest descent are: (i) poor descent procedure for surface which curve more sharply in some directions than others and (ii) high level interference in learning with different patterns. We would like to take large steps where the surface is gently sloping and small step where the surface is steep. We need the learning rate to be different in different directions which are precisely ruled out in steepest descent procedure. So the learning will be very slow and inefficient. To solve the problem we proposed following algorithm.

2.5 Proposed Combined Approach Algorithm

In the proposed algorithm, the learning parameters μ and λ are initialized according to the convergence of the problem. Here network is initially trained with steepest descent method then the corresponding change of weight for both hidden and output layer is determined using linear and nonlinear error. Its mean square error is calculated and compared with threshold value. Based on the comparison, the network is declared as trained. Now the test data is passed to the trained network to get the desired output.

Step1 Initialize the learning parameters μ, λ to some random values.
Step2 Assign Threshold value to a fixed value based on the sigmoid function.
Step3 Linear output is calculated using Eq. (2).
Step4 Nonlinear output is calculated using sigmoid function as in the Eq. (3).
Step5 If number of iterations is less than or equal to 1000, then calculate the weight change from steepest descent method as in the Eq. (10).

If the number of iterations is greater than 1,000, then calculate the weight change for hidden layer is calculated using the Eq. (8) and weight change for output layer is calculated using the Eq. (7).

Step6 Calculate the mean square error using the Eq. (9).
Step7 The above calculated mean square error value is compared with threshold value.
Step8 If the mean square error value is greater than threshold value, then the above steps from 3 to 6 is repeated.
Step9 If the mean square error value is less than threshold value, then declare that the network is trained.
Step10 Now the test data is passed to trained network to produce the desired airfoil profile.

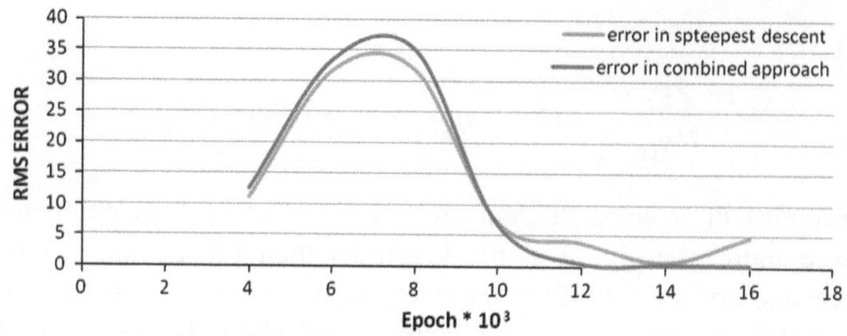

Fig. 4 Convergence comparison

3 Results and Discussion

In our investigation of neural network models for inverse design of airfoil sections, we found that satisfactory results were obtained by using the combined approach of steepest descent and linear and nonlinear error constraints.

In our case, we considered ANN with ten hidden neurons. We have a database comprised of 26 upper and lower-surface *x*- and *y*-coordinates, together with the corresponding coefficient of lift (C_L) and the coefficient of drag (C_D). There were 1,422 patterns in total. The main goal is to determine the airfoil profile for given conditions. This is the "inverse" problem for airfoil design (Fig. 4).

The network was trained with 1,000 patterns to minimize error. Then network was tested with a test set comprising 400 patterns which was not used in the training process. The computed profiles generated during the test process show good agreement with the actual profiles. The new airfoil was tested again for the same flow conditions in XFoil tool to compare C_L, C_D. Six such cases are shown in Fig. 5. Figure 4 shows how the training decreases root mean square (RMS) errors with the epoch. The blue line indicates the error in steepest descent and the red line indicates that of combined approach. From the figure, it is obvious that error in combined approach is less at the converging stage.

Measure of the accuracy of the results obtained can be inferred from examination of error which is defined as:

$$\text{Error} = \frac{y_{i(\text{actual})} - y_{i(\text{computed})}}{\text{airfoil thickness ratio}} * 100$$

where $y_{i(\text{actual})}$ is the actual y-coordinate of the section at location *i*, $y_{i(\text{computed})}$ is the computed y-coordinate. Table 1 shows the maximum error in percentage for the airfoils shown in Fig. 5. From this table, we can conclude that the combined approach predicated comparatively the correct airfoil profiles.

The result obtained from steepest descent and combined approach are compared and tabulated in Table 2. This table clearly proves that the combined approach results in less error and takes less time to predict the airfoil for the given C_L, C_D. The time is denoted by number of clock ticks in the table.

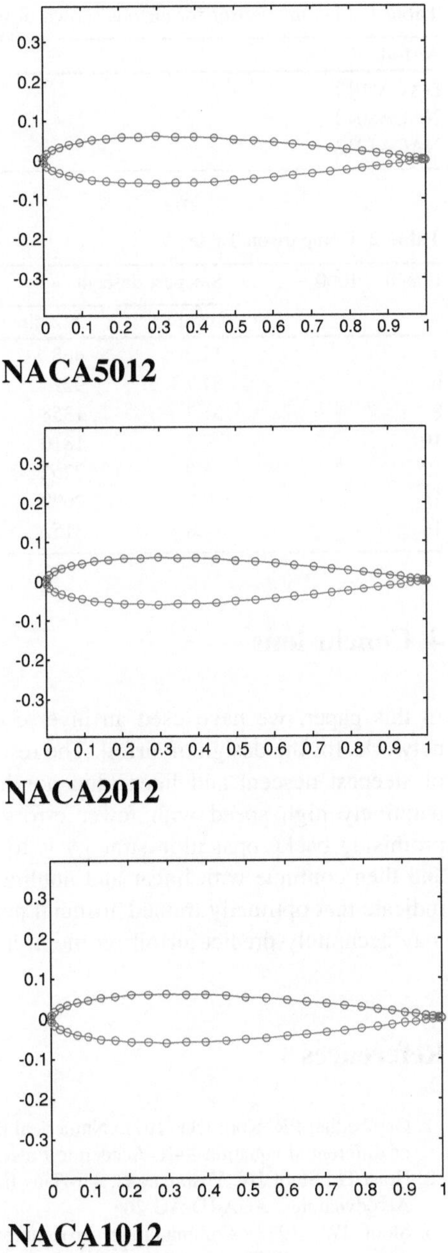

Fig. 5 Airfoil profiles at zero degree angle of attack

Table 1 Maximum error for airfoils shown in Fig. 5

Airfoil	Maximum error (%)
NACA5012	0.09002
NACA2012	0.02677
NACA1012	0.007415

Table 2 Comparison Table

Epoch * 1000	Steepest descent		Combined approach	
	Error	No. clock ticks	Error	No. clock ticks
4	11.2	468	12.6	62
6	31.7	921	33.4	125
8	31.7	1358	34.9	172
10	7.2	1810	6.8	234
12	3.9	2262	0.5	296
14	0.8	2699	0.4	343
16	4.8	3152	0.3	406

4 Conclusions

In this paper, we have used an inverse design methodology in artificial neural networks for the design of airfoil. The results indicate that our combined approach of steepest descent and linear and nonlinear constraints will converge at comparatively high speed with fewer errors. In the proposed algorithm, the most promising backpropagation strategy is to initially use steepest descent algorithm and then continue with linear and nonlinear constraints in the algorithm. Results indicate that optimally trained artificial neural networks using proposed algorithm may accurately predict airfoil profile at a faster pace.

References

1. Garabedian PR, Korn DG (1971) Numerical design of transonic airfoils. Numerical solution of differential equations—II. Academic Press, New York
2. Hoist TL, Sloof JW, Yoshihara H, All-Haus Jr.WFZ (1982) Applied computational Transonic Aerodynamics. AGARD-AG-266
3. Sloof JW (1982) Computational procedures in transonic aerodynamic design. Lecture presented at ITCS Short Course on Computational Methods in Potential Aerodynamics, Amalfi, Italy, 1982 (also NLR MP 82020 U)
4. Ormsbee AI, Chen AW (1972) Multielement airfoils optimized for maximum lift coefficient. AIAA J 10:1620–1624
5. Fray JMJ, Sloof JW (2005) A constrained inverse' method for the aerodynamic design of thick wings with given pressure distribution in subsonic flow. In: AGARD-CP

6. Shankar V, Malmuth ND, Cole JD (1979) Computational transonic design procedure for three-dimensional wings and wing-body combinations. In: AIAA paper 79-0344
7. Garabedian P, McFadden G, Bauer B (1983) The NYU inverse swept wing code. In: NASA CR-3662
8. Shankar V (1981) A full-potential inverse method based on a density linearization scheme for wing design. In: AIAA paper 81-1234
9. Jacobs RA (1988) Increased rates of convergence through learning rate adaptation. Neural Netw 1:295–307
10. Magoulas GD, Plagianakos VP, Vrahatis MN, Van Ooyen A, Nienhuis B (1992) Improving the convergence of the backpropagation algorithm. Neural Netw 5:465–471
11. Yu XH, Chen GA, Cheng SX (1993) Acceleration of backpropagation learning using optimized learning rate and momentum. Electron Lett 29:1288–1289
12. Abid S, Fnaiech F, Najim M (2001) A fast feedforward training algorithm using a modified form of the standard backpropagation algorithm. IEEE Trans Neural Netw 12:424–430
13. Jeong SY, Lee SY (2000) Adaptive learning algorithms to incorporate additional functional
14. Han F, Ling QH, Huang dS (2008) Modified constrained learning algorithms incorporating additional functional constraints into neural networks. Inf Sci 178:907–919
15. Fahlman SE (1988b) Learning variations of backpropagation—An empirical study. In: Proceedings of the 1988 connectionist models summer school, Morgan kanufmann, San Mater, CA
16. Thiang HK, Rendy P (2009) Artificial neural network with steepest descent backpropagation training algorithm for modeling inverse kinematics of manipulator. World Acad Sci Eng Technol 36
17. Sutton RS (1986) Tow problems with backpropagation and other steepest-descent learning procedure for networks. GTE Laboratories, Waltham
18. Hertz J, Krogh A, Palmer RG (1991) Introduction to the theory of neural computation. Addison-Wesley Publishing Co., California
19. Riedmiller M, Braun H (1993) A direct adaptive method for faster backpropagation learning: the RPROP algorithm. In: Proceedings of the IEEE international conference on neural networks, San Francisco, CA. Accessed 28 March–1 April 1993
20. Behera L, Kumar S, Patnaik A (2006) On adaptive learning rate that guarantees convergence in feedforward networks. IEEE Trans Neural Netw 17:1116–1125

Author Biographies

Mr. K. Thinakaran currently working toward the Ph.D. degree at the Anna University. He is currently an Assistant Professor in Computer Science Engineering, Sri Venkateswara College of Engineering and Technology, Thiruvallur, India. His current research interests include Neural Network and Data Mining.

Dr. R. Rajasekar received his doctorate from Department of Aeronautics, Imperial College, London, UK. His aeronautical masters' degree was from IIT, Madras. He is currently working as the Professor and Head of Aeronautical Engineering Department, Kalignar Karunanidhi Institute of Technology, Coimbatore, India. (an affiliated college under Anna University, Chennai). His specialization and research interests are aerodynamics and its applications.

Improvement of Performance and Emission Characteristics of Diesel Engines Using TBC and Emerging Techniques: A Review

G. Sivakumar, R. Nanda Kumaran, M. G. Varun Kumar and S. Senthil Kumar

Abstract This paper reviews an overview concept on insulation in the combustion chamber for a conventional diesel engine and reducing its negative effects on engine performance and emission characteristics using emerging techniques. The main aim is to improve the engine performance and also to reduce the pollutants by using new emerging materials. Several studies have been conducted for various ceramics to increase the engine efficiencies and also decrease the pollutants in diesel engine mainly Particulate Matter (PM). Particulate Matter is one of the main pollutants in Diesel engine because of high sulphur content and it is a carcinogen, recently WHO (World Health Organisation) made a report that 70–80 % of cancers [1] are because of particulate matter. Added to this, another important advantage of ceramic insulation in an engine is the greatest reduction in friction losses due to easy warm up during starting and thus increase the brake power of an engine, simultaneously smoke is reduced at low speed. The main purpose of this study is to evaluate the increase in engine performance and decrease in pollutant level at an extent shown by many researchers.

Keywords Low heat rejection engine · Thermal barrier coating · NOx reduction techniques and engine efficiencies improvement

1 Introduction

Heat transfer plays a major role in compression ignition engine. The heat losses characteristic in the compression ignition engine has a vast field which is shown in Graph 1, major losses are in the form of heat in coolant to radiator and exhaust gas. In order to reduce the heat transfer, metal matrices of ceramics are used.

G. Sivakumar (✉) · R. Nanda Kumaran · M. G. Varun Kumar · S. Senthil Kumar
School of Mechanical Engineering, Vel Tech Dr. RR and Dr. SR Technical University, Chennai, India
e-mail: sec2.veltech@gmail.com

R. P. Bajpai et al. (eds.), *Innovative Design, Analysis and Development Practices in Aerospace and Automotive Engineering*, Lecture Notes in Mechanical Engineering,
DOI: 10.1007/978-81-322-1871-5_37, © Springer India 2014

Graph 1 Energy distribution chart [3]

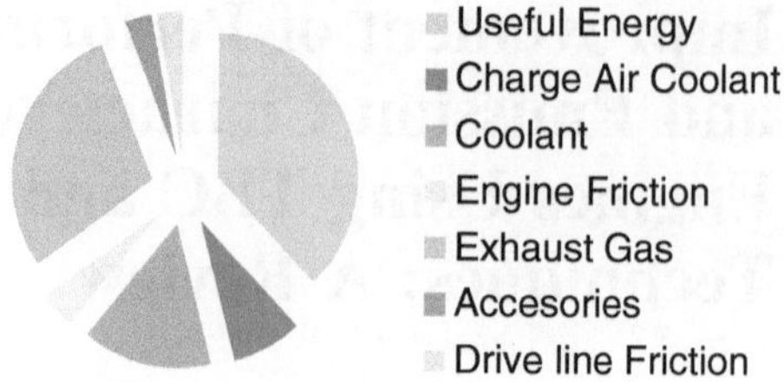

Theoretically, if the heat rejected could be reduced then the thermal efficiency would be improved. Kamo et al. [2] stated that 70 % reduction in heat loss and also Morel et al. [21] stated that 60 % reduction in heat loss of Low Heat Rejection (LHR) engines. The purpose of this paper is to examine the causes for these seemingly contradictory results. An attempt has made to review the previous studies to look into future possibilities of the LHR engine from the viewpoint of combustion, heat transfer and emission. A detail study about engine performance and emission levels using Thermal Barrier Coating (TBC) inside the combustion chamber are described and also various technologies are described to overcome some of the existing problems by using Thermal Barrier Coating.

2 Ceramic Materials

Ceramic materials coating inside the cylinder wall are not new. The research had started since 1970s on the conventional diesel engine into adiabatic engine, i.e. Low Heat Rejection engine. Why ceramics? Ceramics are the composites having a low heat rejecting character (thermal barrier), low heat flux, high mechanical strength and also high lubricating property. The revealed ceramics on the engine applications are Silicon Nitride, Silicon Carbide, Partially Stabilised Zirconia, Alumina, etc. Techniques for coating all these ceramics are by different methods such as Physical Vapour Deposition (PVD), Chemical Vapour Deposition (CVD), Plasma spray technique and Thermal spray technique, etc. Coating thickness is not exceeded by 1 mm otherwise mechanical failure occurs suggested by Rajendra Prasath et al. [4]. Taymaz et al. [5] suggested that 0.5 mm of coating thickness will reduce 5–20 % of Coolant loss.

3 Engine Efficiencies

3.1 Volumetric Efficiency

Volumetric efficiency is the breathing effect of an engine based on pressure difference caused by engine operating conditions. Due to increased temperature inside the combustion chamber, the intake air gets reduced, i.e. small quantity of air is enough to fill in the cylinder. The experimental investigation of Murthy [6] shows that the volumetric efficiency decreases by 8 %. To overcome about this problem various forced air induction technologies [7] are available such as

- Twin Turbo charging
- Wave Supercharging
- Turbo compounding.

Rajendra Prasath et al. [4] showed that the increase in coating thickness results in decrease in volumetric efficiency. By the use of forced air induction system, volumetric efficiency gets increased at the same time exhaust temperature reaches high which causes some secondary pollutants. Thermal barrier coating is good to remove Particulate Matter (PM) through active generation system.

3.2 Thermal Efficiency

Thermal Efficiency is the true indication of the efficiencies, from chemical energy into mechanical power. Thermal efficiency of the low heat rejection engine is increased due to the barring of heat transfer on the cylinder walls. Coatings are also limited to less than 1 mm because, the higher thickness tends to maximise the stress and strain suggested by Samadi and Coyle [8]. Researches of Murthy [6], Shrirao and Pawar [9] found that increase of thermal efficiency by 7 % for Low Heat Rejection engines.

4 Fuel Consumption

Fuel consumption of the Low Heat Rejection engine is likely to be reduced because of thermal barrier and higher thermal efficiency. The volatility of the fuel plays a major role in Low Heat Rejection engine. Even low volatile fuel is enough for Low Heat Rejection engine because higher temperature inside the combustion chamber leads to complete combustion. 2–12 % decrease in fuel consumption is experimented by Shrirao and Pawar [9] and Specific Fuel Consumption is increased by 15–120 % suggested by Samadi and Coyle [8].

4.1 Characteristics of Fuel Injection

Ignition delay plays a major role in the compression ignition engine, which faces both physical delay and chemical delay characteristics. The physical delay is in terms of injection pressure, viscosity, momentum and energy of a fuel; chemical delay is in terms of reaction between fuel and air inside the combustion chamber, in diesel engine physical and chemical delay is high. Viscosity of diesel is high compared to gasoline fuel; viscosity determines the atomization and vaporisation of a fuel. In Low Heat Rejection engines, high injection pressure is not required because the engine itself has more temperature to atomize and vaporise the fuel. So physical delay and chemical delay of mixture get decreased in Low Heat Rejection engines even though the viscosity of diesel is high. At high temperature, the needle lift gets damaged so the alteration is required; Dickey's [10] experimental data showed that the higher temperature in the insulated engine alters both the needle lift. Research of Murthy [6] indicated that the injection timing is found to be 33° before TDC for conventional engine and 32° for Low Heat Rejection engines at injection pressure of 190 bar.

5 Exhaust Emission

5.1 CO and HC Emission

Carbon monoxide and Hydrocarbon are formed due to incomplete combustion. In thermal barrier coating inside the cylinder wall, the temperature gets increased resulting in less production of Carbon monoxide and Hydrocarbon due to complete combustion. Results of Shrirao [11] and Pankaj et al. [9] found that decrease in carbon monoxide and Hydrocarbon level.

5.2 Oxides of Nitrogen (NOx)

NOx is formed from nitrogen and oxygen present in the air at 1,000 °C because of endothermic reaction and Zeldovich mechanism. NOx comprises of 70 % Nitric oxide (NO) and remaining percentage of Nitrogen dioxide (NO_2), Di nitrogen monoxide (N_2O), etc. In diesel engine, NOx formation is high because of lean mixture at elevated temperature; in case of thermal barrier coating NOx level is too high because of high temperature. NOx level is increased by 41 % in Low Heat Rejection engine experimented by Murthy [6]. And also NOx can create smog due to the presence of volatile organic compound in the atmosphere. So NOx reduction in the low heat rejection engine is the major problem.

Lot of technologies have been invented to reduce NOx in Internal Combustion Engines. The technologies are [7]

- In cyclic treatment
- After-treatment devices.

5.2.1 In Cyclic Treatment

The following characteristics or parameters are needed to reduce NOx level in low heat rejection engine:

- High Cetane and volatile fuel
- Reduce peak temperatures
- Reduce residence time in peak temperature zones
- Reduce O_2 content in primary flame zone
- Low excess air
- Phases of Injection
- Staged combustion
- Exhaust Gas Recirculation
- Reduce air preheat.

5.2.2 After Treatment Devices

The after treatment devices are additional support to an engine to reduce pollutants, the possible ways are listed below for NOx reduction in diesel engine [7],

- NSR
- DeNOx
- SCR
- Water injection treatment
- Oxygenated Air Induction.

NOx Storage Reduction

NSR is NOx storage reduction having ceramic monolithic substrate and wash coat material used in NSR is BaO, MgO, etc. Rhodium as a catalyst is used to remove NOx by reduction reaction.

The reaction of NSR is [7]

$$NO + O_2 \rightarrow NO_2$$
$$NO_2 + BaO \rightarrow BaO \cdot NO_2$$
$$BaO \cdot NO_2 + HC \rightarrow BaO + N_2 + H_2O + CO_2$$

The main drawback of NSR is costly because of having precious metals, i.e. Rhodium is three times costlier than platinum, even platinum is costlier.

DeNOx

The lean NOx catalyst washcoat incorporates a zeolite catalyst that acts to adsorb hydrocarbons from the exhaust stream. Once adsorbed on the zeolite, the hydrocarbons will oxidise and create a locally oxygen poor region that is more conducive to reduce NOx. To promote hydrocarbon oxidation at lower temperatures, the wash coat can incorporate platinum or other precious metals. The platinum also helps to eliminate the emission of unburned hydrocarbons that can occur if too much fluid is injected referred to as 'hydrocarbon slip'. With platinum, the NOx conversion can take place at low exhaust temperatures that are typical for diesel engines. The reactions of DeNOx catalytic converter are,

$$NO \rightarrow NO_2$$

$$HC + NO_2 \rightarrow N_2 + CO_2 + H_2O$$

Active lean NOx catalysts have been shown to provide up to 30 % NOx reduction under limited steady state conditions [12]. Kawanami et al. [13] showed NOx reduction 12 % due to excursions outside the optimum NOx reduction efficiency temperature range for these devices.

The 'passive' lean NOx catalyst uses no fluid injection. Therefore, the passive lean NOx catalyst is even more limited in its ability to reduce NOx because the exhaust gases normally contain very few hydrocarbons.

Selective Catalytic Reduction

SCR is selective catalytic reduction uses Diesel Exhaust Fluid (DEF) to control the NOx level, urea is the most commonly used exhaust fluid which concentrates 30–40 % with water solution, Vanadium and Titanium oxide mixture coated on a ceramic substrate in which the exhaust fluid is sprayed.

The reactions of SCR as shown [7]

Hydrolysis

$$(NH_2)_2.CO + H_2O \rightarrow 2NH_3 + CO_2$$

SCR Catalyst

$$4NH_3 + 4NO + O_2 \rightarrow 4N_2 + 6H_2O$$
$$2NH_3 + NO + NO_2 \rightarrow 2N_2 + 3H_2$$
$$8NH_3 + 6NO_2 \rightarrow 7N_2 + 2H_2O$$

Oxidation Catalyst

$$4NH_3 + 3O_2 \rightarrow 2N_2 + 6H_2O$$

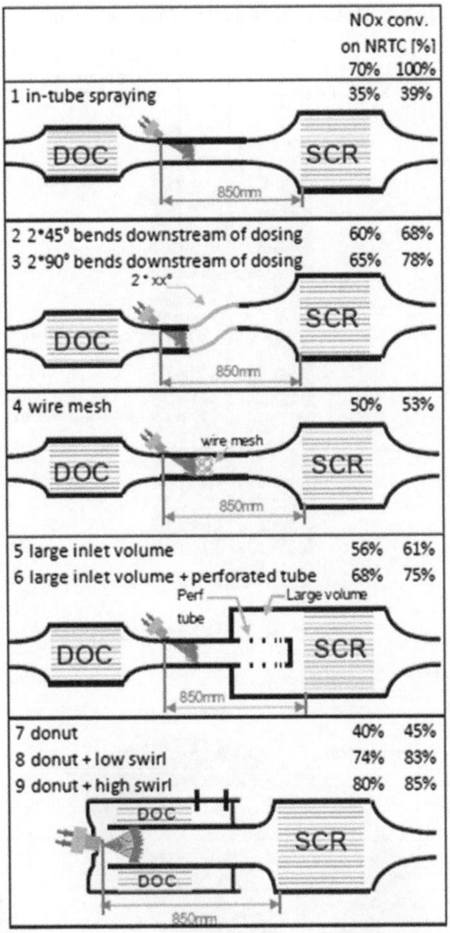

Fig. 1 SCR with different arrangements [14]

De Rudder [14] made an experiment on different SCR mixers and found a result with decrease in NOx level shown in Fig. 1 with various arrangements in first pipe, i.e. pipe immediately after exhaust manifold.

Requirement of separate fluid feed system and Ammonia slip is the main problem for Selective Catalytic Reduction. Various Diesel Exhaust Fluids used for SCR operation as followed by [7] are Urea (liquid), Ammonia (liquid), Urea (solid)-Tenneeco, Ammonia Carbamate-Navistar and Metal ammine based delivery system (strontium chloride ammine)-Amminex technology.

Water Injection Treatment

Water injection treatment reduces the combustion temperature to reduce NOx but on the other side it reduces the Hydro Carbon, Particulate Matter and other

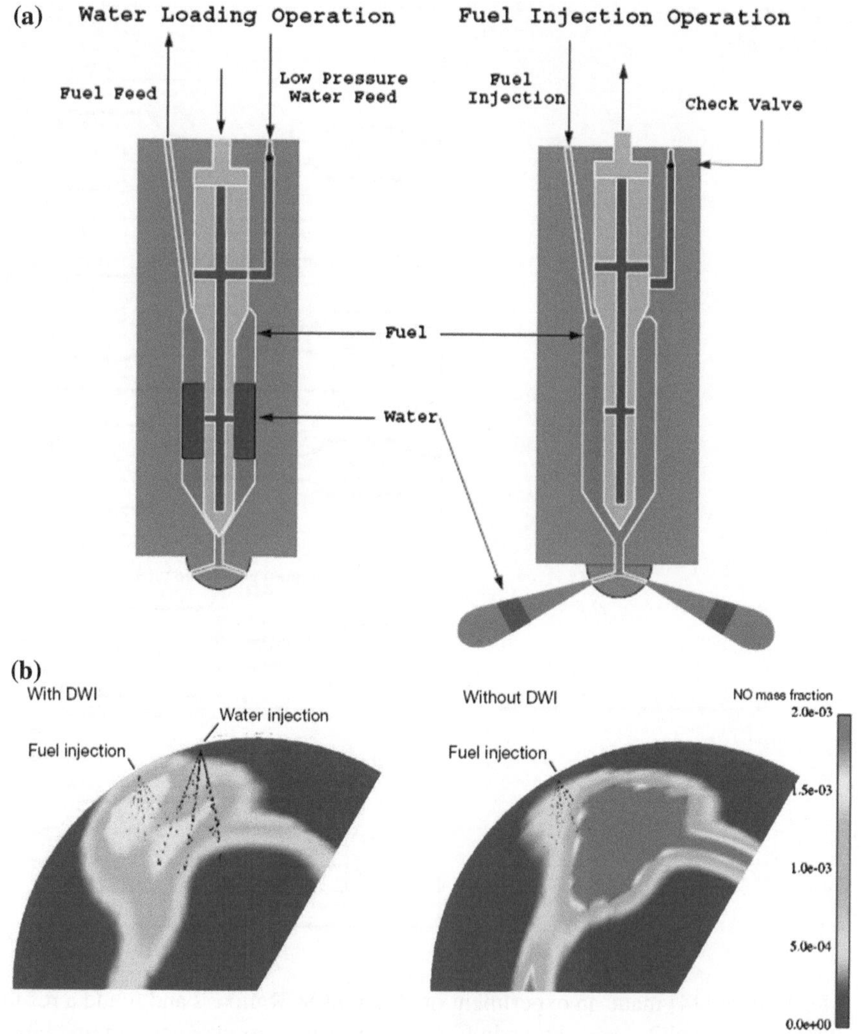

Fig. 2 **a** Combined fuel and water induction arrangement in an Injector [15]. **b** Temperature distribution of Direct Water Injection (DWI) and Without Direct Water Injection [15]

pollutants too because the water forms as a steam tries to improvise the mixture flow motion to achieve complete combustion. Toh Ai Pei Jannell [15] made an experimental study about water injection treatment in a marine diesel engine equipped with a combined injection valve and nozzle that allows injection of water and fuel into the combustion chamber separately as shown in Fig. 2a. Water injection takes place before fuel injection, resulting in cool combustion space and lower NOx emissions shown in Fig. 2b. Water injection stops before fuel injection into the combustion chamber, so the ignition nucleus formation in the Lean Flame

Region (LFR) during the swirl flow motion and combustion process is not disturbed. Water is fed into the cylinder head at high pressure of about 210–400 bar.

The following results are obtained from the experiment [15],

- NOx emissions are reduced by 50–60 %.
- No negative impact on the engine components.
- Ratio of injected water to injected fuel is typically 0.4–0.7.

Oxygenated Air Induction

Diesel Engine operating at lean mixture, i.e. fuel-air ratio, $\lambda > 1$ the excess nitrogen and oxygen in the air form mechanical NOx at elevated temperature. Also in air only oxygen is required for combustion, so the oxygen is to be extracted from the air and supplied to the combustion chamber which leads to produce zero mechanical NOx and complete combustion.

Rajkumar and Govindarajan [16] experimented a single cylinder, 5 HP diesel engine for Oxygenated Air Induction (OAI) and shows increase in NOx emissions because of thermal NOx not from the inducted air, but this problem can be solved by using of Exhaust gas recirculation and lower levels of oxygen enrichment Injection timings can be retarded to reduce NOx emissions without increasing the specific fuel consumption.

Oxygen separation process is quite difficult, one method is to extract from our fossil fuel which again creates an indirect pollution and another method by electrolysis process which costs high, but oxygen enrichment will reduce the NOx as well as CO emission.

5.3 Particulate Matter

Particulate matter consists of Solid Fraction, Soluble Organic Fraction and Sulphur [7]. Solid fraction is a solid pollutant consists of carbon and ashes, etc. and soluble organic fraction is a liquid pollutant because of vapour presence which directly causes harm to human beings. Sulphur is a liquid pollutant and can cause catalytic poison, i.e. failure of catalytic converter like DOC, NSR, etc. and also harm human beings indirectly. Sulphur content present in the fuel is the main factor to increase pollutant so it is reduced from 350 ppm (Euro 3) to 50 ppm (Euro 4) [17]. Particulate Matter is less in Low Heat Rejection engine is reported by various researchers [4, 6]. Particulate matters are harm to humans and causes 70–80 % of cancer reported by WHO [1]. Regeneration of Particulate matters is also ease in Low Heat Rejection engines because high temperature carbon gets burned in the filter.

5.4 Smoke

Smoke is nothing but a soot particle; smoke in a diesel is high when compare to other fuels in an Internal Combustion Engine due to complicated bond structure in the fuel and is difficult to break, so incomplete combustion occurs that's why smoke is formed and also do not have sufficient warm up temperature at low end speeds. But in Low Heat Rejection Engines, smoke is too low because of high temperature and pressure tends to warm up the engine easily and energy gets released from the bond. Hasimoglu [18] stated that smoke emission reduced by 39.8 % at 1,200 rpm and 20.3 % reduced at 2,200 rpm. Murthy [6] showed a result of 6 % decrease in smoke level.

6 Tribology

Low Heat Rejection Engine causes easy warm up which reduces the friction during engine start up. Lubricant used in the engine decomposes at a temperature of 350 °C. Corrosion in the thermal barrier coating is low patented by Heimann and Lehmann [19]. But in Low Heat Rejection engine the temperature will be high, so proper lubricant is required. James [20] stated that the classification of lubricants is based on the required ceramics, i.e. solid lubricant, dislocation of carbonaceous gases, gas lubricant.

7 Conclusion

The effects of thermal barrier coating on engine performance and emission characteristics are discussed. The results of Low Heat Rejection Engine performances taken through reviews are mentioned below:

- increase in thermal efficiency,
- decrease in fuel consumption,
- decrease in CO and HC level,
- increase in NOx level,
- increase in power/weight ratio,
- decrease in volumetric efficiency.

Some emerging techniques are also outlined, which could be used to reduce the negative effects on engine performance and emission characteristics. It is concluded that Thermal Barrier Coating is helpful to boost the performances of an engine to some extent but at the same time more research is needed to overcome some practical problems to benchmark Low Heat Rejection Engines.

References

1. www.greencarcongress.com/2012/06/iarc-20120613.html
2. Kamo R, Bryzik W (1978) Adiabatic Turbo compound engine performance prediction. SAE paper: 780068
3. Wiedemann B (2009) IAV GmbH · 10/2009 · International Experiences V6 final
4. Rajendra Prasath B, Tamilporai P, Shabir MF (2010) Theoretical modelling and experimental study of combustion and performance characteristics of biodiesel in turbocharged low heat rejection D.I diesel engine. World Acad Sci Eng Technol 61 2010
5. Taymaz, Cakir K, Gur M, Mimaroglu A (2003) Experimental investigation of heat losses in a ceramic coated diesel engine. Surf Coat Technol 169–170:168–170
6. Murthy PVK (2010) Performance evaluation of low heat rejection diesel engine with pure diesel. Int J Appl Eng Res 1(3)
7. www.dieselnet.com
8. Samadi H, Coyle TW (2009) Alternative thermal barrier coatings for diesel engine. Centre for Advanced Coating Technologies, University of Toronto, Toronto, Ontario, Canada
9. Shrirao PN, Pawar AN (2011) An overview on Thermal Barrier Coating (TBC) materials and its effect on engine performance and emission. Int J Appl Res Mech Eng 1(2)
10. Dickey DW (1992) The effect of insulated combustion chamber surfaces on Diesel engine performance emission and combustion", SAE paper No: 890292
11. Shrirao PN (2011) Evaluation of performance and emission characteristic of turbocharged diesel engine with mullite on thermal barrier coating. Int J Eng Technol 3(3):256–262
12. Diesel Emission Control-Sulfur Effects Program, Phase I Interim Data Report No. 1, August, 1999. www.ott.doe.gov/decse Copy available in Air Docket A-2001-28
13. Kawanami M (1995) Advanced catalyst studies of diesel NOx reduction for highway trucks. SAE 950154
14. De Rudder K (2012) Tier 4 high efficiency SCR for agricultural applications", SAE Paper: 2012-01-1087
15. Toh Ai Pei Jannell (2006) Direct water injection in marine diesel engine. Singapore Maritime Academy
16. Rajkumar K, Govindarajan P (2011) Impact of oxygen enriched air intake on the exhaust of a single cylinder diesel engine. Am J Environ Sci 7(2):136–140
17. www.epa.gov
18. Hasimoglu C (2008) Exhaust emission characteristics of a low-heat-rejection diesel engine fuelled with 10 per cent ethanol and 90 % diesel fuel mixture. J Automob Eng 222(1):93–100
19. Heimann RB, Lehmann HD (2008) Thermally sprayed coatings for protection against wear and corrosion of engineered Structures. Recent Patents Mater Sci 1:41–55
20. Leidel JA (2012) An optimized low heat rejection engine for automotive use—An inceptive study. SAE paper no: 970068
21. Morel T, Fort EF, Bulumberg PN (1985) Effect of insulation strategy and design parameters on diesel engine heat rejection and performance. SAE Paper 850506

Author Index

A
Ahuja, K., 197
Alexander, J., 207
Anantha Kumar, M., 243
Arankalle, A. R., 269
Augustine, B. S. M., 207

B
Babu, B. H., 253
Babu, J. M., 225, 253, 261
Balasubramanian, E., 117
Bernot, A., 25
Bernot, F., 25
Boller, C., 11, 23
Bragadish, M., 215

C
Chaurasiya, S. M., 269
Cheng, C., 3

D
Dhaka, P., 91

E
Edward, I. A., 129
Egede, P., 13, 235
Esakki, B., 105

F
Falempin, F., 41

G
Ganesh, R. P. V., 171
Gnanavel, C., 189
Goel, V., 91
Gowtham, P., 189
Gowthaman, S., 225, 253, 261

H
Hashim, S. A., 197
Herrmann, C., 13, 235

I
Immanuel Lazur, B., 69

J
Jagadeesh, L., 69

K
Karthik, J., 171
Karthikeyan, B., 69
Karthikeyan, S., 171, 179
Kirpekar, S., 269
Kleemann, S., 13
Kumar, M. R., 253
Kumar, P., 91

M
Madhavan, R., 117
Magin, M., 47
Mallick, P. K., 45

R. P. Bajpai et al. (eds.), *Innovative Design, Analysis and Development Practices in Aerospace and Automotive Engineering*, Lecture Notes in Mechanical Engineering, DOI: 10.1007/978-81-322-1871-5, © Springer India 2014

Manish, N., 197
Manoj Kumar, V., 189
Mariappan, R., 253, 261
Mathiyalagan, P., 225, 253
Milind, M., 79
Mishra, D., 197
Mohammed Ghouse, S., 285
Mugunthan, G., 117
Mukhopadhyay, A., 79

N
Nanda Kumaran, R., 307
Naresh, R., 261, 189
Nehuis, F., 13, 235

P
Pandithurai, R., 129
Pant, R. S., 63, 79
Prabhu, L., 85
Purushothaman, K., 285

R
Rajagopal, V., 105
Rajan, P., 163
Rajasekar, R., 295
Ramakrishnan, V., 285
Rao, A. R., 215

S
Schneider, E., 11, 23
Senthil Kumar, S., 145, 153, 163, 171, 179, 307
Shanmugaraja, M., 69
Sharma, N., 79
Sharma, V., 79, 261
Shivakumar, K., 225
Sing, J., 117
Sivakumar, G., 145, 307
Sivaraj, S., 153
Sreenivasaraja, N., 135
Srihari, R. B., 105
Srinivas, J., 85
Srinivasa Murthy, P., 65

T
Thinakaran, K., 295
Tian, B., 277
Tsau, Y., 277

V
Vannier, J., 25
Varun Kumar, M. G., 307
Venkatachalapathi, N., 215
Vietor, T., 13, 235

Y
Yang, H., 3
Young, W., 277
Yuvaraj, S., 243